Infrared and Raman Spectroscopy of
Biological Molecules

NATO ADVANCED STUDY INSTITUTES SERIES

*Proceedings of the Advanced Study Institute Programme, which aims
at the dissemination of advanced knowledge and
the formation of contacts among scientists from different countries*

The series is published by an international board of publishers in conjunction
with NATO Scientific Affairs Division

A	Life Sciences	Plenum Publishing Corporation
B	Physics	London and New York
C	Mathematical and Physical Sciences	D. Reidel Publishing Company Dordrecht, Boston and London
D	Behavioral and Social Sciences	Sijthoff International Publishing Company Leiden
E	Applied Sciences	Noordhoff International Publishing Leiden

Series C – Mathematical and Physical Sciences

Volume 43 – Infrared and Raman Spectroscopy of Biological Molecules

Infrared and Raman Spectroscopy of Biological Molecules

Proceedings of the NATO Advanced Study Institute
held at Athens, Greece, August 22–31, 1978

edited by

THEO M. THEOPHANIDES
Université de Montréal, Département de Chimie, Montréal, Québec, Canada and
National Hellenic Research Foundation, Athens, Greece

D. Reidel Publishing Company

Dordrecht : Holland / Boston : U.S.A. / London : England

Published in cooperation with NATO Scientific Affairs Division

Library of Congress Cataloging in Publication Data

Nato Advanced Study Institute, Athens, Greece, 1978.
  Infrared and Raman spectroscopy of biological molecules.

  (NATO Advanced Study Institutes series : Series C, Mathematical and physical sciences ; v. 43)
  Bibliography: p.
  Includes index.
     1. Infra-red spectroscopy—Congresses. 2. Raman spectroscopy—Congresses. 3. Molecular
biology—Techniques—Congresses. I. Theophanides, Theo M. II. Title. III. Series.
QH324.9.15N37 1978   574.1'9285   78-31633
ISBN-13: 978-94-009-9414-0   e-ISBN-13: 978-94-009-9412-6
DOI: 10.1007/978-94-009-9412-6

Published by D. Reidel Publishing Company
P.O. Box 17, Dordrecht, Holland

Sold and distributed in the U.S.A., Canada, and Mexico
by D. Reidel Publishing Company, Inc.
Lincoln Building, 160 Old Derby Street, Hingham, Mass. 02043, U.S.A.

CONTENTS

HYDROGEN BONDED SYSTEMS

EXPERIMENTAL TECHNIQUES

PREFACE

For this summer school in Athens, Greece, August 22-21, 1978, I took as my objective the presentation of a timely representative account of the application of infrared and Raman spectroscopy to biological molecules.

A summer school is made up of a number of things – ideas, people, organization international collaboration and sponsorship. The exchange of ideas the student-lecturer interaction in the discussion periods and the tutorials satisfy the urgent need of all the participants to meet and discuss topics of current scientific interest. It seems therefore appropriate to publish this summer school proceedings in order to make it a lasting event and that appreciation be shown to those people and institutions that made it all possible.

The summer school was held under the auspices of the Greek Ministry of Culture and Sciences under the sponsorship of the NATO Scientific Affairs Division in Brussels. In addition, support was provided by the National Hellenic Research Foundation and the Ministry of Culture and Sciences for several social and scientific functions.

Thanks to the use of laser beams as exciting sources in Raman spectroscopy and the introduction of computers and computer analysis of the spectra this technique can actually provide a wealth of information, not only on the small and medium molecules, but also on the very large macromolecular structures which constitute the living matter. It is thus of utmost importance to organize meetings between the spectroscopists and the people who are working with low forms of life i.e., the biologists, chemists and biochemists. Plant physiologists and medical researchers were also among the participants in the program. The presentation included general talks, review and background material for the students and a few reports of original research.

It was hoped that there would be a balance of infrared and Raman lectures and participants. However, Raman spectroscopists were predominant showing the present trend of interest among those applying vibrational spectroscopy.

Thanks to a great many people who helped in various ways either with the school itself or the publication of the manuscripts. First of all, I must thank the authors, for sending in their manuscripts on time. I would also like to thank Professor L. Bernard for secretarial assistance and help in the social events and for advise which was instrumental in arranging the summer school.

National Hellenic Research Foundation, Director
Physical Chemistry and Spectroscopy
Centre B. Constantinou Ave,
Athens 501, Greece, and

Professor
Université de Montréal
Département de Chimie
Montréal, Québec, CANADA

THEO M. THEOPHANIDES

INTRODUCTION AND HISTORICAL SURVEY

INTRODUCTION TO THE RAMAN SPECTROSCOPY OF BIOLOGICAL MOLECULES

Lucien BERNARD

Laboratoire de Recherches Optiques,Université de REIMS
FRANCE

In March 1928 it was published in Nature, by two physicists C.V. Raman and K.S. Krishnan, a new phenomenon which
revolutionized our techniques, commonly used to investigate
matter ; this new technique was called the Raman spectroscopy.

This effect was also observed independently in June
of the same year by the Russians Landsberg and Mandelstam. The
Raman effect was predicted theorically by Smekal in 1923.

The French physicist J. Cabannes had also observed this
phenomenon in experiments with scattered light in high atmosphere
on photographic plates. However, he believed that these new lines
were due to some artefact. Cabannes confirmed immediatly after,
the discovery of the Raman effect.

In this paper we will not be concerned about the
important discovery itself. Raman was largely responsible, because
he was the first one to show this effect conclusively and for
which he was awarded the Nobel Prize in Physics in 1930.

From then onwards, the Raman effect was proved to be
an extraordinary tool for research because ten years later there
were·more than 2000 papers published and today there are tenths
of thousands of them on the same subject.

We shall give a brief description of this phenomenon.
A monochromatic beam of frequency ν_0 falls on a sample
(solid, liquid or gas) the light diffused by the molecules of
the sample contains not only the frequency ν_0, but also the
frequencies $\nu_0 \pm \nu_j$, which are characteristic of the molecular

3

Theo M. Theophanides (ed.), Infrared and Raman Spectroscopy of Biological Molecules, 3–34.

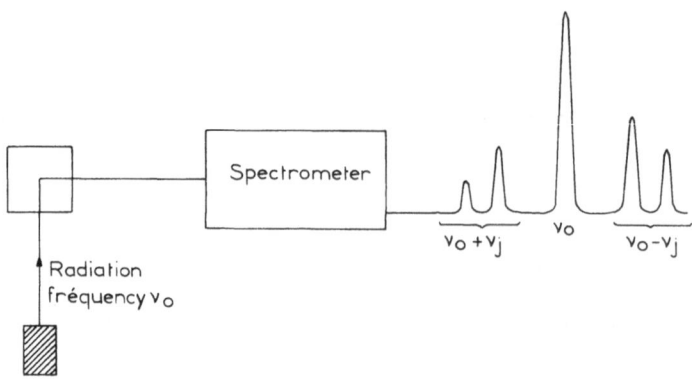

Fig. 1

structure of the sample. The ν_j frequencies constitute the Raman spectra of the sample.

The frequencies $\nu_0 - \nu_j$ are called Stokes Raman lines and the frequencies $\nu_0 + \nu_j$ are called anti-Stokes Raman. The Stokes lines are much stronger than the anti-Stokes Raman lines

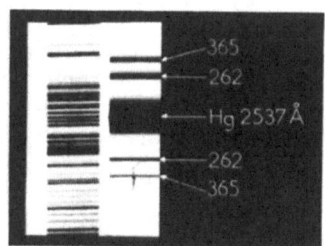

Fig. 2 Raman spectrum of $CHCl_3$ (fundamental frequencies)

If as, the origin of Raman spectroscopy is easy enough to trace back in time, for infrared spectroscopy, on the other hand it's quite a different matter.

Since the end of the last century we know that molecules, made out of constantly moving electric charges are emitting electromagnetic infrared radiations, and infrared spectroscopy has slowly developed as the technique of producing and detecting these radiations.

In this introductory paper, I would like to draw a simple retrospective sketch of origins of the infrared and Raman spectra, so as to emphasize the interest of the two techniques with a particular insistance on their complementarity.

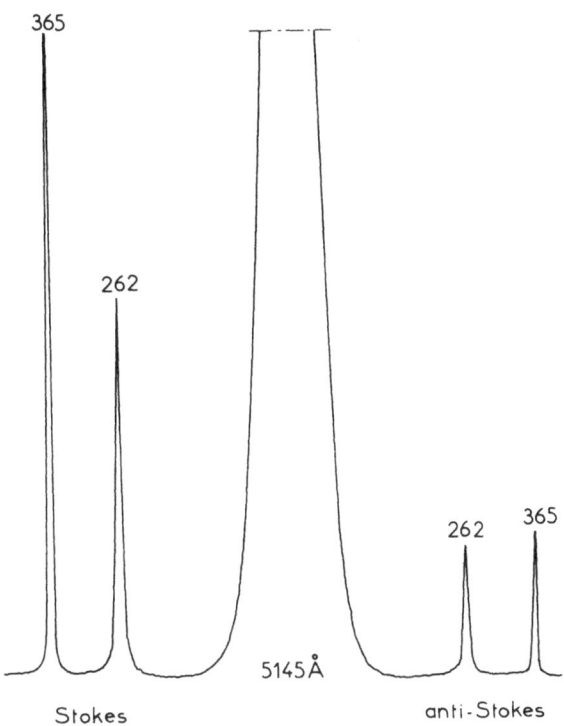

Fig. 3 Raman spectrum of $CHCl_3$ (fundamental frequencies)

I shall proceed to show how important Raman Spectroscopy has become in the last ten years particularly in the study of biological molecules. What I am aiming at in this paper is to help those who are new in the field, to get a better understanding of the topic which will be discussed.

Classical theory of vibration spectra

In the case of small gaseous molecules, we can observe infrared and Raman spectra due to the rotation of molecules. However I will not speak of these spectra because they do not intervene in the spectra of biological molecules.

I'd like to begin with the simple classical theory. It is based on the oscillating electric dipole notion. It is composed from two equal and opposite electric charges at a distance 1 from each other.

a dipole is characterized by its electric moment

$$\vec{p} = q \vec{l} \quad ; \quad \vec{l} = \vec{AB}$$

We know that if the moment p varies according to a sinusoïdal fonction given by :

$$p = p_0 \cos \omega t$$

$\nu = \dfrac{\omega}{2\pi}$; ω is the angular frequency. We get a radiating electromagnetic wave of frequency

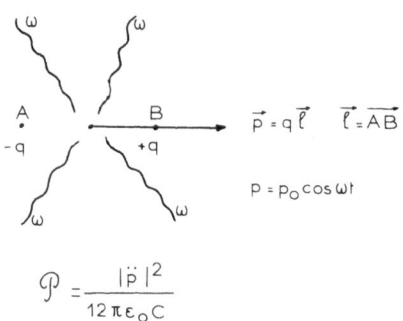

$$\mathcal{P} = \frac{|\ddot{p}|^2}{12\pi\varepsilon_0 C}$$

Fig. 4

The power of radiation in space is given by :

$$= \frac{|\ddot{p}|^2}{12\pi\varepsilon_0 c}$$

where ε_0 is the permittivity of the vacuum and c is the velocity of the light in the vacuum.

Molecules are composed of electrons with a negative charge and of nuclei with a positive charge. If the charges are not symmetrically arranged, the molecule has a permanent electric moment (H Cl, H_2O etc...) and is called a polar molecule. On the other hand, if the charges are symmetrically arranged, the molecule has not a permanent electric moment, and is called a non-polar molecule (CO_2, CH_4, C Cl_4, etc...).

When an electric field $\vec{E}$ is acting on a molecule, there is a shift in charges and there may appear an induced secondary electric moment P. There is a relation between the moment P and the electric field, which can be given by :

$$\vec{P} = \alpha \vec{E}$$

as a first approximation. We can neglect higher terms of very little importance, and they play a role only in the case of very intense field.

Usually, α is a tensor, called the molecular polarizability ; which means that the vectors $\vec{P}$ and $\vec{E}$ are not colinear. When one projects these two vectors onto xyz axes, we have :

$$P_x = \alpha_{xx} E_x + \alpha_{xy} E_y + \alpha_{xz} E_z$$

$$P_y = \alpha_{yx} E_x + \alpha_{yy} E_y + \alpha_{yz} E_z$$

$$P_z = \alpha_{zx} E_x + \alpha_{zy} E_y + \alpha_{zz} E_z$$

where α_{ij} are the 9 components of the tensor.

MOLECULAR VIBRATIONS

The atoms which constitute a molecule vibrate around an equilibrium position ; if the molecule contains N atoms, it constitutes a system of 3N degrees of freedom.

In order to study how these atoms move about within the molecule, the usual methods of classical mechanics are used. Once adequate coordinates have been chosen, the kinetic energy and the deformation potential energy of the molecule can be calculated. After the usual LAGRANGE equations have been applied, we get 3N differentials equations.

In the case of cartesian coordinates, the 3N differential equations are linked together by coupling terms, which are then rather difficult to solve. That is why a very special system, called normal coordinated system is preferable and which gives 3N differential equations independent from one another. Each differential equation being that of an harmonic oscillator, whose the solution is of the following form :

$$Q_K = Q_{oK} \cos (\omega_K t + \phi_K)$$

$$K = 1, 2, 3 \ldots 3N$$

ϕ_K is the angle phase

ω_K's are the specific frequencies of the molecule. It is easy to show that six of these frequencies vanish, since they correspond to the translation and the rotation movements of the whole molecule. What needs to be remembered, is that a molecule with N atoms, can have (3N - 6) specific vibration frequencies.

Example : H_2O has 3 specific frequencies (3.3-6 = 3)

CH$_3$ Cl has 9 specific frequencies (3.5-6 = 9)

ORIGIN OF INFRARED SPECTRA

If a molecular vibration of frequency ω_K, modifies the molecular electric moment according to :

$$\vec{p} = \vec{p}_0 \; \text{Cos} \; (\omega_K t + \phi_K)$$

then there will be an electromagnetic wave with an angular frequency ω_K, and the wave will be in the infrared region :

$$8\mu < \lambda < 200\mu$$

Such frequencies can be observed in absorption, since each atom or molecule can only absorb the radiation that it is able to emit.

Infrared spectroscopy is a simple technique. Using an infrared source which gives a continuous spectrum one sends a beam through the substance and observes the transmitted light which is then analysed by an infrared spectrometer.

Infrared spectroscopy has quickly developped after the second world-war, making the study of numerous structures possible, particularly in the field of organic chemistry.

ORIGIN OF RAMAN SPECTRA

The polarizability tensor of a molecule, which characterizes a change of the molecule by the influence of an electric field, can be expressed in relation with the normal coordinates given by a Taylor series :

$$\alpha_{ij} = (\alpha_{ij})_0 + \sum_K (\frac{\partial \alpha_{ij}}{\partial Q_K})_0 \; Q_K$$

+ higher order terms which can be written.

$$\alpha_k = \alpha_0 + \alpha_K \; Q_K$$

where K is the mode of vibration with :

$$Q_K = Q_{0K} \; \text{Cos} \; (\omega_K t + \phi_K)$$

We assumed that the molecule vibrates with frequency ω_K.

If the molecule is exposed to an electromagnetic radiation, with an electric field given by :

$$E = E_0 \cos \omega_0 t$$

the induced electric moment is expressed by :

$$P_K = \alpha_K E$$

using

$$P_K = \alpha_0 E_0 \cos \omega_0 t + \alpha'_K Q_{0K} \cos(\omega_K t + \phi_K) E_0 \cos \omega_0 t$$

with the well known relation :

$$\cos a \cos b = \frac{1}{2} \cos(a-b) + \cos(a+b)$$

We get the expression :

$$P_K = P_0 \cos \omega_0 t + P_1 \cos (\omega_0 - \omega_K)t - \phi_K$$
$$+ P_2 \cos (\omega_0 + \omega_K)t + \phi_K$$

which means that the molecule emits 3 frequencies :

ω_0 the frequency of the incident wave, that is the Rayleigh scattering, which is also called the elastic scattering without any change in frequency.

$\omega_0 - \omega_K$ Stokes Raman lines and

$\omega_0 + \omega_K$ anti-Stokes Raman lines.

This traditional and very simple theory gives a very qualitative interpretation of the phenomenon. The incident wave, frequency ω_0 is modulated by the molecular vibrations of frequency ω_K. But traditional mechanics cannot give a quantitative interpretation of the phenomenon and more precisely they cannot account for the considerable differences which occur in the intensities of the various scattered lines.

QUANTUM ASPECT OF THE RAMAN SCATTERING

The energy of vibration is in fact quantified, which means that this energy can only have some very specific discontinuous values. The same applies to the electronic energy due to the changes in the electronic motion. Such energies will be represented by horizontal lines.

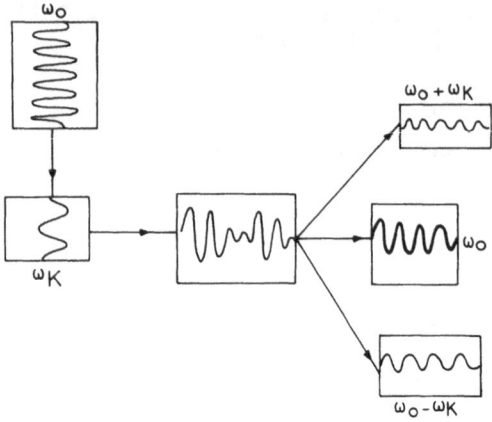

Fig. 5

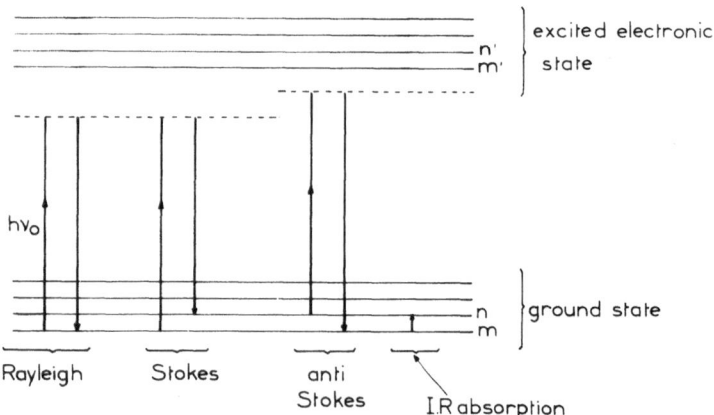

Fig. 6

In the lowest electronic state, which is called the ground state, we will represent different vibrating states of the molecule : m, n, ...

If the molecule in its ground state absorbs one $h\nu_0$ photon, it reaches an excited level which is not a molecular state and which is in fact very unstable so the molecule can either go back to the initial state (Rayleigh line) or to an excited vibrating state by emitting a Stokes Raman line with a frequency : $\nu_s = \nu_0 - \nu_{mn}$.

If the molecule which absorbs the $h\nu_0$ photon is in an excited vibrating level n, it can come back to the vibrating states m, thus emitting an anti-Stokes Raman lines of frequency $\nu_{as} = \nu_0 + \nu_{mn}$.

The Stokes lines are more intense than the anti-Stokes, because in the case of usual temperatures close to 20°C, the number of molecules in the m state is much larger than in the n state and it is more likely that we will get a transition leading to an anti-Stokes line.

DIFFERENT TYPES OF RAMAN EFFECT

In his quantum theory published in 1934, PLACZEK had shown that the intensity of a Raman line depended on the exciting frequency ν_0, and that this intensity must grow when ν_0 is close to ν_a, ν_a being an electronic absorption frequency.

Such explanation has been experimentally tested in 1955 or so by different physicists : BEHRINGER, BRANDMULLER, SCHORYGIN and HARRAND.

We must then distinguish several types of Raman effects according to the value of ν_0 in relation to ν_a

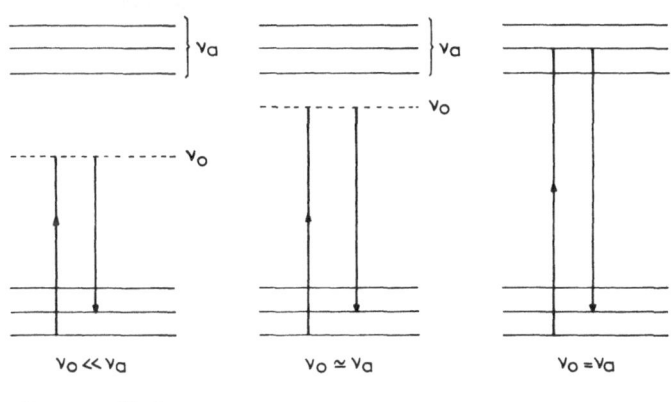

$$\nu_0 \ll \nu_a \qquad \nu_0 \simeq \nu_a \qquad \nu_0 = \nu_a$$

ordinary Raman effect   preresonance   resonance

Fig. 7

$\nu_0 \ll \nu_a$: ordinary Raman effect

$\nu_0 \simeq \nu_a$: preresonance Raman effect

$\nu_0 = \nu_a$: resonance Raman effect

Since 1955, numerous theorical and experimental studies have been conducted on the resonance Raman effect (SCHORYGIN, BEHRINGER, ALBRECHT, JACON, BERJOT, PETITCOLAS, etc...) and we shall come back later on the importance of the resonance Raman effect in the study of biological molecules.

ON THE COMPLEMENTARITY OF INFRARED AND RAMAN SPECTROSCOPY

When one comes back to the traditional interpretation, one can see that an infrared radiation is due to periodical variations of the electric moment of the molecule, whereas the Raman scattering effect takes into account the variations of the induced moment, that is to say the polarizability variations.

The emission of the frequency ω_k line will be linked with $\frac{\partial p}{\partial Q_k}$ in infrared, and with $\frac{\partial \alpha}{\partial Q_k}$ in the Raman effect.

If these expressions vanish, the ω_k frequency will not be visible, and one can already foresee that some frequencies

a) can be observed both in Raman and in infrared,

b) can be observed in Raman but not in infrared,

c) can be observed in infrared but not in Raman,

d) cannot be observed neither in Raman nor in infrared.

Hence, the complementarity of these two techniques.

Let's take some very simple cases.

The following figures are extracted from the excellent book "The Raman Spectroscopy" by D.A. LONG.

First, one must insist on some particularly important characteristic of the type ABA molecule (CO_2, H_2O ...).

The examination of the Raman and Infrared spectra allow to choose between the linear form which has :

1 active vibration in Raman, and

2 active vibrations in Infrared;

and the triangular form which has :

3 active vibrations in Raman and in Infrared.

Second, one must point out that a linear molecule which has a center of symmetry has active vibrations in Raman which are inactive in Infrared and vice versa. This remark can be applied to all molecules with a center of symetry.

As early as 1932, J. CABANNES has established selection rules based on geometrical considerations.

	(a) A—A	(b) A—B
Molecule		
Mode of vibration		
Variation of polarizability with normal coordinate (schematic)		
Polarizability derivative	$\neq 0$	$\neq 0$
Raman activity	Yes	Yes
Variation of dipole moment with normal coordinate (schematic)		
Dipole moment derivative	$= 0$	$\neq 0$
Infrared activity	No	Yes

Fig. 8 Comparison of polarizability and dipole moment variations in the neighbourhood of the equilibrium position and vibrational Raman and infrared activities for (a) an A_2 and (b) an AB molecule.

Molecule	o—O—o		
Mode of vibration			
Variation of polarizability with normal coordinate (schematic)			
Polarizability derivative	$\neq 0$	$= 0$	$= 0$
Raman activity	Yes	No	No
Variation of dipole moment with normal coordinate (schematic)			
Dipole moment derivative	$= 0$	$\neq 0$	$\neq 0$
Infrared activity	No	Yes	Yes

Fig. 9 Polarizability and dipole moment variations in the neighbourhood of the equilibrium position and vibrational Raman and infrared activities for a linear ABA molecule.

Molecule	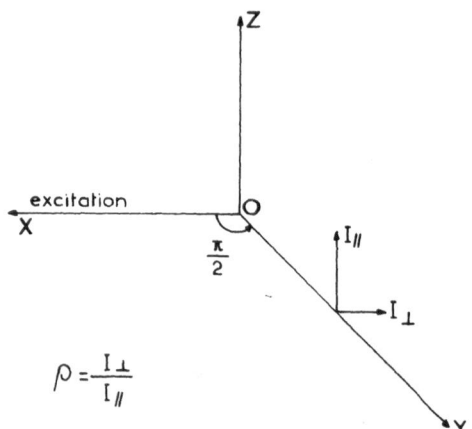	x out of plane	
Mode of vibration			
Variation of polarizability with normal coordinate(schematic)			
Polarizability derivative	$\neq 0$	$\neq 0$	$\neq 0$
Raman activity	Yes	Yes	Yes
Variation of dipole moment with normal coordinate(schematic)			
Dipole moment derivative	$\neq 0$	$\neq 0$	$\neq 0$
Infrared activity	Yes	Yes	Yes

Fig. 10 Polarizability and dipole moment variations in the neighbourhood of the equilibrium position and vibrational Raman and infrared activities for a non-linear A - B - A molecule.

The group theory applied to molecular vibrations enables one to give a classification of the vibrations, each type based on some common characteristic features and to calculate the number of vibrations for each type. Quantum mechanics makes it then possible to define selection rules and to know which vibrations are active and which are inactive in Raman and in Infrared spectroscopy.

$$\rho = \frac{I_\perp}{I_\parallel}$$

Fig. 11

POLARIZATION STATE OF RAMAN LINES

When one observes scattered light at right angles, one defines the depolarization ratio by :

$$\rho = \frac{I_\perp}{I_\parallel} \quad . \quad \left(\text{Fig. 11}\right) \; .$$

The value of ρ depends on the value of the components α_{ij} of the tensor of polarizability, consequently it depends on the symmetry of the molecule. It is important to know the value of the depolarization ratio, because it helps to determine the molecular geometry.

CHARACTERISTICS OF A RAMAN LINE

A Raman line is characterized by the following elements :

its frequency, intensity, depolarization ratio, symmetry and also its profile.

These characteristics will be developped later in this lecture. I will only say a few words about frequency, so as to give a clearer presentation of the various applications of Raman spectroscopy.

A Raman or infrared line frequency is generally expressed by the wavenumber $\sigma_K = \frac{1}{\lambda_K}$, that is to say $\sigma_K = \frac{\nu_K}{c}$ and that number is measured in cm^{-1}. This way of presenting frequencies is interesting in two reasons : the wavenumber is a comparatively small number, furthermore like frequency, it is proportional to the vibration energy.

A Raman or infrared line frequency depends on the mass of the vibrating atom or group of atoms, together with the binding forces which link the atoms or group of atoms to the rest of the molecule. If the molecule is not an isolated one, the frequency depends on the molecular environment (solvent effects, intermolecular forces, etc...). It is still difficult now to calculate all the normal frequencies of a molecule even more so when it is a complicated one. The way computers are developing make us feel hopeful for the future.

In order to define frequencies, we use characteristic frequency tables of atoms depending on their bonds in the molecules.

Gradually, specific terms have been used to name the different types of vibrations :

ν : stretching vibration

σ : bending vibration

τ : twisting vibration

etc...

Some characteristic frequencies are given in the following table

Vibration[a]	Region (cm^{-1})	Intensity[b]	
		Raman	Infrared
$\nu(O - H)$	3650-3000	w	s
$\nu(N - H)$	3500-3300	m	m
$\nu(\equiv C - H)$	3300	w	s
$\nu(= C - H)$	3100-3000	s	m
$\nu(- C - H)$	3000-2800	s	s
$\nu(- S - H)$	2600-2550	s	w
$\nu(C \equiv N)$	2255-2220	m - s	s - O
$\nu(C \equiv C)$	2250-2100	vs	w - O
$\nu(C = O)$	1820-1680	s - w	vs
$\nu(C = C)$	1900-1500	vs - m	O - w
$\nu(C = N)$	1680-1610	s	m
$\nu(N = N)$, aliphatic substituent	1580-1550	m	O
$\nu(N = N)$, aromatic substituent	1440-1410	m	O
$\nu_a((C -)NO_2)$	1590-1530	m	s
$\nu_s((C -)NO_2)$	1380-1340	vs	m
$\nu_a((C -)SO_2(- C))$	1350-1310	w - O	s
$\nu_s((C -)SO_2(- C))$	1160-1120	s	s
$\nu((C -)SO(- C))$	1070-1020	m	s
$\nu(C = S)$	1250-1000	s	w
$\delta(CH_2), \delta_a(CH_3)$	1470-1400	m	m
$\delta_s(CH_3)$	1380	m - w, s, if at C=C	s - m
$\nu(CC)$, aromatics	1600,1580 1500,1450 1000	s - m m - w s (in mono-;m;1,3, 5-derivatives)	m - s m - s O - w

[a] ν stretching vibration, δ bending vibration, ν_s symmetric vibration, ν_a antisymmetric vibration.

[b] ν_s very strong, s strong, m medium, w weak, O very weak or inactive.

Fig. 12 Characteristic wavenumbers and Raman and infrared intensities of groups in organic compounds.

Let's take the example of the double bond carbon-carbon stretching frequency : $\nu(C=C)$. Its frequency varies from 1500 cm^{-1} to 1900 cm^{-1}, depending on the bonds formed by the carbon atoms :

$$H_2C = CH_2 \qquad \nu(C=C) = 1620 \ cm^{-1}$$

$$CH_3CH = CH_2 \qquad \nu(C=C) = 1647 \ cm^{-1}$$

$$(CH_3)_2C = CH_2 \qquad \nu(C=C) = 1655 \ cm^{-1}$$

$$(CH_3)_2C = C(CH_3)_2 \qquad \nu(C=C) = 1672 \ cm^{-1}$$

$$Cl_2C = CCl_2 \qquad \nu(C=C) = 1571 \ cm^{-1}$$

the same thing is observed for the other characteristic vibrations.

EXPERIMENTAL TECHNIQUES

The Raman effect is a very weak phenomenon since the strongest lines are only 10^{-6} of the power in the exciting line. It is thus necessary to use very strong exciting sources, luminous spectrometer and sensitive detectors. The following figure shows the basic equipment used in Raman spectroscopy.

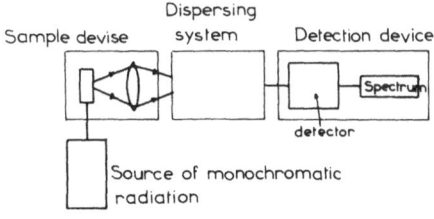

Fig. 13 Block diagram of equipment for observation of Raman spectra.

The first researchers to experiment in this field used mercury arc lamps, prism spectrographers and photographed the spectra of the scattered light.

The mercury arc lamp gives various radiations, which can be visually examined, of which the most intense is λ = 4358 Å, which has been used at large. The mercury vapor spectrum contains another very intense line in the ultraviolet, the resonance line λ = 2537 Å; this line has not be used so much because one needs quartz optics and it causes a rather embarassing photochemical decomposition of the sample.

Until about 1950, the use of prism and photographic plates spectrometers made it necessary to have time exposures which could be as long as 100 or 120 hours ! The prints were of a rather mediocre quality. The frequency and especially the intensity measures were not very precise ; furthermore about ten cm^3 of sample were needed !

In spite of those difficulties a considerable number of experiments had been made, thus giving the chemists a great number of precious results about the molecular structures, particularly in the field of organic chemistry.

Things were made easier for the first time in 1950 when prisms could be replaced by gratings and the photographic plate by a photoelectric recording. However, in spite of such technical changes for the better, Raman spectroscopy was miles behind infrared spectroscopy. Almost all the organic chemistry laboratories had one or even several infrared spectrometers, but very few had a good Raman spectrometer, because it costed much more than the infrared spectrometer.

From 1960 onwards thing have been completely different with the appearance of the LASERS and all the possibilities and hopes that was offered in Raman spectroscopy.

First the pulsed laser has paved the way to the stimulated Raman and the hyper Raman effect, new phenomena which however proved to be not applicable in the study of the molecular structures.

It is the gas lasers which opened new possibilities for Raman spectroscopy.

First the He - Ne laser which gives a very thin radiation of 6328 Å was used. This laser is remarkable for its stability but it is rather weak and gives only one line in the red.

Now, we are using almost exclusively the ion lasers (Ar and Kr), in particular the Argon ion laser, whose stability and fiability are quite acceptable. Their power goes up to 20 watts and they give several lines ; the most intense of which are given in the table :

λ_{air} (Å)	ν_{vac} (cm^{-1})	Relative output powers (mW)			
		He/Ne	Kr	Ar/Kr	Ar
7993.2	12507		30		
7931.4	12605		10		
7525.5	13285		100		
6764.4	14779		120	20	
6470.9	15450		500	200	
6328.2	15798	70			
5681.9	17595		150	80	
5308.7	18832		200	80	
5208.3	19195		70	20	
5145.3	19430			200	800
5017.2	19926			20	140
4965.1	20135			50	300
4879.9	20487			200	700
4825.2	20719		30	10	
4764.9	20981			60	300
4762.4	20992		50		
4726.9	21150				60
4657.9	21463				50
4579.4	21831			20	150
4545.1	21996				20
3637.9	27481				20
3511.1	28473				
3564.2	28049		40		
3507.4	28503				

Fig. 14 Wavelengths (and wave numbers) available from typical gas lasers.

By using a frequency doubler one can get quite acceptable light intensity powers until 2500 Å.

Raman spectroscopy, particularly the resonance Raman effect, has also benefited a great deal by the tunable dye lasers ; a fluorescent substance excited by radiations from an argon laser gives a continuous spectrum out of which we can select the proper radiation to the substance we want to study.

Recently, the N_2 pulsed laser gives good results in the ultraviolet region.

The laser beam produces very narrow fine lines. The very small divergence of the beam allows, after focussing, to obtain a great energy density and therefore it makes it possible to study very small sample, $\approx 10^{-6}$cm^3 for liquids and 10^{-9} for solids. Furthermore, if we are under resonant conditions, the intensity of the bands is much greater than in the ordinary Raman effect , thus much smaller samples can be used.

The use of the laser beam has made it possible to solve
almost completely the question of the exciting sources in a much
more satisfying way than the spectroscopist could have imagined
some twenty years ago.

There have been other technological improvements in the
course of these last ten years, which have made Raman spectroscopy become one of the most powerful techniques in the investigation of the structures of matter. Various physical states are
considered (pressure, temperature, weak dilutions, etc...).

From all technical improvements, which Professor
DELHAYE will talk about later, I will first insist of the
quality of the gratings

- improved techniques of engraving and printing,

- the perfecting of holographic gratings which solve
  the problem of diffused light, thus furnishing the
  spectroscopist with almost perfect dispersion devices.

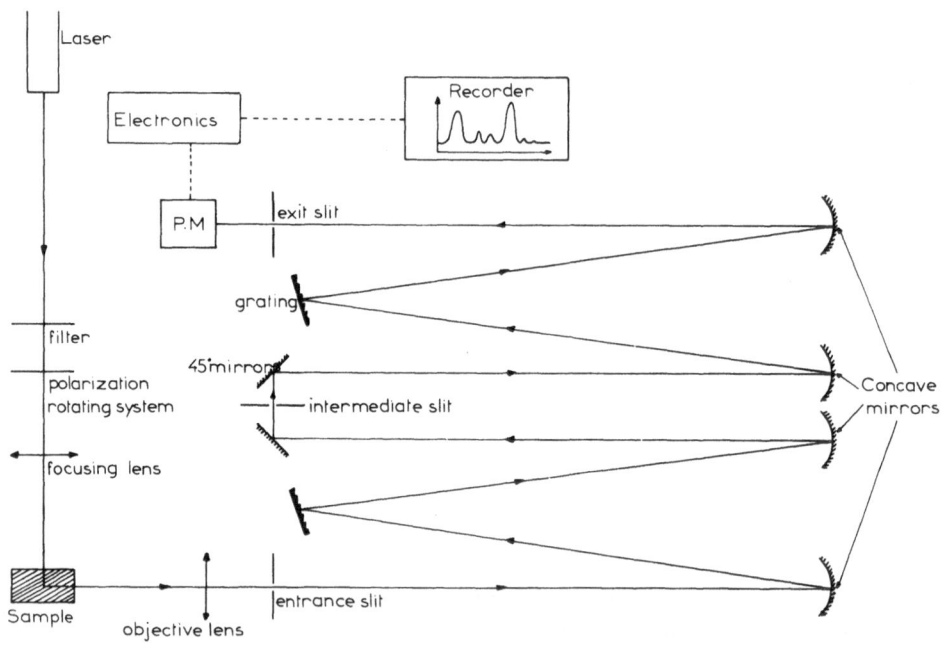

Fig. 15 Laser Raman equipment.

The use of double or triple monochromators makes it
possible to come very close (a few cm^{-1}) to the exciting line,

thus enabling us to observe vibrations of very low frequencies, which is of particular interest in the case of crystals and biological molecules.

Finally, the considerable progress that has been made in electronics has also largely contributed to the success of Raman spectroscopy.

- great improvement of the photomultipliers performance,

- the developpement of computer technics to extract information from the observed spectra.

APPLICATIONS OF RAMAN SPECTROSCOPY

For half a century Raman spectroscopy together with infrared spectroscopy have proved to be very useful in Chemistry, Physics, Astrophysics, Biology, Medecine. I will only give a brief summary of these applications.

QUALITATIVE ANALYSIS

Each molecule has a specific Raman and infrared spectrum, something like its finger-print which is a means of knowing and identifying it. So this technique can be used in the analysis of oils and other complex natural substance.

STRUCTURE OF MOLECULES

Each bond in a molecule is characterized by several Raman and infrared frequencies according to as which vibrations : stretching, bending, twisting, are considered. These frequencies are slightly influenced by the environment of the group of atoms which is responsible for them.

The following example is particularly simple.

In spite of such variations in frequencies, when we can observe a Raman or infrared band between 1500 and 1900 cm^{-1}, one may conclude that the vibration is a double bond, $C = C$, $C = O$, $C = N$ or even $N = N$ with aliphatic substituent in the same way a vibration between 2800 and 3100 cm^{-1} points out a simple bond $C - H$.

Thus it is possible to known from the spectra with different types of bonding and of structure are responsible for a given spectrum.

Raman and infrared spectroscopy can also help us to

be specific on the symmetry of molecules and to choose between
various isomeric forms. Let's take the simple example of cis-
trans isomery.

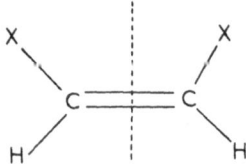
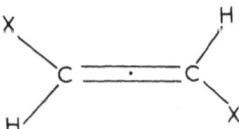

cis (C$_{2v}$ symmetry)

10 Raman and Infra-red active vibrations

2 Raman active and Infra-red

inactive vibrations.

trans (C$_{2h}$ symmetry)

6 Raman active and Infra-red

inactive vibrations.

6 Infra-red active and Raman

inactive vibrations.

Fig. 16

The cis form has :

- 10 Raman and infrared active vibrations

- 2 Raman active vibrations which are forbiden in IR

whereas the trans form, which has a center of symmetry has :

- 6 Raman active vibrations, no active in I.R.,

- 6 I.R. active vibrations no active in Raman.

The measurement of the depolarization ratio allows one
to give the exact symmetry of the vibrational mode.

With the previous example we have shown how complementary are the two techniques. The only way to do a complete study
of a structure is to apply both Raman and infrared spectroscopy
for the study of a molecule.

INTERMOLECULAR BONDING

The vibration frequencies of an atom in a molecule A
can be more or less modified when the atom is bound with another molecule B. The spectrum of the isolated molecule (gaseous
state or matrix) can be different from that of the liquid state
spectrum or the spectrum of the sample in solution. Such molecular interactions are particularly important in water and in
alcohols, where there is hydrogen bonding between the molecules.

It is certainly necessary to insist here on the great importance of hydrogen-bond in the intra and inter-molecular configurations of biological molecules.

Till now, most of the molecular spectroscopy of the hydrogen bond has been done by infrared techniques. However, laser Raman spectroscopy has been shown to be a very valuable tool in this field ; this is due to its intrinsic peculiarities and especially to the gain of information due to polarization measurements. This has been clearly established in the case of liquid alcohols by J.P. PERCHARD (PARIS) and in our Laboratory by J.L. BEAUDOIN.

The vibrations of the fonctional group (OH, NH_2, etc...) are very sensitive to the nature and to the strength of the hydrogen bond. The vibrational stretching ν_{OH}, ν_{OD}, ν_{NH} are the most commonly used "probes".

In the case of weak bonds, the ν_{OH} frequency is strongly dependent on the strength of the H-bond. Similar trends are observed for N H groups. As a consequence the width of the ν_{OH} can be used to estimate the degree of order in the intermolecular arrangements. The following example shows the $_{OH}$ bands of liquid and super-cooled liquid glycerol, and the corresponding bands of an oriented single crystal of the same substance. These spectra have been obtained in our Laboratory by J.L. BEAUDOIN.

Type de liaison	ν_{OH} cm^{-1}	$R_{O...O}$ Å	H kcal/mole	Exemples
OH libre (gaz)	$\sim$ 3650			Alcools
OH libre (liquide)	$\sim$ 3620			T- butanol
liaison H faible	3500 - 3200	$\geqslant 2,7$	$\leqslant 5$	eau, alcools
liaison H moyenne	3100 - 2800	$> 2,6$	6 - 8	R-COOH (acides)
liaison H forte	2700 - 700	2,6 - 2,4	> 8	sels acides

$$\frac{d(\nu_{OH})}{d(R_{O...O})} \simeq 1500 \, cm^{-1}/Å$$

in the case of weak hydrogen bonding.

Fig. 17

Of course when aqueous solutions are used, the regions of interest may be more or less obscured by the band of water ; however, the substracting techniques can be helpful in these cases.

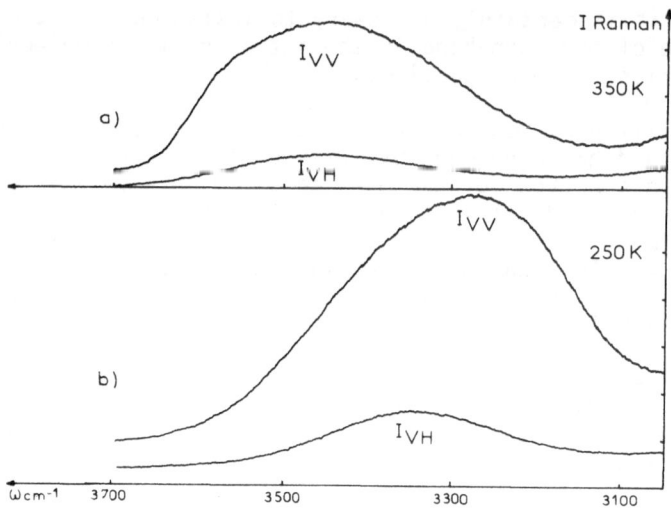

Fig. 18 Raman bands ν_{OH} of liquid glycerol.

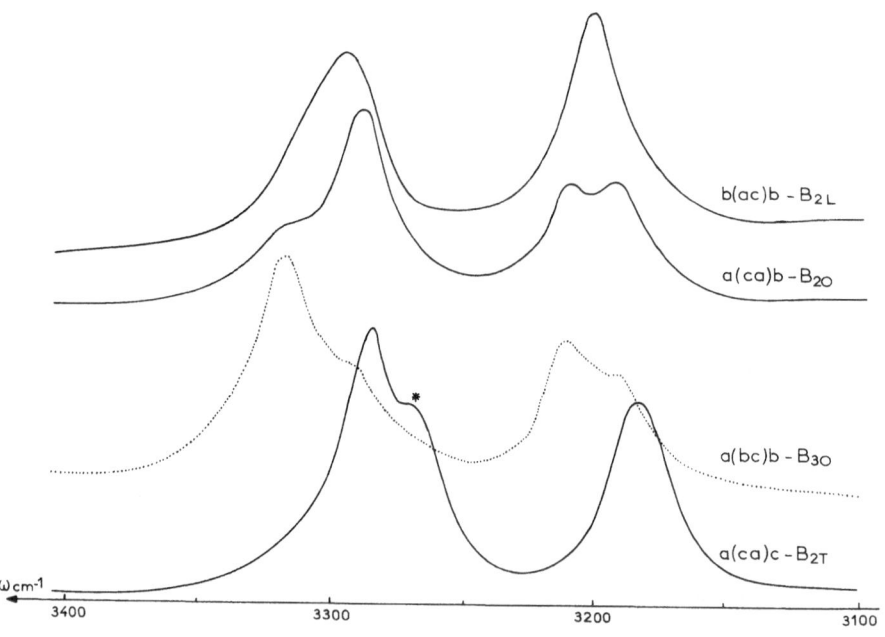

Fig. 19 Example of the polarities in the Raman spectrum of a glycerol crystal: bands and 'supernumerary bands'.
* Fermi resonance.

Some work of that kind is in progress in our Laboratory about fibrinogen and fibrin.

STUDY OF BIOLOGICAL MOLECULES

The use of Raman spectroscopy applied to biological molecules has been developed in the last ten years only. Here, infrared spectroscopy is rather difficult to use, since water being a natural biological medium has a great capacity for absorbing infrared radiations, whereas the Raman spectrum of water is not very strong and can be easily eliminated by the substraction methods.

Raman spectroscopy gives information on the detailed fine structure of molecules, in contrast to many other optical techniques which give information only on the overall gross structure ; the power of vibrational Raman spectroscopy as a structural diagnostic tool stems in large measure from the fact that each band in the spectrum is characterized not only by the frequency shift but also by the band profile, the band intensity and the depolarization ratio. With such a wealth of information available it is possible to investigate all manner of quite subtle structural changes in biological systems resulting from effects of temperature, pH, ionic strenght, uptake of oxygen, interaction with enzymes, interaction with virus, antibodies, various drugs (for instance action of the antitumour drugs)and in selected cases in vivo studies can be made.

Moreover, as we have already pointed out, Raman spectroscopy only needs a very small quantity of low concentration sample, especially if the resonance Raman effect is used.

However, there may be some difficulties in using this technique, which can be difficult to overcome.

One difficulty lies in the fact that the molecule is very complex with spectra very difficult to interprete, since a molecule with N - atoms has 3N - 6 normal vibrations; hence 3N -6 different frequencies depending on the nature of atoms and on how they are arranged in space. A good thing is that a great number of identical atoms or group of atoms are contained in a biological molecule and they have similar frequencies whose intensities add up. This applies to C-C, C-S-, S-S bonding and also to the amide group :

$$
\begin{array}{c}
O \\
\parallel \\
/^{C}\diagdown_{N}\diagup^{H} \\
\mid
\end{array}
$$

whose vibrations are called Amide I, Amide II, Amide III, etc.

Their frequencies range between 1630 and 1700 cm^{-1} for Amide I and between 1220 and 1300 cm^{-1} for Amide III, and they can give important precisions about the secondary structure of the proteins.

The characteristic vibrations of some amino acids are easily indentified.

In nucleotides and nucleic acids one can easily distinguish some bands which are characteristic of the purine and pyrimidine bases. The study of the evolution of spectra under the influence of various physical or chemical factors makes it possible to investigate changes in the molecular structure.

In the case of biological molecules, the resonance Raman effect can be very useful. When a molecule presents an absorption band in the visible spectrum or near the ultra-violet, this band is due either to a group of atoms, called the chromophore group, or to a metal atom in the molecule, with a specific biological function (Hemoglobine, Cytochrom, etc...).

If one excites the molecule with a radiation, of frequency ν_0 coinciding with the absorption frequencies, the Raman lines due to the chromophore vibrations are enhanced and their intensity can be hundred times stronger than that of bands in the ordinary Raman effect. Then, it is possible to use very low concentrations, ranging from 10^{-3} to 10^{-6} moles/liter.

Moreover resonance Raman effect theories show that selection rules and depolarisation ratios are different from that of the ordinary Raman effect, which makes it possible sometimes to get interesting precisions about the structure and the way these chromophore groups are working.

I will not give here any precise example on all these applications of Raman spectroscopy in the study of biological molecules. Many specialists will discuss at length such this important topic, which is the main reason for this Congress.

The biomolecules are difficult to study but there are other difficulties with them.

Since the Raman spectra are often overlapped by fluorescence, a much more intense phenomenon than Raman effect, it can obstruct the observation of the Raman effect.

The fluorescence may be due to impurities contained in the substance, in which case an adequate exhaustive purification is necessary. It frequently occurs that fluorescence is

due to the studied substance, especially when molecules are
excited in the absorption band that is in the case of the
resonance Raman effect. It is possible then to remove fluorescence altogether taking into account the ·fact that the excited
state life time of the molecule, which produces the Raman effect,
is much shorter than that of the excited state which brings about
the fluorescence so the molecule must be excited by very short
pulses and the Raman effect must be recorded only during the
excited pulses.

        Finally one can also come across another difficulty
the photochemical decomposition, a particularly impeding phenomenon where the molecule is excited in its absorption band. It
is then necessary to control the intensity of the lines, so
as to control the evolution of the phenomenon. Such a decomposition may be the result of an overheating of the scattering
substance which is more particularly frequent in the case of
resonance Raman effect. Very precise devices can be used to cope
with such difficulties.

WORK ACTUALLY IN PROGRESS IN EUROPE

        It is from 1970 on that biological molecular spectros-
copy has started to develop. A colloquium organized by Société
de Chimie Physique at the CNRS in THIAIS (FRANCE) in 1975 has
brought together about seventy scientists (physicians, chemists
and biologists).

        I shall summarize some of these works done in several
countries of Europe.

        In West Germany, there is Professor SCHMID in the
Institut für Physikalische Chemie der Universität FREIBURG, who
is directing a group of researchers in this field with the
following research project :

- Raman studies on ribonucleoside monophosphates in
  aqueous solution.

- Estimation of Poly A secondary structure from Raman
  scattering in aqueous solution.

- Influence of hydrogen bonding on the streching vibration of various nucleic acid derivatives.

- Raman studies on dinucleoside monophosphates.

- Raman investigation into the self-association of
  5'G M P in neutral aqueous solution.

In the University of Dortmund, B. SCHRADER and
R. SCHNEIDER study the vibrations of molecules in the free
and cristalline state.

In Austria, in the University of Graz, B. PALETTA,
R. MULLER, M. LIPPITSH work on measurements of cancer cell
metabolism.

In Italy, in the University of Bologna, A. BERTOLLUZA,
C. FAGNANO, M.A. MORELLI and R. TOZY study Raman spectra of
acetamide and aqueous acid solutions.

In the University of Milano, G. MASSETI, G. DELLEPIANE
and G. ZERBI study the resonance Raman spectra of free and
cristalline state and the Resonance Raman spectra of azoaldolase
and modeles molecules.

In England, Professor D.A. LONG supervises at the
School of Chemistry, University of BRADFORD, important works,
specially on the interaction between D N A and the oncolytic
agent bleomycin in collaboration with D.P. FAIRCLOUGH, V. FAWCETT,
L.H. TAYLOR and R.L. TURNER.

In France, in the Laboratoire Curie - Institut du
Radium and Institut de Biologie Moléculaire of PARIS VII, impor-
tant works are in progress on resonance Raman effect in visible
and ultraviolet on the nucleic acid complexes with proteins,
D N A and D N A actinomycine D, by P.Y. TURPIN, L. CHINSKY and
M. DUQUESNE.

In the Laboratoire de Spectroscopie Infra-rouge in the
University of BORDEAUX, there is the work of P.V. HUONG on
resonance Raman spectra and excitation profiles of vitamine B_{12}
derivatives and the work of J. LASCOMBES, P. COMBELAS and
C. GARIGOU-LAGRANGE, who has studied by infrared spectroscopy,
the transformation helix-random-coil of the peptides and those
of Anne Marie BELLOCQ, J.C. BOILOT, E. DUPART and M. DUBIEN on
the conformational analysis of the hypotalamic hormones by
Raman spectroscopy.

In the Laboratory of C.E.N. in SACLAY, the works of
M. LUTZ :
  - Resonance Raman spectra of chlorophyll proteins
    complexes.

  - Resonance Raman spectra of cation radical of bacterio
    chlorophyll in solution and in reaction centers of
    photosynthetic bacteria.

- Resonance Raman spectra of cis conformers of C - 4 carotenoïds. Evidence for cis carotenoïds in bacterial reaction centers.

The Laboratoire du CNRS de Spectrochimie Infra-rouge et Raman de THIAIS et de LILLE, supervised by M. DELHAYE ; there are the works of C. MERLIN and A. DUPAIX on resonance Raman spectroscopy studies of trypsin inhibitor interactions, and the studies of the bending of 2 - (4' hydroxy phenylazo) benzoic acid on avidin and bovine serum albumin.

- M. COPPEY, C. de LOZE, A. BURNEAU on the resonance Raman effect of cooperative and non cooperative hemoglobin.

- Lately, the work of C. de LOZE, M.H. BARON and F. FILLAUX on the non planar structure of the peptide group.

In the Centre de Technologie Biomedicale (I.N.S.E.R.M.) a recent work of G. VERGOTTEN, Y. MOSCHETTO, G. LEROY and J.M. DEVYNCK, has appeared on the use of Raman microprobe and microscope on the study of powder heterogeneity in compounds with therapeutic interest.

In the Laboratoire de Recherches Optiques of REIMS, we study :

- The transformation of the fibrinogen as a function of temperature and the transformation fibrinogen into fibrin in collaboration with F. CAPET of the University of MONTREAL and G. HUDRY-CLERGEON of GRENOBLE.

We have developed a method to measure the ratio of secondary structures of the proteins.

- In collaboration with the Center of Physical Chemistry and Spectroscopy in ATHENES supervised by Th. THEOPHANIDES, specialized in the study of platinum complexes, we have studied several antitumor agents (cis - $Pt(NH_3)_2 Cl_2$ adriamycin, daunamycin) and their mechanism of action.

REFERENCES

88128462 Chemabs No. 18 journal
A relation between Raman resonance frequencies and the ring core
radius of metalloporphyrins
  Pham Van Huong: Pommier, Jean Claude
  Lab. Spectrosc. Infrarouge Univ. Bordeaux I Talence Fr.
  C. R. Hebd. Seances Acad. Sci., Ser. C; (77) P 519-22;
  Vol 285;

No 16; In Fr; Coden CHDCA ISSN 0567654

88046734 Chemabs No. 07 conference
  Vibrational spectra of cholorophylls a and b labeled with
magnesium-26 and Nitrogen-15
  Lutz, Marc; Kleo, Jacques; Gilet, Roland; Henry, Monique; Plus,
Remi; Leicknam, Jean Pierre
  Dep. Biol. CEN Saclay Gif-sur-Yvette Fr.
  Proc. Int. Conf. Stable Isot., 2nd; (76) P 462-9: In Eng; Avail.
Klein, E. Roseland; Klein, Peter D; Coden 37EVA Date; 75 Publisher
NTIS, Springfield, Va

88023506 Chemabs No. 04 conference
  Resonance Raman Studies of conjugated macromolecules
  Bloor, D.; Batchelder, D. N.
  Phys. Inst. Univ. Stuttgart Stuttgart Ger.
  Proc. Int. Conf. Raman Spectrosc., 5th; (76) P 475-84; In Eng;
Avail. Schmid, Eduard D.; Brandmueller, J.: Kiefer, W;
Coden 36RDA

Publisher Hans Ferdinand Schulz Verlag, Freiburg/Br., Ger.

88007283 Chemabs No. 01 conference
  Resonance Raman spectra and excitation profiles of vitamin B12
derivatives
  Pham Van Huong
  Lab. Spectrosc. Infrarouge Univ. Bordeaux I Talence Fr.
· Proc. Int. Conf. Raman Spectrosc., 5th; (76) P 204-5; In Eng;
Avail. Schmid, Eduard D.; Brandmueller, J.; Kiefer, W; Coden
36RDA

Publisher Hans Ferdinand Schulz Verlag, Freiburg/Br., Ger.

88001820 Chemabs No. 01 conference
   The uses of high resolution resonance Raman excitation profiles
to study electronic structure of heme proteins
   Friedman, J.M.: Rousseau, D.L.
   Bell Lab. Murray Hill N.J.
   Proc. Int. Conf. Raman Spectrosc., 5th: (76) P 324-5; In Eng:
Avail. Schmid, Eduard D.; Brandmueller, J.; Kiefer, W; Coden
36RDA

Publisher Hans Ferdinand Schulz Verlag, Freiburg/Br., Ger.

88001818 Chemabs No. 01 conference
   Effect of peripheral substituents on resonance Raman spectra
of metalloporphyrins and heme proteins
   Verma, A. L.
   Lab. Phys. Chem. Univ. Amsterdam Amsterdam Neth.
   Proc. Int. Conf. Raman Spectros., 5th; (76) P 198-9; In Eng;
Avail. Schmid, Eduard D: Brandmueller, J.; Kiefer, W.; Coden
36RDA

Publisher Hans Ferdinand Schulz Verlag, Freiburg/Br. Ger.

88001761 Chemabs No. 01 conference
   Raman intensity measurements on nucleotides
   Herbeck, R.; Schlenker, P.; Gramlich, V.; Schmid, Eduard D.
   Inst. Phys. Chem. Univ. Freiburg Freiburg/Br. Ger.
   Proc. Int. Conf. Raman Spectrosc., 5th: (76) P 192-3; In Eng;
Avail. Schmid, Eduard D.; Brandmueller, J.; Kiefer, W.; Coden
36RDA

Publisher Hans Ferdinand Schulz Verlag, Freiburg/Br., Ger.

87197202 Chemabs No. conference
   Resonance Raman scattering of photosynthetic pigments in vivo.
Light harvesting structures and reaction centers
   Lutz, M.
   Dep. Biol. CEN Saclay Gif-sur-Yvette Fr.
   Proc. Int. Conf. Raman Spectrosc., 5th; (76) P 200-1; In Eng;
Avail. Schmid, Eduard D.; Brandmueller, J.; Kiefer, W.; Coden
36RDA

Publisher Hans Ferdinand Schulz Verlag, Freiburg/Br., Ger.

87195707 Chemabs No. 25 conference
  Resonance Raman spectra of highly fluorescent molecules: free
and DNA-bonded ethidium bromide
  Chinsky, L.; Turpin, P.V.; Duquesne, M.
  Lab. Curie Paris Fr.
  Proc. Int. Conf. Raman Spectrosc., 5th; (76) P 196-7; In Eng;
Avail. Schmid, Eduard D.; Brandmueller, J.; Kiefer, W.; Coden
36RDA

Publisher Hans Ferdinand Schulz Verlag, Freiburg/Br., Ger.

87191466 Chemabs No. 24 conference
  Effects of excitation light frequency on resonance Raman spectra
of platinum hematoporphyrin complexes
  Berjot, M.; Theophanides, T.
  Fac. Sci. Univ. Reims Reims Fr.
  Proc. Int. Conf. Raman Spectrosc., 5th; (76) P 202-3; In Eng;
Avail. Schmid, Eduard D.; Brandmueller, J.; Kieffer, W.; Coden
36RDA

Publisher Hans Ferdinand Schulz Verlag, Freiburg/Br., Ger.

87136241 Chemabs No. 17 journal
  Raman studies of platinum-nucleoside complexes
  Theophanides, T.; Hadjiliadis, N.; Berjot, M.; Manfait, M.;
Bernard, L.
  Dep. Chem. Univ. Montreal Montreal Que.
  J. Raman Spectrosc.; (76) P 315-23; Vol 5; No 4; In Eng; Coden
JRSPA

87118097 Chemabs No. 16 journal
  Raman polarization techniques in the study of macromolecules
  Shepherd, I. W.
Phys. Dep. Manchester Univ. Manchester Engl.
Adv. Infrared Raman Spectrosc. ; (77) P 127-66; Vol 3; In Eng;
Coden AIRSD

87080040 Chemabs No. 11 journal
  Fast Raman spectroscopy of cytochrome c using intracavity
resonance Raman amplification
  Fabiann H. ; Lau, A. ; Werncke, W. ; Lenz, K.

Zentralinst. Molekularbiol. DAW Berlin-Buch E. Ger.
Chem. Phys. Lett. ; (77) P 607-10; Vol 48 ; No 3; In Eng;
Coden CHPLB

87050265 Chemabs No. 07 journal
  Antenna chlorophyll in photosynthetic membranes. A study by
Resonance. Raman spectroscopy
  Lutz, Marc
  Dep. Biol. CEN Saclay Fr.
  Biochim. Biophys. Acta; (77) P 408-30; Vol 460 ; No 3; In Eng;
Coden BBACA

87006364 Chemabs No. 01 journal
  Magnetic resonance Raman optical activity of ferrocytochrome c
at different laser wavelengths
  Barron, L.D.
  Chem. Dep. Univ. Glasgow Glasgow Scot.
  Chem. Phys. Lett. ; (77) P 579-81; Vol 46 No 3 In Eng; Coden
CHPLB

86184711 Chemabs No. 25 journal
  Structural investigation of poly d (BrU-A) by ultraviolet
resonance
Raman spectroscopy
  Chinsky, L.; Turpin, P.V. ; Duquesne, M. ; Brahmes, J.
  Sect. Phys. Chim. Inst. Radium Paris Fr.
  Biochem. Biophys. R s. Commun.; (77) P 766-71;Vol 75; No 3
In Eng; Coden BBRCA

86023907 Chemabs No. 04 journal
  Raman resonant effect on a cobalt histidine complex, a model
of a metalloprotein
  Chottard, G.; Bolard, J.
  CNRS Univ. Pierre et Marie Curie Paris Fr.
  Inorg. Chim. Acta; (76) P L17-L19; Vol 20; No 2; In Eng;
Coden ICHAA

85143394 Chemabs No.19 journal
  Carbon-13 magnetic resonance spectra of C-nucleosides. 3.

Tautomerism in formycin and formycin B and certain
pyrazolo (4,3-d) pyrimidines
  Chenon, Marie T.; Panzica, Raymond P.; Smith, James C.;
  Pugmire, Ronald J.; Grant, David M.; Townsend, Leroy B.
  Serv. Spectrochim. Infrarouge Raman CNRS Thiais Fr.
  J. Am. Chem. Soc.; (76) P 4736-45; Vol 98; No 16; In Eng;
Coden JACSA

        Many figures are extracted from Raman Spectroscopy
by D.A. LONG - Mc GRAW-HILL International Book Company.

HISTORICAL SURVEY OF THE INFRARED AND RAMAN SPECTROSCOPIC STUDY OF BIOLOGICAL MOLECULES

J. R. Durig and D. J. Gerson

Department of Chemistry
University of South Carolina
Columbia, S.C. 29208 U.S.A.

1. INTRODUCTION

In the survey which follows there is no attempt to reference all the work in the area. Representative papers are referenced but because of the restricted length of this article many excellent studies could not be included. We have rather arbitrarily divided the survey into decades for convenience. We have pointed out papers which appear to be milestones but in some cases there may be a certain amount of personal prejudice involved. Omission of some important papers simply reflects the limitations of the authors and not a judgement that they are less important than the referenced works.

2. 1930's

Within the decade after the discovery of the Raman effect by Sir C. V. Raman in 1928, several investigators began examining the Raman spectra of simple biological molecules. Amongst the first of these early studies was an investigation of some amino acids by N. Wright and W. C. Lee[1]. In their letter to _Nature_, Wright and Lee reported the measurement of Raman frequencies for several amino acids in solution. Although they did not give any vibrational or group frequency assignments, they did show that the Raman technique was applicable to biological molecules. These workers also reported the observation of fluorescence which could be "burnt out" by extended periods of exposure to the excitation source or quenched by the addition of small amounts of potassium iodide to the sample.

Theo M. Theophanides (ed.), Infrared and Raman Spectroscopy of Biological Molecules, 35–43.
All Rights Reserved. Copyright © 1979 by D. Reidel Publishing Company, Dordrecht, Holland.

In another early study, Edsall[2] reported the Raman spectra
and vibrational assignments for guanidine and urea derivatives.
These compounds have long been models for biological compounds and
at the time of Edsall's study, compounds whose structure in solution was uncertain. From the vibrational assignments given by
Edsall on the basis of Wigner's molecular symmetry it was concluded
that the C_{2v} structure for urea and the D_{3h} structure for guanidine were present in solution and not a C_{2v} immonium ion structure
(see Structures I & II).

This paper by Edsall marked a major cornerstone in the application
of Raman spectroscopy to biological molecules: the use of the
Raman spectra of a molecule to determine its structure in solution,
a technique which has been refined in recent years (see section on
1970's).

During the remainder of the 1930's, Edsall[3-7] continued to
expand the library of Raman spectra of amino acids and their derivatives. These papers included vibrational assignments on the
basis of symmetry and group frequencies.

The experimental techniques used by Wright, Edsall and others
were based upon modification of the readily available atomic spec-
trographs. These modifications included low dispersion optics,
220 Å/mm, the use of a mercury arc as the excitation source and
photographic detection. Although primitive by today's standards,
this basic design was used until the early 1960's with the availability of the Cary 81 spectrometer.

3. 1940's

In the 1940's Raman spectroscopy of biological molecules took
a back seat to the infrared study of biological molecules. This
was partly due to the second World War, but mainly due to the development at Harvard University in 1938 of an automatic recording
infrared spectrometer[9]. This work included the infrared spectra
of gelatins by Ellis[65], of chlorophyll and plant pigments by
Coblentz and coworker[66] and carboxylic acids by Edsall and Wilson[67].

In 1940 Loofbourow[8] reviewed the uses and the potential of infrared and Raman spectroscopy as a physical technique in biochemistry, as well as a philosophical approach which is as applicable today as it was 38 years ago. It was this philosophical approach that Lord[10], and others at the Spectroscopy Laboratory of MIT would use in the investigation of biological molecules in the 1960's and 1970's.

After the end of the second World War, several articles appeared in the literature by M. Renard[11], H. Deslandres[12] and others on the application of molecular spectroscopy to the structural analysis of proteins and enzymatic reaction products. In his 1945 paper, H. Deslandres[12] proposed using the symmetry analysis as developed by Wigner, Herzberg and others to assign the spectra of proteins. He continued by applying this analysis to some biological molecules, mainly amino acids, and although his analysis was later shown to be incomplete by Edsall et al.[13] and Lord et al.[14], it showed excellent foresight.

4. 1950's

During the 1950's vibrational spectroscopy of biological molecules entered a stage of empirical correlation between peptide structure and vibrational frequency. In 1950, Elliot and Ambrose[15] proposed empirical correlations between peptide structure and the amide I and amide II bands. Three years later, these workers showed[16] that for a polypeptide in a helical conformation the amide I band is at 1655 cm^{-1} and the amide II band is at 1540 cm^{-1}, whereas for an extended or β-structure the bands are at 1630 and 1520 cm^{-1}, respectively. Throughout the remainder of the decade Elliot and coworkers[17] continued their examination of various peptide structures, showing that their correlation did not apply to unoriented poly-L-glutamate films, poly-L-glycine and some proteins[17].

In 1958, Miyazawa and coworkers[18] reported the results of an extensive normal coordinate analysis of the amide I, II and III bands at 1650, 1550 and 1250 cm^{-1}, respectively. Their results showed that these vibrational motions are localized in the peptide group, as seen below.

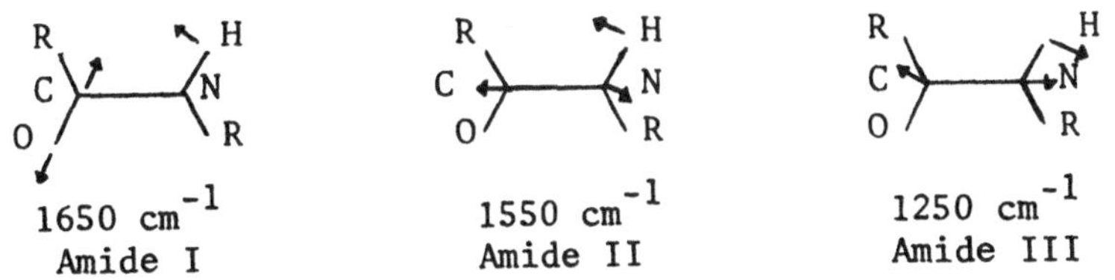

1650 cm^{-1}
Amide I

1550 cm^{-1}
Amide II

1250 cm^{-1}
Amide III

It was this normal coordinate analysis that Blout, Krimm and others would use as a basis for the development of more sophisticated structural correlations in the 1960's.

Also during the 1950's, Raman spectra were continuing to be reported, mainly by Edsall, of several different classes of biological molecules. In the early 1950's Edsall, et al.[19] reported the Raman spectra and vibrational assignments of several amino acids including glycine, cysteine and cystine. In 1958, with the marketing of the Cary 81 Raman spectrometer, Edsall reported the spectra of histidine[20], deutero-amino acids[21,23], protein degradation products[24] and lysozyme[22]. Of these, the Raman spectra of lysozyme[22] was a major cornerstone. This was the first Raman spectra of a protein, taken in aqueous solution, to have a relatively complete vibrational assignment. Edsall et al.[22] reported 14 faint Raman lines of which 6 were assigned and later shown by Lord[14] to be correct.

5. 1960's

The 1960's marked the beginning of a new era in the application of vibrational spectroscopy. With the availability of new detectors, new electronics and high speed digital computers, more experimental data were collected and theoretical correlations made in the 1960's than in the previous 30 years. In the early 1960's Miyazawa, Shimanouchi, Faseman and Blout[25,18,68] conducted extensive work in both normal coordinate calculations and structural correlations of polypeptide structure and observed infrared spectra. As an example, their 1961 paper[25] entitled "The infrared spectra of polypeptides in various conformations: Amide I and II bands" provides an excellent window to view "the state of the art" as of that date. Miyazawa and Blout, using the theoretical calculations of Shimanouchi[18], assigned specific frequencies to several conformations. For example, they found that an α-helical conformation will give rise to amide I and II bands at 1656 and 1635 cm^{-1}, respectively, where as an antiparallel sheet or β-structure will give rise to bands at 1685 and 1530 cm^{-1} for these same vibrations. The potential energy distribution over these two modes was shown by Miyazawa[25,26] to depend upon the displacements along the polymer chain of the amide linkage, with the particular frequency highly dependent upon intrachain hydrogen bonding. In 1962, Krimm[27] presented a method, based on that of Miyazawa and Blout[25], for correlating protein structure to the observed frequencies of the amide I and II bands, along with determining the relative contributions of the various normal modes[26] to each vibrational frequency. This paper more or less defines "the state of the art" in infrared spectroscopy of polypeptides until the mid 1970's with the arrival of commercially available Fourier Transform infrared spectrometers.

Also in the 1960's, nucleic acids were beginning to be studied by infrared spectroscopy. In the laboratories of Lord at M.I.T., infrared studies[28-30] were conducted in the early 1960's into the structural effects of water on DNA. Later, in the mid 1960's, Lord and coworkers[31-33] conducted an extensive investigation into the structural effects of substitution and non-aqueous solvation on hydrogen bonding in uracil and adenine derivatives. These studies were later used as guidelines by Lord and Thomas in assigning the Raman spectra of ribonucleic acid[34-36].

In 1967, Lord and Thomas[34] reported a complete vibrational analysis of 20 nucleic acid or nucleic acid derivatives. This paper, one of the most ambitious of its time, is still referred to as a standard despite the use of pre-laser excitation spectra. In the conclusions to this landmark paper, Lord and Thomas proposed (1) group frequency assignments for both the nitrogenous bases and the ribose-nitrogen linkage, (2) electronic and structural correlations for both pH and solid state effects on the observed spectra and (3) group frequencies for base pair interactions. Although Lord and Thomas did report two other studies[34-35] on nucleic acids, most of the research in this area did not begin until the 1970's with the availability of the laser.

In 1969 Peticolas and coworkers[37] reported one of the first laser-Raman spectra of polypeptides and polynucleotides. In their paper, Peticolas, et al. reported a group-theoretical analysis of the Raman tensor for helical polymers, selection rules for the Pauling α-helix, factor group $\bar{C}$ 18/s = spectroscopic group C 18, and a vibrational assignment based upon polarization studies of polymer fibers. Although this study was one of the last of the decade, it gave a taste of things to come in the 1970's.

6. 1970's

The 1970's marks the decade where vibrational spectroscopy comes into its own as another physical technique in biochemistry. It is at this point in time, the words of J. R. Loofbourow[8] come to mind:

> "When the physicist becomes better acquainted
> with biological and biochemical problems and
> the biological worker learns more of the
> terminology and techniques of the physicist,
> one may expect rapid progress."

This then is the 1970's, a time of rapid experimental and theoretical development in the vibrational spectroscopy of biological molecules. There are several reasons for this rapid development,

the two major reasons being stable, commercially available gas
lasers and advanced monochromator design as seen in the Cary 82
and Spex 1400 series monochromators for use in Raman spectroscopy
and Fourier Transform interferometers interfaced to high speed
digital computers for infrared spectroscopy. Recent developments
in the experimental design of both infrared[38] and Raman[39] Spectroscopy have recently been reviewed.

In the early 1970's, extensive work was conducted into the
structure of enzymes by Raman spectroscopy[40-47] and reviewed extensively by Koenig[48] and Thomas[49]. Lord and coworkers[40-44] conducted a series of investigations into the use of group frequencies
derived from amino acids in the assignment of the Raman spectra of
proteins. In one of the earliest of these studies[40], they showed
that the Raman spectra of an enzyme, lysozyme, could be predicted
by obtaining the Raman spectra of its constituent amino acids in
the correct mole fraction. It was further shown that the resultant spectra correlated well to that of the native enzyme except
at the amide I and III bands which are conformationally dependent.
In a latter paper[43], Lord and coworkers reported a crude conformation amide band frequency correlation scheme, which was summarized
by Frushour and Koenig[48].

	amide I	amide III
α helix	$1655 - 1659$ cm^{-1} (s)	$1250 - 1280$ (w)
antiparallel β sheet	$1667 - 1672$ cm^{-1} (s)	$1229 - 1240$ (s)
Random coil	$1665 - 1675$ cm^{-1} (s)	$1243 - 1265$ cm^{-1} (m)

Among the enzymes studied in the early 1970's were lysozyme, ribo-
nuclease, α-chymotrypsin, insulin, tropomyosin, α-casein, serum
albumn, β-lactoglobin, keratins, collagen and elastin, all of
which have been reviewed by Frushour and Koenig[48].

In 1976 Lippert, et al.[50] reported a method of determining
the relative percentage of α-helix, β-structure and random coil in
proteins whose Raman spectra have been recorded. Using empirical
methods similar to that used by Krimm[27] in infrared spectroscopy,
they correlated relative intensities of the amide bands of the
protein to that of homopolypeptides to give a structural determination that agrees to within 10% with circular dichroism
analysis[50,51].

In 1977, N. T. Yu published an extensive review of work on
the Raman spectra of proteins covering work through early 1977.
In 1977 and early 1978 several studies have been reported on the
Raman spectra of various biopolymers including histones[52], lens
proteins[53], and thymidylate synthetase[54] among others.

Also during the 1970's the Raman spectra of ribonucleic acids developed tremendously. This field has been reviewed by Thomas[49] and Peticolas[55] in 1975. Peticolas[55] and coworkers have developed structural correlations for simple polynucleic acids as well as examination of both Raman hyperchromism and hypochromism in the spectra. They and Thomas[49] later concluded that hyperchromic activity is restricted to bands above 1450 cm-1 and hypochromic activity to bands below 1300 cm-1.

In a series of papers in the mid 1970's Thomas and coworkers[56] (reviewed[57] in late 1976) investigated the structure of viruses both in situ and without protein coat. These workers concluded that the use of UV lasers would allow for pre-resonance enhancement of nucleic acid bands while visible lasers would allow observation of Raman bands due to the protein coat. In a very recent article, Shic, Dobrov and Tikchonenko[58] examined the Raman spectra of two related viruses; they concluded that although the protein coats differed in structure, their nucleic acids showed similar spectra.

Infrared spectroscopy of biopolymers during the late 1960's and early 1970's took a back seat to Raman spectroscopy. However, with the development of Fourier Transform infrared spectrometers, the field should open up. In a recent review, Koenig[59] presents both the experimental methods and analysis of spectra taken by this technique, as well as review of past work in the field.

Another experimental technique which has developed into a field unto itself is Resonance Raman Spectroscopy. In the early 1970's with the development of tunable dye lasers, resonance enhancement of biological molecules, usually porphyrin containing, was achieved. In an excellent series of papers, Spiro[60,61] reported an extensive and ongoing investigation into the electronic structure of the metallo-porphyrin complex. In 1976, Spiro[62] reviewed progress in the field to date and also reported on two new techniques: vidicon detection and resonance labels. Vidicon detection[63] is a new method using a silicon intensified target tube, a holographic grating monochromator and computerized control and data storage. This technique has been applied to the spectra of cytochrome c. Resonance Raman labeling is a technique, developed by Carey, Schneider and Bernstein for examining the structure of the active site of enzymes using chromophoric substrates. The progress in this technique has recently been reviewed by Carey[64].

REFERENCES

1. N. Wright and W. C. Lee, Nature 136, 300 (1935).
2. J. T. Edsall, J. Phys. Chem. 41, 135 (1937).
3. J. T. Edsall, J. Chem. Phys. 4, 1 (1936).

4.   J. T. Edsall, J. Chem. Phys. 5, 225 (1937).
5.   J. T. Edsall, J. Chem. Phys. 5, 508 (1937).
6.   J. T. Edsall, Cold Spring Harbor Symposium 6, 40 (1938).
7.   J. T. Edsall and H. Scheinberg, J. Chem. Phys. 8, 520 (1940).
8.   J. R. Loofbourow, Rev. Mod. Phys. 12, 267 (1940).
9.   H. Gershinowitz and E. B. Wilson, Jr., J. Chem. Phys. 6, 197
     (1938).
10. R. C. Lord, Appl. Spect. 31, 187 (1977).
11. M. Renard, Mem. Roy. Soc. Sci. Liege 7, 196 (1945).
12. H. Deslandres, Compt. Rend. 221, 193 (1945).
13. J. T. Edsall and D. Garfinkel, J. Am. Chem. Soc. 80, 3827
     (1958) and papers cited within.
14. R. C. Lord and N. T. Yu, J. Biol. Chem. 51, 203 (1970).
15. A. Elliot and E. J. Ambrose, Nature 165, 921 (1950).
16. A. Elliot and E. J. Ambrose, Proc. 3rd Int. Cong. Biol. Chem.
     (1956).
17. A. Elliot, W. E. Hanby and B. R. Malcom, Disc. Faraday Soc.
     25, 167 (1958) and papers cited within.
18. T. Miyazawa, T. Shimanouchi, S. Mizushima, J. Chem. Phys. 29,
     611 (1958).
19. J. T. Edsall, J. W. Otvos and A. Rich, J. Am. Chem. Soc. 72,
     474 (1950).
20. D. Garfinkel and J. T. Edsall, J. Am. Chem. Soc. 80, 3807
     (1958).
21. M. Takeda, R. E. S. Iavazzo, D. Garfinkel, I. H. Scheinberg
     and J. T. Edsall, J. Am. Chem. Soc. 80, 3813 (1958).
22. D. Garfinkel and J. T. Edsall, J. Am. Chem. Soc. 80, 3818
     (1958).
23. D. Garfinkel and J. T. Edsall, J. Am. Chem. Soc. 80, 3823
     (1958).
24. D. Garfinkel, J. Am. Chem. Soc. 80, 3827 (1958).
25. T. Miyazawa and E. R. Blout, J. Am. Chem. Soc. 83, 742 (1961)
     and papers cited within.
26. T. Miyazawa, J. Chem. Phys. 33, 1647 (1961).
27. S. Krimm, J. Mol. Biol. 4, 528 (1962).
28. R. C. Lord, M. Falk and K. A. Hartman, Jr., J. Am. Chem. Soc.
     84, 3842 (1962).
29. R. C. Lord, M. Falk and K. A. Hartman, Jr., J. Am. Chem. Soc.
     85, 387 (1963).
30. R. C. Lord, M. Falk and K. A. Hartman, Jr., J. Am. Chem. Soc.
     85, 391 (1963).
31. R. C. Lord, R. M. Hamlin and A. Rich, Science 148, 1734 (1965).
32. R. C. Lord, Y. Kyogoku and A. Rich, J. Am. Chem. Soc. 89, 496
     (1967).
33. R. C. Lord, Y. Kyogoku and A. Rich, Proc. Nat. Acad. Sci.,
     USA 57, 250 (1967).
34. R. C. Lord and G. J. Thomas, Spectrochim. Acta 23A, 2551
     (1967).
35. R. C. Lord and G. J. Thomas, Biochim. Biophys. Acta 142, 1
     (1967).

36. R. C. Lord and G. J. Thomas, Dev. Appl. Spectrosc. 6, (1968).
37. B. Fanconi, B. Tomlinson, L. Nafie, W. Small and W. Peticolas, J. Chem. Phys. 51, 3993 (1969).
38. P. Griffiths, Chemical Infrared Fourier Transform Spectrosc., Wiley, New York, N.Y., 1976.
39. N. T. Yu, Critical Rev. in Biochem., 229 (1977).
40. R. C. Lord and N. T. Yu, J. Mol. Biol. 50, 509 (1970).
41. R. C. Lord and N. T. Yu, J. Mol. Biol. 51, 203 (1970).
42. R. C. Lord, Pure Appl. Chem. 28, (1971).
43. R. C. Lord, A. M. Bellocq and R. Mendelsohn, Biochim. Biophys. Acta 257, 280 (1972).
44. R. C. Lord, M. C. Chen and R. Mendelsohn, Biochim. Biophys. Acta 328, 252 (1973).
45. J. L. Koenig and B. G. Frushour, Biopolymers 11, 2505 (1972).
46. N. T. Yu, C. S. Culver and D. C. O'Shea, Biochim. Biophys. Acta 263, 1 (1972).
47. N. T. Yu and C. S. Liu, J. Am. Chem. Soc. 94, 3250 (1972).
48. B. G. Frushour and J. L. Koenig, Advances in IR and Raman Spect. Vol. I, Heyden, New York, N.Y., 1975.
49. G. J. Thomas, Vibrational Spectra and Structure Vol. III, Marcel Dekker, ed. by J. R. Durig, New York, N.Y., 1974.
50. J. L. Lippert, D. Tyminski and P. J. Desmeules, J. Am. Chem. Soc. 98, 7075 (1976).
51. N. Greenfield and G. D. Fasman, Biochemistry 8, 4108 (1969).
52. J. G. Guillot, M. Pezolet and D. Pallota, Biochim. Biophys. Acta 491, 423 (1977).
53. N. T. Yu, E. J. East and R. C. C. Chang, Exp. Eye Res. 24, 321 (1977).
54. R. K. Sharma, R. L. Kisliuk, S. P. Verma and D. F. H. Wallach, Biochim. Biophys. Acta 391, 19 (1975).
55. W. L. Peticolas, Biochimie 57, 417 (1975).
56. T. A. Turano, K. A. Hartman and G. J. Thomas, J. Phys. Chem. 80, 1157 (1976) and papers cited within.
57. G. J. Thomas, Appl. Spectrosc. 30, 483 (1976).
58. M. Shie, E. N. Dobrov, T. I. Tikchonenko, Biochem. Biophys. Res. Communic. 81, 907 (1978).
59. L. D. Esposito and J. L. Koenig, Fourier Transform IR Spectroscopy Vol. I, Acad. Press, N.Y., 1978.
60. T. G. Spiro and T. C. Strekas, J. Am. Chem. Soc. 96, 338 (1974) and papers cited within.
61. T. G. Spiro, Acc. Chem. Res. 7, 339 (1974).
62. T. G. Spiro, Vibrational Spectra and Structure Vol. V, Elsevier, ed. by J. R. Durig, Amsterdam and New York, 1976.
63. W. H. Woodruff and G. H. Atkinson, Anal. Chem. 48, 186 (1976).
64. P. R. Carey and H. Schneider, Acc. Chem. Res. 11, 122 (1978).
65. J. W. Ellis and J. Bath, J. Chem. Phys. 8, 632 (1940).
66. L. Stair and W. W. Coblentz, Bur. Stand. J. Res. 11, 703 (1943).
67. J. T. Edsall and E. B. Wilson, J. Chem. Phys. 8, 124 (1940).
68. G. D. Fasman, M. Idelson, and E. R. Blout, J. Amer. Chem. Soc. 83, 709 (1961).

CLASSICAL THEORY OF VIBRATIONS AND GROUP FREQUENCIES

CLASSICAL THEORY OF MOLECULAR VIBRATIONS

J. R. Durig and W. J. Natter

Department of Chemistry, University of South Carolina,
Columbia, S.C. 29208 U.S.A.

1. Introduction

Normal vibrations are characterized as having particle motions
which are all either in-phase or out-of-phase at the same instant
in time, in other words, for a given normal vibration each atom in
the molecule oscillates along a line at the same frequency. For
a diatomic molecule only one vibration occurs, only one vibrational
energy level pattern exists, and only one fundamental transition
occurs. The frequency of this transition is proportional to the
force constant and the reduced mass of the vibration. For a polyatomic molecule there are $3N-6$ normal modes of vibration (or $3N-5$
for a linear molecule) and the vibrational energy level diagram
can be very complicated, and overtones, combination and difference
bands may be observed. It is necessary then to see how the frequencies of these transitions can be used to determine the force
constants and provide an understanding of how the molecule
vibrates.

The treatment of polyatomic molecules will be simplified by
the use of the Lagrange equation. It will then be necessary to
deduce an expression for both the kinetic and potential energies
of the vibrating molecule. After a discussion of the normal modes
involved, the selection rules for infrared and Raman activity and
finally a systematic method developed by Wilson for performing
normal coordinate analysis will be outlined.

2. Normal Modes of Vibration

In order to generate these normal modes of vibration easily,

Theo M. Theophanides (ed.), Infrared and Raman Spectroscopy of Biological Molecules, 47–57.

we place the center of mass of the molecule under consideration at $(x, y, z) = (0, 0, 0)$ in the standard x, y, z coordinate system. Since it is important that each normal motion be a true vibration and not a translation, the following relations must hold:

$$\sum_{i=1}^{N} m_i \dot{x}_i = 0, \quad \sum_{i=1}^{N} m_i \dot{y}_i = 0, \quad \sum_{i=1}^{N} m_i \dot{z}_i = 0$$

for all N atoms with coordinates (x_i, y_i, z_i) where $i = 1, 2, \ldots, N$; and $\dot{x}, \dot{y},$ and $\dot{z}$ are the velocity components. These insure no movement of the center of gravity. Further restrictions are imposed to rule out any rotations. Here the relations

$$\sum_{j=1}^{N} m_j (x_j \dot{y}_j - y_j \dot{x}_j) = 0, \quad \sum_{j=1}^{N} m_j (z_j \dot{x}_j - x_j \dot{z}_j) = 0,$$

$$\sum_{j=1}^{N} m_j (y_j \dot{x}_j - x_j \dot{y}_j) = 0 \quad j = 1, 2, \ldots, N$$

require that no angular momentum is given to the molecule by the velocity components. Applying these restraints leaves 3N-6 (or 3N-5) independent displacement coordinates for the polyatomic molecule. It is easiest if all 3N coordinates are treated and then remove the translations and rotations at the appropriate point.

The kinetic energy, T, of the system can be expressed in terms of displacement coordinates for any atom k as $\Delta x_k = x_k - x_{ke}$, $\Delta y_k = y_k - y_{ke}$ and $\Delta z_k = z_k - z_{ke}$ where (x_{ke}, y_{ke}, z_{ke}) is the equilibrium position of atom k. In this case

$$T = \frac{1}{2} \sum_{k=1}^{N} m_k [\dot{x}_k^2 + \dot{y}_k^2 + \dot{z}_k^2]$$

or transformed to mass weighted cartesian coordinates $q_1, q_2, \ldots, q_{3N}$ for simplicity

$$T = \frac{1}{2} \sum_{i=1}^{3N} \dot{q}_i^2$$

where $q_1 = \sqrt{m_1} \Delta x_1$, $q_2 = \sqrt{m_1} \Delta y_1$, $q_3 = \sqrt{m_1} \Delta z_1$, $q_4 = \sqrt{m_2} \Delta x_2, \ldots,$
$q_{3n-2} = \sqrt{m_n} \Delta x_n$, $q_{3n-1} = \sqrt{m_n} \Delta y_n$, $q_{3n} = \sqrt{m_n} \Delta z_n$

The potential energy for small displacement, about the equilibrium nuclear positions is

$$U(q) = \frac{1}{2} \sum_{i=1}^{3N} \sum_{j=1}^{3N} \frac{\delta^2 U}{\delta q_i \delta q_j}\bigg|_e q_i q_j = \frac{1}{2} \sum_{i=1}^{3N} \sum_{j=1}^{3N} f_{ij} q_i q_j$$

where $f_{ij} = \dfrac{\delta^2 U}{\delta q_i \delta q_j}\bigg|_e$ and naturally $f_{ij} = f_{ji}$.

Now by substituting into Lagranges equation of motion in mass-weighted coordinates q_i

$$\frac{d}{dt} \frac{\delta T}{\delta \dot{q}_j} + \frac{\delta U}{\delta q_j} = 0 \qquad\qquad j = 1, 2, \ldots, 3N$$

we have

$$\ddot{q}_j + \sum_{i=1}^{3N} f_{ij} q_i = 0$$

There are 3N second order differential equations having solutions q_i. One form of these solutions, for a simple harmonic motion is

$$q_i = A_i \cos (\lambda^{\frac{1}{2}} t + \phi)$$

where A_i is the amplitude, λ is a constant, t is time, and ϕ is a phase constant. The first and second time derivatives of the q_i's are readily obtainable for the i = j solutions.

$$\dot{q}_j = A_j \lambda^{\frac{1}{2}} \sin (\lambda^{\frac{1}{2}} t + \phi) \text{ and}$$

$$\ddot{q}_j = -A_j \lambda \cos (\lambda^{\frac{1}{2}} t + \phi)$$

These can now be substituted into the Lagrange equation in mass-weighted coordinates to yield

$$-A_j \lambda + \sum_{i=1}^{3N} f_{ij} A_i = 0$$

which rearranges to

$$\sum_{i=1}^{3N} (f_{ij} - \delta_{ij} \lambda) A_i = 0 \qquad j = 1, 2, \ldots, 3N$$

where δ_{ij} is the familiar Kronecker delta;

$$\delta_{ij} = 0 \qquad \text{for } i \neq j$$

$$\delta_{ij} = 1 \qquad \text{for } i = j$$

so that the λ term is non-zero only for $i = j$. With the above form of the equation we can form the so-called secular determinant of the amplitudes A_i which must be equal to zero for non-trivial solutions.

$$
\begin{vmatrix}
f_{11} - \lambda & f_{12} & f_{1,3N} \\
f_{21} & f_{22} - \lambda & f_{2,3N} \\
\vdots & \vdots & \vdots \\
f_{3N,1} & f_{3N,2} \cdots & f_{3N,3\lambda}
\end{vmatrix} = 0
$$

This equation is 3N degree in λ and therefore has 3N roots, six of which are zero (five for a linear molecule) because of the initial restrictions placed on the system to rule out translations and rotations. The remaining 3N-6 roots are genuine vibrational degrees of freedom. The constant $\lambda = 4\Pi^2 \nu^2$ for each of the 3N-6 vibrations can be found once we find a unique set of A_{im}'s such that

$$A_{im} = K_m \ell_{im} \qquad i = 1, 2, \ldots, 3N-6$$

where K_m is a proportionality constant relating the amplitudes A_{im} and ℓ_{im}. The ℓ_{im}'s are normalized for convenience by requiring

$$\sum_{i=1}^{3N-6} \ell_{im}^2 = 1$$

Thus we can now obtain unique solutions of the form

$$q_i = K_m \, \ell_{im} \cos (\lambda_m^{\frac{1}{2}} t + \phi_m) \qquad i = 1, 2, \ldots, 3N-6$$

for the amplitude in each direction of each particle. The net amplitude of these is used to construct net vectors, proportional to the mass-weighted coordinates, describing the amplitude and direction of each particle displaced during the normal motion.

It is important to remember the physical significance of these conclusions. For a given normal vibration each atom oscillates in only one direction with a frequency of $\sqrt{\lambda_m} / 2\Pi$. Furthermore all atoms reach their minimum and maximum positions at the same time and hence only the amplitudes of this harmonic motion are different. The amplitudes of course are determined in part by the mass of the particle.

The normal modes of HCN illustrated in Fig. 1 below demonstrate the use of these vectors and their amplitudes. The actual displacements of the atoms are indicated by the vector length. The two identical roots among the 3N-5 for a linear molecule correspond to the degenerate bending motion.

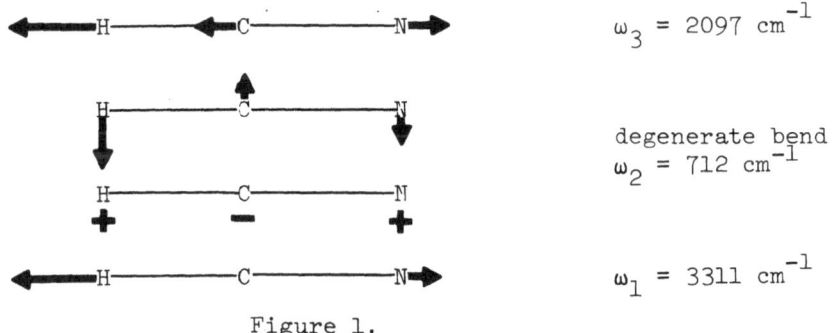

$$\omega_3 = 2097 \text{ cm}^{-1}$$

degenerate bend
$$\omega_2 = 712 \text{ cm}^{-1}$$

$$\omega_1 = 3311 \text{ cm}^{-1}$$

Figure 1.

3. Vibrational Energy of a Polyatomic Molecule

In order to solve the multidimensional Hamiltonian by the method of separation of variables it will first be necessary to set up kinetic and potential energy expressions which contain no cross terms $q_i \, q_j$. A more convenient set of mass-weighted coordinates can be obtained by the following transformation.

$$Q_k = \sum_{i=1}^{3N-6} \ell_{ik} \, q_i \qquad k = 1, 2, 3, \ldots, 3N-6$$

It will be more convenient to refer to these in the matrix form

$$\bar{Q} = \bar{\ell} \; \bar{q}$$

The Q_k's or column vectors of $\bar{Q}$ are the normal coordinates and the ℓ's are chosen in such a way that the kinetic and potential energies are proportional to the sum of the squared terms:

$$2T = \bar{\dot{Q}}' \; \bar{\dot{Q}} \qquad\qquad 2U = \bar{Q}' \; \bar{\lambda} \; \bar{Q}$$

where $\bar{Q}'$ and $\bar{\dot{Q}}'$ are the transpose matrices of $\bar{Q}$ and $\bar{\dot{Q}}$ respectively. This transposition is equivalent to substituting q_i for the inverse transformation

$$q_i = \sum_{k=1}^{3N-6} (\ell_{ik})^{-1} \; Q_k$$

$$\text{or } \bar{q} = \bar{\ell}^{-1} \; \bar{Q}.$$

It follows from this and our previous equation for kinetic and potential energies in q that

$$2T = \bar{\dot{q}}' \; \bar{\dot{q}} = (\bar{\ell}^{-1} \; \bar{\dot{Q}})' \; (\bar{\ell}^{-1} \; \bar{\dot{Q}}) = \bar{\dot{Q}}' \; (\bar{\ell}')' \; (\bar{\ell}^{-1}) \; \bar{\dot{Q}} = \bar{\dot{Q}}' \; \bar{\dot{Q}}$$

where $(\bar{\ell}')' \; \bar{\ell}^{-1} = \bar{\ell} \; \bar{\ell}^{-1} = E$ the identity matrix and

$$2U = \bar{q}' \; \bar{F}\bar{q} = (\bar{\ell}^{-1} \; \bar{Q})' \; \bar{F}(\bar{\ell}^{-1} \; \bar{Q}) = \bar{Q}' \; (\bar{\ell} \; \bar{F} \; \bar{\ell}^{-1}) \; \bar{Q} = \bar{Q}' \; \bar{\lambda} \; \bar{Q}$$

where $\bar{\ell} \; \bar{F} \; \bar{\ell}^{-1} = \bar{\lambda}$ and $\bar{F}$ is the 3N-6 by 3N-6 symmetric matrix of f_{ij}'s.

The Hamiltonian can now be written from the rows of Q

$$H = T + U = \tfrac{1}{2} \sum_{k=1}^{3N-6} \hat{\dot{Q}}_k^{\,2} + \tfrac{1}{2} \sum_{k=1}^{3N-6} \lambda_k \; \hat{Q}_k^{2}$$

where the operators $\hat{\dot{Q}}$ and $\hat{Q}$ are defined as $\hat{\dot{Q}}_k = \dfrac{h}{2\Pi i} \dfrac{d}{dQ}$ and $\hat{Q}_k = Q_k$. With these substitutions the Schrödinger equation becomes

$$\frac{-h^2}{8\Pi^2} \sum_{k=1}^{3N-6} \frac{\delta^2 \Psi_v}{\delta Q_k^2} + \frac{1}{2} \sum_{k=1}^{3N-6} \lambda_k Q_k^2 \Psi_v = E_v \Psi_v$$

where the Ψ_v are the product of all the Ψ_v's for non-interacting modes of vibration:

$$\Psi_v = \Psi_{v_1}(Q_1) \, \Psi_{v_2}(Q_2) \, \cdots \, \Psi_{v_{3N-6}}(Q_{3N-6}).$$

The Schrödinger equation is then separable into equations of the form:

$$\frac{-h^2}{8\Pi^2} \frac{\delta^2}{\delta Q_k} \Psi_{v_k}(Q_k) + \frac{1}{2} \lambda Q_k^2 \Psi_{v_k}(Q_k) = E_k \Psi_{v_k}.$$

Solutions to this equation are analogous to those of the familiar one-dimensional harmonic oscillator with their general form being

$$\Psi_{v_k}(Q_k) = H_{v_k}(\gamma_k^{\frac{1}{2}} Q_k) N_{v_k} \exp\left(-\frac{1}{2}\gamma_k Q_k^2\right)$$

where γ_k is a normalizing constant of the Hermite polynomial H_{v_k}. The total vibrational energy corresponding to this equation is the sum

$$E_v = (\bar{v}_1 + \tfrac{1}{2})\omega_1 + (\bar{v}_2 + \tfrac{1}{2})\omega_2 + \cdots + (\bar{v}_{3N-6} + \tfrac{1}{2})\omega_{3N-6}$$

of each fundamental frequency ω_k multiplied by $\frac{1}{2}$ plus its vibrational state quantum number $\bar{v}_k$.

For the H_2S molecule we have the following normal modes of vibration with their characteristic frequencies

$\omega_1 = 2615$ cm^{-1} $\omega_2 = 1183$ cm^{-1} $\omega_3 = 2626$ cm^{-1}

The zero point energy of the H_2S molecule is the total vibrational energy in the ground vibrational states and is equal to (i.e. $v_k = 0$ for all $k = 1, 2, 3, \ldots, 3N-6$)

$$E_v = \tfrac{1}{2}\,(2615) + \tfrac{1}{2}\,(1183) + \tfrac{1}{2}\,(2626) = 3212\ cm^{-1}$$

On the vibrational energy level diagram shown below this vibrational state has quantum numbers $(v_1, v_2, v_3) = (0, 0, 0)$.
Fundamental transitions occur between the zero energy level and
each of the levels $(1, 0, 0)$, $(0, 1, 0)$ and $(0, 0, 1)$. Overtones
of these fundamentals occur when more than one quanta of energy
is absorbed and the molecule is excited to the second vibrational
state of a given mode, for example, to the $(2, 0, 0)$ level. Combination and difference bands may also occur between energy levels
of different vibrations. These are shown for the combination of
the symmetric stretch v_1 and the bend v_2. A difference band
between these two fundamentals is also shown.

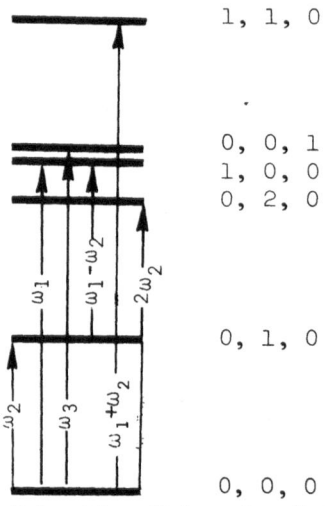

4. Selection Rules for Infrared Transitions

    To determine the selection rules which govern the electric-
dipole for rotation-vibration spectra of molecules, we must ascertain the restrictions that allow us to obtain nonzero values
of the various integrals and terms of the transition moment
matrix.

$$\int \Psi_m^{*}\,\hat{\mu}\,\Psi_k\ d\tau = \int \Psi_m^{*}\,\mu_x\,\Psi_k\ d\tau + \int \Psi_m^{*}\,\mu_y\,\Psi_k\ d\tau +$$

$$\int \Psi_m^{*}\,\mu_z\,\Psi_k\ d\tau$$

For a given electronic state, the volume elements $d\tau$ and the wavefunctions Ψ are the products of rotational and vibrational parts.

$$d\tau = d\tau_r \, d\tau_v \text{ and } \Psi = \Psi_r \, \Psi_v$$

We will also denote the initial and final states as double prime (") and single prime ('), respectively. It will also be convenient to denote the space-fixed axes whose origin is located at the center of mass as X, Y, Z. This coordinate system translates with the molecule. The fixed axes of the molecule are made up from the three rotational axes x, y, and z. A relationship between the two coordinates exists so the transition moment matrix can be express-ed in terms of the space fixed coordinates

$$\mu_z = \mu_x \, \phi_{Zx} + \mu_y \, \phi_{Zy} + \mu_z \, \phi_{Zz}$$

where the ϕ's are the appropriate direction cosine matrix elements showing the angle between each of the axes of the two coordinate systems. Similar dipole components along the X and Y axes can be generated. Substituting back into the Z moment integral

$$\int \Psi_{r'}^{*} \, \phi_{Zx} \, \Psi_{r''} \, d\tau_r \int \Psi_{v'}^{*} \, \mu_x \, \Psi_{v''} \, d\tau_v +$$

$$\int \Psi_{r'}^{*} \, \phi_{Zy} \, \Psi_{r''} \, d\tau_r \int \Psi_{v'}^{*} \, \mu_y \, \Psi_{v''} \, d\tau_v +$$

$$\int \Psi_{r'}^{*} \, \phi_{Zz} \, \Psi_{r''} \, d\tau_r \int \Psi_{v'}^{*} \, \mu_z \, \Psi_{v''} \, d\tau_v.$$

Substitutions for X and Y will yield similar integrals.

It is important to remember that all the nuclei and electrons are in motion so we must consider an average effective change. This can be approximated by a Taylor series expansion of the normal coordinates

$$\mu_z = \mu_z^{e} + \sum_{k=1}^{3N-6} \left(\frac{\delta\mu_z}{\delta Q_k}\right)_e Q_k + \tfrac{1}{2} \sum_{k=1}^{3N-6} \sum_{k=1}^{3N-6} \left(\frac{\delta^2\mu_z}{\delta Q_k \, \delta Q_j}\right) Q_k \, Q_j + \cdots$$

Where μ_z^{e} is the z component of the equilibrium dipole moment. We continue the derivation of Z_z and μ_z integrals for which we have similar components in x and y. Now, substituting this last series approximation into the integral for μ_z gives

$$\mu_z^{e} \int \Psi_{r'}^{*} \, \phi_{Zz} \, \Psi_{r''} \, d\tau_r \int \Psi_{v'}^{*} \, \Psi_{v''} \, d\tau_v +$$

$$\int \Psi_{r'}^{*} \phi_{Zz} \Psi_{r''} d\tau_r \sum_{k=1}^{3N-6} (\frac{\delta\mu_z}{\delta Q_k})_e \int \Psi_{v'}^{*} Q_k \Psi_{v''} d\tau_v$$

when we assume all further terms in the series to be negligible
to a first approximation. It was shown in the previous section
that the wavefunction for a polyatomic molecule is the product of
an orthonormal set

$$\int \Psi_{v'}^{*} \Psi_{v''} d\tau_v = \prod_{k=1}^{3N-6} \int \Psi_{v'_k}^{*} (Q_k) \Psi_{v''_k} (Q_k) dQ_k$$

where the integral to the right is equal to one for $v'_k = v''_k$ and
equal to zero for $v'_k \neq v''_k$ Thus the first term of the expanded
equation for moments about z reduces to terms of the form

$$\mu_z^e \int \Psi_{r'}^{*} \phi_{Zz} \Psi_{r''} d\tau_r$$

which can be used to derive the selection rules for pure rotational
motion. The rotational integral is nonzero for $\Delta J = \pm 1$ and
$\Delta K = 0$ where z is the unique axis or figure axis. For an oblate
top, M_z is M_c^e whereas for a prolate top M_z is M_a^e. Similar
selection rules are obtained for ϕ_{Xx} and ϕ_{Yy} along M_z.

    Substitution into the second term of the expanded equation
for the moments of z yields

$$\int \Psi_{r'}^{*} \phi_{Zz} \Psi_{r''} d\tau_r (\frac{\delta\mu_z}{\delta Q_k'})_e \int \Psi_{v'_k}^{*}(Q_k') Q_k' \Psi_{v''_k} (Q_k') dQ_k' \text{ x}$$

$$\prod_{k=1}^{3N-7} \int \Psi_{v'_k}^{*}(Q_k) \Psi_{v''_k} (Q_k) dQ_k$$

The first term gives rise to rotational selection rules for vibrations where

$$\frac{\delta\mu_z}{\delta Q_k'}$$

is nonzero. These are analogous to the pure rotational selection
rules derived above. The next term in the first part of this
equation Q_k' remains nonzero only for $\Delta v_k' = \pm 1$. The last term
which is a product of all the wavefunctions in Q_k' is nonzero only

when all Δv_k's are zero. Thus, for harmonic oscillators only fundamental transitions are allowed. The rotational transitions depend on the symmetry of the molecule. In linear and symmetric top molecules, vibrations which cause atom displacements parallel to the dipole moment vector (μ_z only) are denoted as parallel bands. Those vibrations with atom displacements resulting in

$$(\delta\mu_y/\delta Q_k)_e \neq 0 \text{ and } (\delta\mu_x/\delta Q_k)_e \neq 0$$

that is, with some motion not along the z axis, are referred to as perpendicular bands. The selection rules for rotation-vibration motions are summarized below:

Linear	parallel	$\Delta J = \pm 1$	$\Delta v = \pm 1$
Linear	perpendicular	$\Delta J = 0, \pm 1$	$\Delta v = \pm 1$
Spherical top	-	$\Delta J = 0, \pm 1$	$\Delta v \pm 1$
Symmetric top	parallel	$\Delta J = 0, \pm 1$ when $K \neq 0$	$\Delta K = 0$ $\Delta v = \pm 1$
		$\Delta J = \pm 1$ when $K = 0$	$\Delta K = 0$ $\Delta v = \pm 1$
Symmetric top	perpendicular	$\Delta J = 0, \pm 1$	$\Delta K = \pm 1$ $\Delta v = \pm 1$
Asymmetric top	-	$\Delta J = 0, \pm 1$	$\Delta v = \pm 1$

These, of course, are the rotational-vibrational selection rules for infrared transitions of molecules in the gaseous phase. In the liquid phase we would expect to see a single band lacking rotation-vibration fine structure at approximately the same frequency as in the gas phase spectrum. For a more detailed discussion of the classical theory of molecular vibrations one should consult one of the references [1-3] listed below.

REFERENCES

1. Guillory, W. A., Introduction to Molecular Structure and Spectroscopy, Allyn and Bacon, Inc., Boston, MA, 1977.
2. Herzberg, G., **Molecular** Spectra and Molecular Structure, Vol. II. Infrared and Raman Spectra of Polyatomic Molecules, Van Nostrand Co., Princeton, N.J., 1945.
3. Wilson, E. B., Jr., Decius, J. C., and Cross, P. C., Molecular Vibrations, McGraw-Hill Book Co., New York, N.Y., 1955.

CORRELATION OF GROUP FREQUENCIES WITH MOLECULAR CONFORMATION

J. R. Durig and W. J. Natter

Department of Chemistry, University of South Carolina,
Columbia, S.C. 29208 U.S.A.

1. INTRODUCTION

Since biological molecules are often of high molecular weight
with many unknown structural features, the study of smaller molecular weight organic molecules having known structures of variable
geometry can be very useful. Several examples of correlations of
this type are known and are usually directly applicable to the
larger systems in providing the desired relationship. Some of the
strategy and tactics in Raman spectroscopy of biological molecules
was presented in a recent article by Lord [1]. Group frequencies
provide the vast amount of Raman information used to correlate
the observed frequency with molecular conformation. Two examples
of potentially useful correlations of this type will be given in
the section which follows immediately. The first shows the dependence of the frequency of the carbonyl stretch on the C-C-C
angle at the carbonyl carbon. The second relates frequency of
the OH-vibration in strongly hydrogen bonded crystals with the
O-O distance determined by x-ray diffraction. Subsequent sections
will include correlations of other vibrational frequencies of
proteins with molecular geometries as illustrated by group frequencies of amides, sulfides, and phosphates. This will provide
a basis for the comtemporary work on the interpretation of Raman
spectra of biological molecules.

2. GROUP FREQUENCIES AND TRENDS IN SMALL MOLECULES

A carbonyl group situated between two methylene groups represents the simplest case of a ketone C=O stretch. The examination
of the carbonyl stretching frequencies of several aliphatic

Theo M. Theophanides (ed.), Infrared and Raman Spectroscopy of Biological Molecules, 59–67.
All Rights Reserved. Copyright © 1979 by D. Reidel Publishing Company, Dordrecht, Holland.

cyclic ketones as summarized by Bellamy [2] along with the internal ring angles at the carbonyl position determined by microwave
spectroscopy [3], allow the following trend to be established.
If one plots the observed C=O stretching frequency versus the
C-C-C angle at the carbonyl position for several alicyclic ketones,
one immediately sees a correlation between the interior angle and
the observed frequency. Such a plot is shown below in Fig. 1a.
It is not surprising that as the C-C-C angle becomes larger
the carbonyl stretch becomes of larger amplitude and the observed
frequency shifts downward. Even though the data plotted is for
ring compounds, the point for acetone falls on the curve very near
to that of cycloheptanone.

A second example gives the dependence of the OH-vibrational
frequency in hydrogen bonded crystals on the oxygen-oxygen distance in the crystal. A decrease in the difference between the
OH stretching frequency in the vapor and solid phases denoted as
$\Delta \nu = \nu_{vapor} - \nu_{solid}$ versus the O-O distance in the solid is
shown in Fig. 1b. This correlation was noted in 1953 by several
authors [4].

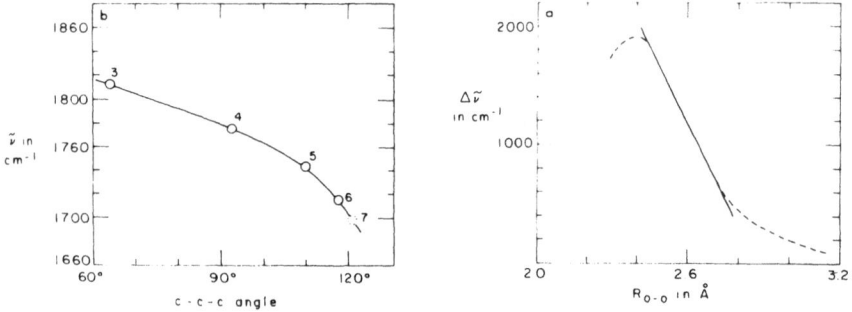

Fig. 1.   Correlation of the CCC angle and the O-O distance with
          frequency. Used by permission [1].

Clearly, if we can measure the carbonyl (C=O) or OH-stretching
frequency of the type in Fig. 1, in a complicated biological
molecule, it is a straightforward matter to approximate the
C-C-C angle or the O-O distance to a reasonable degree of accura-
cy. This type of correlation is of potential importance in all
types of measurements of vibrational frequencies which follow.

3. DEPENDENCE OF THE AMIDE MODES ON CONFORMATION

Proper characterization of the vibrational spectra of polypeptides and proteins requires an understanding of the amide group
frequencies. The motion of the peptide group has associated with
it 9 definitive frequencies which were initially described in a
normal coordinate treatment of the model peptide containing molecule N-methylacetamide. Among the most important to a conforma-

tional study are the amide I, II, and III with their respective
frequencies of 1653, 1567, and 1299 cm^{-1}. These have been the
objects of extensive studies over the past decade using both infrared and Raman instrumentation [5]. The amide I vibration is
found to be about 70% C=O stretching and about 16% C-N stretching.
The amide II and III bands, on the other hand, result from a coupling of the C-N stretching and the N-H in-plane bending motions.
The amide II motion shows most of its activity in the infrared
spectrum and is seldom seen in the nonresonance situation. However
if one uses excitation sources near the $\pi \rightarrow \pi^*$ transition of the
peptide chromophore (about 190 nm), enhancement of this and other
amide fundamentals occurs. This happens because the $\pi \rightarrow \pi^*$ transition causes the C-N distance to shorten [6]. The atomic displacements in each of these modes of vibration is shown in Fig. 2
below.

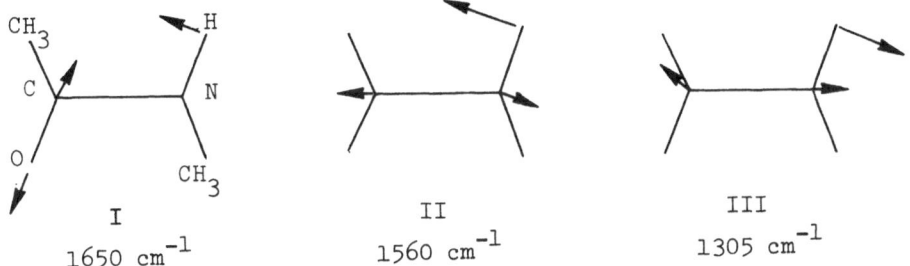

Fig. 2. Atomic displacements for the Amide I, II and III modes.

The most important factors in determining the amide I, II,
and III frequencies are the ϕ and Ψ dihedral angles in the peptide
chain, hydrogen bonding, and interaction effects. The effect of
the ϕ and Ψ angles on the vibrations of the amide I was recently
studied using the model compound N-acetyl-L-alanine-N-methylamide
[7]. The geometrics of this molecule varied from α-helical to
β-pleated sheet conformations and the corresponding amide I frequencies varied from 1653.5 to 1664 cm^{-1}. The line-width of measured
frequencies was as good as 2 cm^{-1} depending on the degree of order
of the structures. For the disordered structure the line-width
was greater because of a wider distribution of ϕ and Ψ angles.
The frequencies of the amide III region are more sensitive to
changes in ϕ and Ψ angles. In the α-helical region, a change of
10° in Ψ causes a shift by 11 cm^{-1} in this frequency, whereas in
the β-sheet region, a 10° change in ϕ or Ψ causes a shift of only
6 cm^{-1}. These results show the sensitivity of the amide bands to
changes in the peptide dihedral angles and underscores the importance of these vibrations as a probe for protein conformational
change.

Hydrogen bonding in peptides and proteins has the effect of

decreasing the amide I frequency but tends to increase the frequency of the amide II and III bands. The shifts of amide I for
phenylacetamidoacetate as it is changed from a hydrophobic chloroform solution to a solid is about 30 cm^{-1}. The amide II band
under the same conditions increases by about 40 cm^{-1} [8]. A similar shift in frequencies of the amide I, II, and III bands was
noted for N-methylacetamide. In going from the gaseous to liquid
phase the amide I frequency was observed to shift from 1715 to
1650 cm^{-1}, the amide II from 1494 to 1563 cm^{-1}, and the amide III
from 1263 to 1300 cm^{-1} [9]. Clearly the amide frequencies are a
good indicator of H-bonding effects.

The interaction of vibrations among the peptide units has
received considerable attention. In perturbation theory a localized amide frequency can be given by

$$\nu\,(\delta,\,\delta') = \nu_0 + \sum_i (D_i \cos i\delta + D_i' \cos \delta')$$

where

ν_0 is the unperturbed frequency

D_i are the interaction constants between the i-th neighbors
in the chain

D_i' are the interactions of adjacent peptide groups

$\delta,\,\delta'$ are the phase angles between the vibrations as shown
in Fig. 3 below.

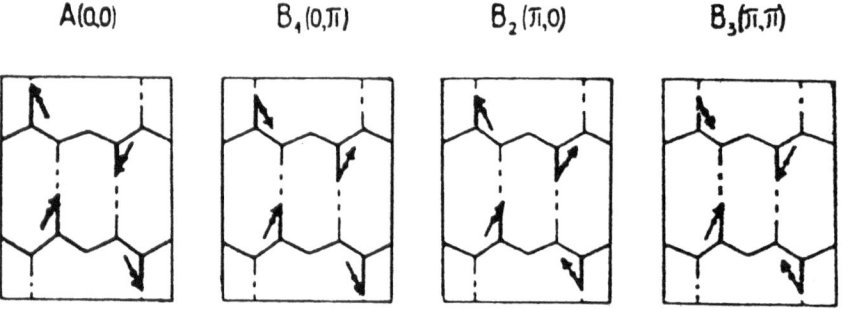

Fig. 3.   Interaction of the vibrations among peptine units.   Used
          by permission [16].

For the anti-parallel β-sheet structure shown in Fig. 3 only
the D_1 and D_1' are assumed to be important and consequently the
above equation reduces to

$$\nu(\delta, \delta') = \nu_0 + D_1 \cos \delta + D_1' \cos \delta'$$

The four modes shown are represented as follows for the different phase combinations

$$\nu(0, 0) = \nu_0 + D_1 + D_1' \qquad A_1 \text{ mode}$$

$$\nu(0, \pi) = \nu_0 + D_1 - D_1' \qquad B_1 \text{ mode}$$

$$\nu(\pi, 0) = \nu_0 - D_1 + D_1' \qquad B_2 \text{ mode}$$

$$\nu(\pi, \pi) = \nu_0 - D_1 - D_1' \qquad B_3 \text{ mode}$$

Only the totally symmetric A_1 mode is Raman polarized and would be expected to show measurable intensity in the non-resonance case. The others are infrared active.

Observation of the $\nu(0, 0)$ mode at 1674 cm^{-1} [5,10] in the Raman spectrum of polyglycine caused considerable controversy in subsequent applications of the theory. In order to provide a better understanding of the perturbation expression a new term, D_{11}, was added to include transition dipole coupling.

$$\nu(\delta, \delta') = \nu_0 + D_1 \cos \delta_1 + D_1' \cos \delta_1' + D_{11} \cos \delta \cos \delta'$$

Assuming $D_1 = 0$ the following expressions are obtained

$$\nu(0, 0) = 1674 = \nu_0 + D_1' + D_{11}$$

$$\nu(0, \pi) = 1685 = \nu_0 - D_1' - D_{11}$$

$$\nu(\pi, 0) = 1636 = \nu_0 + D_1' - D_{11}$$

$$\nu(\pi, \pi) = [1723] = \nu_0 - D_1' + D_{11} \text{ (predicted)}$$

Differences in the structure of β-polypeptides are reflected in the similarities and differences of the parameters ν_0, D_1, D_1', and D_{11} obtained from the observed amide frequencies.

For a polypeptide in an α-helical conformation the interchain D_1' can be omitted. D_3 terms are usually introduced in the expression to include interaction with third neighbors.

4. DEPENDENCE OF THE S-S AND S-C STRETCHING FREQUENCIES ON THE GEOMETRY OF DISULFIDE BONDS

In the Raman spectra of disulfides the S-S and C-S stretching motions occur in the region from 500 to 750 cm^{-1}. A relation-

ship between the ν (S-S) and the CS-SC dihedral angle was presented by van Wart et al. [11]. It was later shown by Nogami et
al. [12] that the relationship is actually among the three torsional angles about the C-S bonds in C-C-S-S-C-C. The frequencies of the two characteristic modes depend on whether the two
torsional angles about the S-S bond are both _gauche_, both _trans_,
or one _gauche_ and one _trans_, assuming that the dihedral of the
S-S bond is fixed at 90° (gauche). The S-S stretching frequencies
are 510 cm^{-1} for the GGG, 525 cm^{-1} for the GGT, and 540 cm^{-1} for
the TGT [5]. Conformations of this type are shown in Fig. 4.

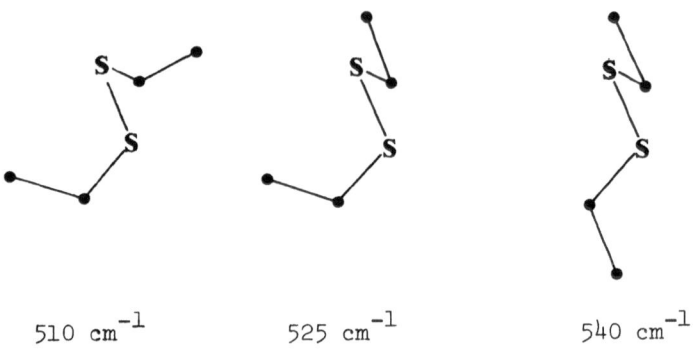

510 cm^{-1} 525 cm^{-1} 540 cm^{-1}

Fig. 4.   Torsional angles around the S-S bond and their respective
          frequencies.

The frequency of ν (C-S) has been shown to be dependent upon
the atom X substituted on the primary disulfide

$$X-\overset{|}{\underset{|}{C}}-CH_2-S-S$$

When the atom is H the C-S stretching vibration is between 630
and 670 cm^{-1}. When it is C the ν (C-S) is about 720 cm^{-1} and when
the atom is N the band is observed at about 700 cm^{-1}.

The correlation between the C-S frequency and the torsional
angles around the C-S and C-C bonds was made by Nogami et al.
[12]. The two C-S frequencies are expected at 760 and 719 cm^{-1}
for the _trans-trans_ form, at 746 and 697 cm^{-1} for the _trans-gauche_
form, at 667 cm^{-1} for the _gauche-trans_ form, and at 723 and 645
cm^{-1} for the _gauche-gauche_ form. The above considerations have
been applied to the Raman data of several protein molecules [5].

5. GROUP FREQUENCIES AND CORRELATION OF PHOSPHATES

One of the most active areas of biological Raman spectroscopy has been the examination of nucleic acids and viruses. Much of the information gained from Raman spectra of DNA and RNA comes from the regions of the phosphate stretches. It has been shown that the sugar phosphate vibrations occur in the region of 750 to 850 cm^{-1} and that these are sensitive to the conformation of the DNA fibers. Based on normal coordinate calculations of dimethyl phosphate, the symmetric -O-P-O- stretch should appear at 814 cm^{-1} Extension of the normal coordinate calculation to larger molecules like a ribose phosphate chain, indicates that this mode will be significantly coupled to the vibrations of the attached sugar. In fact, the frequency of the symmetric phosphate stretch is sensitive to conformations of the ribose-phosphate backbone. For A-type DNA, the phosphate groups are at the C_3' endo position of the ribose ring and the corresponding phosphate frequency occurs at 807 cm^{-1}. For B-type DNA, the phosphates are at the C_3' exo position and the frequency of this same mode is shifted to 787 cm^{-1}. For polyribo-adenylic acid, a synthetic form of RNA, one observes the phosphate stretch at 814 cm^{-1}. Hence, ribonucleic acids appear to be in the A form. Other weaker bands occur in the Raman spectra at about 830 cm^{-1} and are attributed to the antisymmetric -O-P-O- stretches [13].

The other symmetric phosphate motion, the dioxy stretch $O=P-O^-$, appears in the spectrum at about 1100 cm^{-1} and does not seem to be very conformation dependent. Since one of the advantages of Raman spectroscopy is that it can be used as an analytical tool, the 1100 cm^{-1} band serves as an internal standard. One thus calculates intensity ratios of I_{814}/I_{1100} to obtain an idea of the relative order or A-form character of a particular sample of nucleic acid under investigation [14].

6. THE EFFECT OF ENVIRONMENT ON THE TYROSINE DOUBLET

Raman spectroscopy has also been shown to be a valuable probe in determining the environment of hydrophobic groups such as tyrosine. The doublet at 830 to 850 cm^{-1} in the spectra of tyrosine containing polypeptides and proteins is indicative of the number of such residues "buried" in the interior of the molecule. In Fig. 5 below, Lord, Shimanouchi, and coworkers [15] have depicted the way in which hydrogen bonding affects the relative intensity of the two bands which make up the Fermi doublet. If the tyrosine acts as a strong hydrogen bond donor, the ratio of I_{830}/I_{850} is between 0.3 and 0.5. Hydrogen bonding in which the phenolic hydrogen acts as an acceptor results in a much higher ratio, sometimes greater than 1.0. A quantitative estimate of the number of "buried" tyrosines can be made on this basis. For carbonic anhydrase-B, the number of "buried" and "exposed" tyrosines was found to agree quite favorably with the results of the x-ray diffraction studies [15].

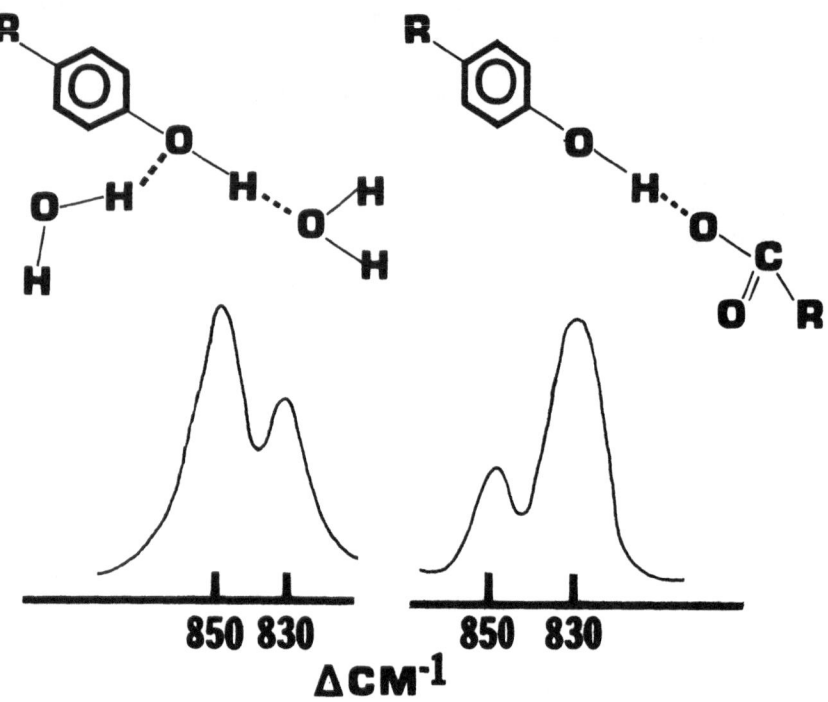

Fig. 5.   Effect of H-bonding and relative intensities of the bands in the tyrosine Fermi doublet. Used by permission [17].

REFERENCES

1.   R. C. Lord, Appl. Spectrosc. 31, 187 (1977).
2.   L. J. Bellamy, Infrared Spectra of Complex Molecules, 3rd ed. (John Wiley and Sons, New York, NY, 1975).
3.   D. Coffey, Jr. and T. E. Hooker, J. Mol. Spectrosc. 40, 158 (1971).
4.   R. C. Lord and R. E. Merrifield, J. Chem. Phys. 21, 166 (1953). (see also E. R. Lippincott and R. Schroeder, J. Chem. Phys. 23, 1099 (1955).
5.   N. T. Yu, Critical Reviews in Biochemistry January, (1977).
6.   I. Harada, Y. Sygawara, H. Matsurra, and T. Shimanouchi, J. Raman Spectrosc. 4, 91 (1975).
7.   S. L. Hsu, W. H. Moore, and S. Krimm, Biopoly. 15, 1513 (1976).
8.   R. E. Richards and H. W. Thompson, J. Chem. Soc. 1248, (1947).

9. See references cited in reference 5 above.
10. E. W. Small, B. Fanconi, and W. L. Peticolas, _J. Chem. Phys._ 52, 4369 (1970).
11. H. E. van Wart, A. Lewis, H. A. Scheraga and F. D. Saeva, _Proc. Nat. Acad. Sci._, USA 70, 2619 (1973).
12. N. Nogami, H. Sugeta, and T. Miyazawa, _Bull. Chem. Soc. Jpn._ 48, 2417 (1975).
13. E. B. Brown and W. L. Peticolas, _Biopolymers_ 14, 1259 (1975).
14. G. J. Thomas, Jr., _Vibrational Spectra and Structure_, Vol. 3, J. R. Durig, ed., Marcel Dekkar, New York (1975).
15. M. E. Siamwiza, R. C. Lord, M. C. Chen, T. Takamatsu, I. Harada, and T. Shimanouchi, _Biochem._ 14, 4870 (1975).
16. Y. N. Chirqadze and N. A. Nevskaya, _Biopolymers_ 15, 607 (1976).
17. B. P. Gaber, _Amer. Lab._ 9, 15 (1977).

VIBRATIONAL SPECTRA OF SOLIDS

J. R. Durig

Department of Chemistry, University of South Carolina,
Columbia, S.C. 29208 U.S.A.

1. Introduction

The Bravais space cell is used by molecular spectroscopists
to obtain the irreducible representation for the lattice vibrations. The crystallographic unit cell may be identical with the
Bravais cell or it may be larger by a multiple of two, three or
four. This information can be obtained from the capital letter in
the x-ray symbol which is used to designate the crystal symmetry.
For all crystal structures designated by a symbol P (primitive),
the crystallographic unit cell and the Bravais unit cell are
identical. Crystal structures designated with capital letters A,
B, C, or I are doubly primitive and thus the crystallographic
unit cells contain two Bravais cells. Crystal structures designated with capital letters R or F are triply and quadrupoly
primitive, respectively, and the crystallographic unit cells contain three and four Bravais cells, respectively. Thus, to obtain
the desired Bravais space cell from the crystallographic unit cell,
one simply divides the number of molecules in the crystallographic
unit cell by the number of lattice points. In summary, the number
of molecules in the Bravais space cell Z^B is the number of molecules in the crystallographic unit cell (Z) divided by the number
of lattice points (LP): $Z^B = Z/LP$ and this reduction is summarized.

Crystals can be classified into six systems as indicated in
Table II. In the earlier literature seven crystal systems were
recognized and some authors still use seven, but it is now general
practice to place the rhombohedral crystals into the hexagonal
system. The crystal systems are commonly defined by reference to

69

Theo M. Theophanides (ed.), Infrared and Raman Spectroscopy of Biological Molecules, 69–80.
All Rights Reserved. Copyright © 1979 by D. Reidel Publishing Company, Dordrecht, Holland.

Table I. The number of lattice points which reduces the
crystallographic unit cell to the Bravais space cell

Type of crystal structure	Number of lattice points
P	1
A	2
B	2
C	2
I	2
R	3[a]
F	4

[a]For this crystal structure, the crystallographic group may have
already decreased by three, so care must be used when working
with rhombohedral crystals.

the shapes of their unit cells but the minimum symmetry required
by the system is more consistent. There are only ten possible
point-symmetry operations in crystallography: 1, 2, 3, 4, 6, $\bar{1}$,
$\bar{2}$, $\bar{3}$, $\bar{4}$ and $\bar{6}$. The bar indicates an inversion axis. In the
Hermann-Mauguin symbols the center of symmetry is represented by
its equivalent, $\bar{1}$, but a plane is usually symbolized by m (mirror)
though $\bar{2}$ is just as logical. There are 32 ways of combining ten
operations which gives the 32 crystallographic point-groups. An
infinite planar array of points, all of them identical in their
environment, is known as a plane net. In three dimensions the
problem is to combine an infinite number of plane nets together,
while maintaining the condition that no point can be distinguished
from any other in its environment. Obviously the nets must be
parallel to one another, and they must be equidistantly spaced.
In 1848 Bravais proved that there are only 14 possibilities and
these give the 14 Bravais space lattices. However, in repeat
patterns, there are various translational symmetry operations,
besides the ten types of point-symmetry operations, and these
additional operations increase the number of possibilities from
the 32 point groups to the 230 space groups.

     Apart from the operation of simple translation, which gives
rise to the lattices, the translational operations in three
dimensions are glide-planes and screw-axes. A glide-plane involves
the combined operation of reflection across a plane followed by
translation parallel to the plane and through a distance equal to
a simple fraction of a primitive translation, nearly always

Table II.   The Crystal Systems

System	Lattice Types	Minimum Symmetry	Number of Point Groups	Number of Space Groups
Triclinic	P	1	(2) C_1, C_i	2
Monoclinic	P, C (or A)	one 2 (or $\bar{2}$)	(3) C_s, C_2, C_{2h}	13
Orthorhombic	P, C (or A or B) I, F	three 2 (or $\bar{2}$)	(3) C_{2v}, D_2, D_{2h}	59
Tetragonal	P, I	one 4 (or $\bar{4}$)	(7) C_4, S_4, C_{4v}, C_{4h}, D_{2d}, D_4, D_{4h}	68
Hexagonal or Rhombohedral	P, R	one 3 (or $\bar{3}$) or one 6 (or $\bar{6}$)	(12) C_3, C_{3i}, C_{3v}, C_{3h}, D_3, D_{3d}, D_{3h}, C_6, C_{6v}, C_{6h}, D_6, D_{6h}	52
Cubic	P, I, F	four 3	(5) T, T_h, T_d, O, O_h	36
Totals	14		32	230

one-half.   The glide-plane is symbolized by a, b, or c accordingly as the fractional translation is parallel to these axes or by n (certain circumstances by d) when the fractional translation is along a diagonal of the unit cell.   A screw-axis involves the combined operation of rotation about an axis of order 2, 3, 4, or 6 and translation in the direction of an axis through fractions which must be multiples of 1/2, 1/3, 1/4, or 1/6 of the primitive translation.   Thus when the translational symmetry operations are added to the ten crystallographic point-symmetry operations, there

are 230 self consistent space groups. These are summarized in
Table II. For a more in-depth discussion of the crystal systems
one should consult reference 1.

Space groups were originally symbolized by the Schoenflies
symbol for the corresponding point group with a numerical superscript - e.g. C_s^4 for one of the monoclinic space groups. Crystal-
lographers, following a recommendation of their International
Union, now use symbols based on the Hermann-Mauguin system, though
the Schoenflies symbol is sometimes added as well. The molecular
spectroscopist continues to use the Schoenflies notation.

The potential energy of a crystal can be considered to be made
up of the following terms:

$$V_{Total} = \sum_j V_j + \sum_i \sum_j V_{ij} + V_L + V_{Lj}$$

where V_j is the potential due to the internal coordinates, V_{ij} is
the potential due to the correlation field, V_L is the potential
for the external degrees of freedom and V_{Lj} represents the potential
due to the interaction of the internal modes with the lattice
modes. There are three symmetries which must be considered when
studying the vibrational spectrum of a crystal sample and these are
the molecular symmetry, the site symmetry, and the factor group
symmetry. It should be pointed out that the factor group is not
a point group but it is isomorphous with the space group which
means there is a one to one correspondence between the two. It
will now be illustrated with an example how these various symmetries
must be considered.

The various site symmetries for the triclinic and monoclinic
crystal systems are listed in Table III. These listings are to be
used with the correlation method. The site symmetries are arranged
in alphabetical order reading from left to right. The number in
the parentheses after the site symmetry is the occupation number
and it gives the number of equivalent atoms which sit on the site.
The number preceding the site symmetry indicates the number of
nonequivalent distinct sets. In some cases there are an infinite
number of distinct sets and these are listed with the infinity
symbol ∞. For a complete set of site symmetry tables for all
Bravais space cells the reader should consult the article by
Fateley et al[2,3,4].

2. Ordered Crystals

At this point, it should be instructive to consider some

examples. Crystallographic studies[5,6] have established the structures of oxamide ($CONH_2$) and dithiooxamide ($CSNH_2$) as having molecular symmetry <u>trans</u> C_{2h} with space group of $P\overline{1}$. Oxamide has one
molecule per unit cell, whereas the sulfur analog contains two.
The molecules occupy C_i sites. In Table IV, the group theory for
these molecules having these molecular, site, and factor group
symmetries is given.

For the oxamide molecule the twenty-four internal fundamentals
span the representations, $9A_g + 4A_u + 3B_g + 8B_u$, with the A_g and
B_g being Raman active and the A_u and B_u being infrared active.
The site symmetry does not change the number or activity: $12 A_g$
$+ 12 A_u$. However, for the dithiooxamide molecule correlation field
splitting is possible since there are two molecules per primite
cell. The only indication of such splitting was observed in ν_{12},
the SCN out-of-plane bend, which exhibited[7] a splitting of 13 cm^{-1}.

Oxamide should exhibit three Raman active librations whereas
six Raman librations and three infrared translations are predicted
for dithiooxamide. For oxamide, the three lattice bands are very
prominent features in the Raman spectrum, occuring at 106, 134,
and 157 cm^{-1} with corresponding bands at 100, 137 and 156 cm^{-1} in
the d_4 compound. In agreement with the group theory no lattice
bands were observed in the far infrared spectrum. It was not
possible to associate these librational frequencies with a particular librational axis because the predicted shift factors for all
three moments were very small and nearly the same value (1.0452,
1.0568 and 1.0510 for I_A, I_B and I_C, respectively). An estimated
range for the force constant of the librations can be calculated
using an equation of the form $F_L = \lambda I/4r^2$, where an arbitrary r
of 1Å is used for consistent units. By associating the highest
librational frequency with the largest moment of inertia and the
lowest frequency with the smallest moment, the widest possible
range for F_L is found to be 0.14 - 0.79 mdyn/A. These values are
relatively large for librational force constants and probably
reflect the high degree of hydrogen bonding in this compound.

For the dithiooxamide molecule, the three infrared optical
translations were observed at 79, 92 and 99 cm^{-1}. The six Raman
active librations apparently occur as doublets with three strong
lines observed at 74, 98 and 112 cm^{-1} and two weaker shoulders at
67 and 93 cm^{-1}. Presumably the remaining librational mode is contained in the unsplit line of highest frequency. Librational force
constants calculated for this molecule fall in the range 0.089 -
0.75 mdyne/A. The fact that the observed frequencies for dithiooxamide are shifted from those of oxamide by about 1.4 suggests
that the forces are about the same in the two crystals and the
difference in the number of molecules per unit cell has little
effect.

Table III. Site symmetries for the
triclinic and monoclinic Bravais space cells

Space Group Number[a]	Hermann-Mauguin Symbol	Schoenflies Symbol	Site Symmetries
1	P1	C_1^1	$\infty C_1(1)$
2	P$\bar{1}$	C_i^1	$8C_i(1)$; $\infty C_1(2)$
3	P2	C_2^1	$\infty 4C_2(1)$; $\infty C_1(2)$
4	P2$_1$	C_2^2	$\infty C_1(2)$
5	B$_2$ or C$_2$	C_2^3	$\infty 2C_2(1)$; $\infty C_1(2)$
6	Pm	C_s^1	$\infty 2C_s(1)$; $C_1(2)$
7	Pb or Pc	C_s^2	$\infty C_1(2)$
8	Bm or Cm	C_s^3	$\infty C_1(1)$; $\infty C_1(2)$
9	Bb or Cc	C_s^4	$\infty C_1(2)$
10	P2/m	C_{2h}^1	$8C_{2h}(1)$; $\infty 4C_2(2)$; $\infty 2C_s(2)$; $\infty C_1(4)$
11	P2$_1$/m	C_{2h}^2	$4C_i(2)$; $\infty C_s(2)$; $\infty C_1(4)$
12	B2/m or C2/m	C_{2h}^3	$4C_{2h}(1)$; $2C_i(2)$; $\infty 2C_2(2)$; $\infty C_s(2)$; $\infty C_1(4)$
13	P2/b or P2/c	C_{2h}^4	$4C_i(2)$; $\infty 2C_2(2)$; $\infty C_1(4)$
14	P2$_1$/b or P2$_1$/c	C_{2h}^5	$4C_i(2)$; $C_1(4)$
15	B2/b or C2/c	C_{2h}^6	$4C_i(2)$; $\infty C_2(2)$; $C_i(4)$

[a] Space groups one and two are triclinic and the others are monoclinic.

Table IV. Correlation of the external and
internal modes from the factor group analysis

Molecular symmetry C_{2h}	Site symmetry C_i	Crystal symmetry $P\bar{1} - C_i^1$
$(R_z)\ A_g$		
	A_g	$A_g\ (R_x,\ R_y,\ R_z)$
$(R_x)(R_y)B_g$		
$T_z\ A_u$		
	A_u	$A_u\ (T_x,\ T_y,\ T_z)$
$(T_x)(T_y)B_u$		

Oxamide (1 molecule per unit cell)

    Γ (acoustical translations) = $3A_u$ (inactive)

    Γ (optical translations) = 0

    Γ (optical librations) = $3A_g$ (Raman active)

Dithiooxamide (2 molecules per primitive cell)

    Γ (acoustical translations) = $3A_u$ (inactive)

    Γ (optical translations) = $3A_u$ (infrared active)

    Γ (optical librations) = $6A_g$ (Raman active)

In carrying out the group theory predictions for the lattice
modes of solids, it is important to keep the accounting straight.
For example the number of translational modes will be three times
the number of molecules (or ions) in the primitive cell and these
will always include the three acoustical modes. The number of
librational modes will also be three times the number of molecules
(or ions except when the ion is an atom, i.e. Cl^-) in the primitive
cell. The correlation method can be used for ionic crystals as
well as molecular crystals. For ordered crystal systems the site
group must be a subgroup of both the molecular and factor group.
However for disordered crystals, the site symmetry may be "higher"
than the molecular symmetry. Since overtone or combination bands
(two phonon bands) of lattice modes need not obey the k = 0 selection rule, their wavenumbers may be somewhat different than the sum
of the bands from which they arise.

Table V. Factor group analysis of phase II
$(CH_3)_3CCl$

Molecular symmetry	Site symmetry	Crystal symmetry
C_{3v}		$D_{4h}^7 - P4/nmm \ (Z = 2)$

	$(T_z) \ A_1$	A_{1g}
		A_{2u}
	$(R_z) \ A_2$	A_{2g}
		A_{1u}
	$(T_x \ T_y \ R_x \ R_y) \ E$	E_g
		E_u

Γ (acoustical translations) $= A_{2u} + E_u$ (inactive)

Γ (optical translations) $= A_{1g} + E_g$ (both Raman active)

Γ (optical librations) $= A_{2g} + A_{1u} + E_g + E_u$ (E_u infrared active
   and E_g Raman active whereas others are inactive)

3. Disordered Crystals

Frequently for molecular crystals where the molecules are
fairly spherical there will be a high temperature crystal phase
where the site symmetry is higher than the molecular symmetry. As
an example let us consider the tertiary-butyl chloride molecule.
Rudman and Post[8] have reported the phase II (-90 to -53.6°C)
crystal structure for the molecule to be tetragonal with P4/nmm
(D_{4h}^7) symmetry and two molecules per primitive cell. Since the
molecules occupy sites of C_{4v} symmetry the site group is not a
subgroup of the molecular point group. In such cases it is thought
that dynamic disorder must result in a time-average infinite-fold
rotation axis along the C-Cl bond. It is still possible to predict the activity of the lattice modes of this disordered crystal
and this is illustrated in Table V.

Bertie and Whalley[9] treated the problem of lattice vibrations of orientationally disordered solids and their results indicated that such disorder should lead to a broadening of the vibrational bands and all modes may be active in both the infrared and Raman spectra. Spectra of this type are referred to as density-of-states spectra since the band peaks correspond to the flat points in the dispersion curve. The vibrational spectrum[10] of this molecule in this high-temperature phase was consistent with their predictions. Raman bands were observed at 60 and 46 cm^{-1} in the light molecule and at 57.4 and 45 cm^{-1} in the heavy compound for phase III of $(CH_3)_3CCl$. The shift factors indicate that these are the optical translations predicted in Table V. These bands shift to lower frequency and broaden at higher temperatures. It is not possible to assign the symmetry species of the Raman-active lattice bands on the basis of the available data.

4. Line-Group Analysis

For molecules with chainlike or polymeric structures the line-group analysis of Tobin[11] is appropriate for predicting the symmetry and number of the fundamental modes. The line group is essentially a one-dimensional space group which will contain, in addition to the familiar point covering operations, translations and possibly glide planes and screw axes. Like factor groups of the space groups, the factor groups of line groups are isomorphous with one of the point groups. Because of the existence of inter-molecular forces between polymer chains and the possibility of more than one chain in the unit cell, one expects to observe in the spectra of polymers, as in the case of organic crystals, factor group splittings, correlation field splittings and external lattice modes. All of these factors have actually been observed for a few simple polymers. Because of the flexibility of the polymer chain the distinction between internal and external modes in the low frequency spectral region is sometimes very difficult because of strong coupling between these motions. Although it is frequently possible to distinguish between internal and external modes of the same symmetry species for rigid organic molecules, it is generally impossible to do so for many polymers since the internal modes may fall at equal or lower frequencies than those for the external modes. The splitting of the intramolecular (line-group) modes because of intermolecular (space-group) interactions are generally very small, and are not observed for all of the modes even when symmetry allowed. The motions of the groups of atoms which perform comparatively large amplitude motions and which are located at the exterior of the polymer chain are generally candidates for exhibiting factor group splittings. Interchain distances are also an important factor and the larger the chain-chain distance, the smaller should be the factor group splitting.

As an example of the utility of line-group analysis, let us consider the methanol molecule. The crystal structure of methanol may be considered as a zig-zag chain of hydrogen-bonded molecules similar in structure to polyethylene[12]. The methanol molecules occupy positions in the polyethylene-like chain with the oxygens in the positions of the methylene groups and a hydrogen bond connecting each oxygen atom. However, the hydrogen bonds are not nearly as strong as the covalent carbon-carbon bonds in polyethylene. If the methyl groups are considered as point masses, 18 normal modes are predicted (3N) for the two-molecule repeat unit of the chain. Of these 18, 3 are acoustical translations corresponding to movement of the chain as a whole in the three directions. One mode corresponds to rotation about the chain axis which is also an acoustic mode if interaction between the chains is neglected. Of the remaining 14 modes, 6 arise from the in-phase and out-of-phase intramolecular normal modes of the isolated methanol molecule. Five modes arising from rotations and three modes arising from translations of the two molecules of the repeat unit account for the remaining eight internal modes of the chain which are essentially hydrogen-bonded modes. These motions are the in-phase and out-of-phase motions of the O-H torsional mode, ν_t; the O-H---O stretching mode, ν_σ; and the O-H----O in-plane and out-of-plane bending modes, ν_β and ν_α, respectively. The symmetry of the line-group factor group depends on the positions of the methyl groups and the hydroxyl hydrogens. If the methyl groups were in the plane, the line group would be D_{2h} or C_{2v} depending upon whether or not the hydroxyl hydrogens are midway between the two oxygens of the hydrogen bond. Both of these symmetries have been ruled out on the bases of both the infrared[13] and x-ray data[14]. With the methyl groups out of the plane, there are again two line-group symmetries, C_2 and C_{2h}, which must be considered depending on whether the hydrogen bond is symmetrical. In Table VI are listed the predicted and observed activity for these two line-group symmetries. For the low-temperature α phase there are four far infrared bands which cannot be assigned to multiphonon processes. Since only two bands are predicted by the C_{2h} line-group symmetry, the effective symmetry of the chain must be C_2. This means that the hydroxyl hydrogens are "significantly" displaced from the symmetric positions between the oxygen atoms of the chain.

The assignments[12,13] given for the eight hydrogen-bonded modes are as follows. The two O-H torsional modes, ν_t, were assigned at 790 and 685. The O-H$\cdots$O stretch, ν_σ, was assigned to a band at 350 cm^{-1} on the bases of its intensity and isotopic shift factor. This band did not appear to be split. The C-O$\cdots$H in-plane bending modes, ν_β, were assigned to bands at 196 and 176 cm^{-1} and the out-of-plane motions, ν_α, to bands at 109 and 57 cm^{-1}. Only one of these bands, 196 cm^{-1}, was observed in the Raman effect whereas

INFRARED AND RAMAN SPECTRA OF METHANOL

Table VI. Predicted fundamentals for solid methanol
of C_{2h} and C_2 line-group symmetries

| Predicted | | Observed | | Approximate |
C_{2h}	C_2	β phase	α phase	description
A_g, B_u	A, B	1 ir, 1 R	2 ir, 2 R	C-O stretch
A_u, B_u	A, B	2 ir, No R	2 ir, No R	O-H in-plane bend
A_u, B_u	A, B	2 ir, No R	2 ir, No R	O-H stretch
A_u, B_u	A, B	2 ir, No R	2 ir, No R	O-H out-of-plane bend
A_g, B_g	A, B	1 ir, No R	1 ir, No R	O-H···O stretch
A_u, B_g	A, B	1 ir, 1 R	1 ir, 1 R	C-O···H in-plane bend
A_g, B_u	A, B	2 ir, No R	2 ir, No R	C-O···H out-of-plane bend

the other bands were observed in the infrared spectrum.

Line-group analysis has been applied to several polymers. For example polyethylene which has a three atom repeat unit, $(CH_2)_n$, can be treated as a perfect one-dimensional crystal (single chain polymer) in which case there will be two acoustical branches and seven optical branches. The normal coordinates of an infinite polyethylene chain with planar zig-zag conformation may be characterized by the phase difference (δ) between adjacent methylene groups. The dynamical matrices (G and F) of infinite order may be factored into a set of matrices of finite order of 5 for the in-plane modes and 4 for the out-of-plane modes. Calculations for various δ values between 0 and π have been carried out and the assignments for the nine branches (dispersion curves) given[15]. The in-plane modes can be characterized as the CH_2 symmetric stretch, v_1, the CH_2 scissors, v_2, the CH_2 wagging, v_3, the CC stretch - CCC bend, v_4, and the CCC bend - CC stretch, v_5. The out-of-plane motions can be characterized as the CH_2 antisymmetric stretch, v_6, the CH_2 rock - CH_2 twist, v_7, the CH_2 twist - CH_2 rock, v_8, and the torsion, v_9. The modes v_5 and v_9 are the acoustical modes. The observed frequencies of n-alkanes as well as those of polyethylene are broadly in agreement with the calculated curves[15]. It has been found that v_5, the longitudinal

acoustical mode (accordion vibration after the similarity of its
motion with that of an accordion), gives rise to a relatively
strong Raman band in n-alkanes and an experimental formula was
derived by Schaufele and Shimanouchi[16] for correlating the chain
length with the frequency of this fundamental as well as with its
overtone frequencies. A more detailed description of the vibrations of chain molecules will follow in the next chapter.

REFERENCES

1. J. C. D. Brand and J. C. Speakman, Molecular Structure: The
    Physical Approach, Edward Arnold, London, (1960).
2. W. G. Fateley, N. T. McDevitt and F. F. Bentley, Appl.
    Spectrosc. 25, 155 (1971).
3. W. G. Fateley, Appl. Spectrosc. 27, 395 (1973).
4. W. G. Fateley, F. R. Dollish, N. T. McDevitt and F. F. Bentley,
    Infrared and Raman Selection Rules for Molecular and Lattice
    Vibrations: The Correlation Method, Wiley-Interscience, New
    York, (1972).
5. E. M. Ayerst and J. R. C. Duke, Acta Cryst. 7, 588 (1954).
6. P. J. Wheatley, J. Chem. Soc. 396, (1965).
7. J. R. Durig, S. C. Brown and S. E. Hannum, Mol. Cryst. Liq.
    Cryst. 14, 129 (1971).
8. R. Rudman and B. Post, Mol. Cryst. 5, 95 (1968).
9. J. E. Bertie and E. Whalley, J. Chem. Phys. 46, 1264 (1967).
10. J. R. Durig, S. M. Craven and J. Bragin, J. Chem. Phys. 51,
    5663 (1969).
11. M. C. Tobin, J. Chem. Phys. 23, 891 (1955).
12. J. R. Durig, C. B. Pate, Y. S. Li and D. J. Antion, J. Chem.
    Phys. 54, 4863 (1971).
13. M. Falk and E. Whalley, J. Chem. Phys. 34, 1554 (1961).
14. K. J. Tauer and W. N. Lipscomb, Acta Cryst. 5, 606 (1952).
15. M. Tasumi, T. Shimanouchi and T. Miyazawa, J. Mol. Spectrosc.
    9, 261 (1962).
16. R. F. Schaufele and T. Shimanouchi, J. Chem. Phys. 47, 3605
    (1967).

GROUP FREQUENCIES AND THE CHEMICAL BOND

R.NORMAN JONES

Division of Chemistry, National Research Council
of Canada

Ottawa,Ontario, Canada, K1A OR6

There is a profound dichotomy in the interpretation of infrared and Raman spectra of complex molecules.This has existed for the past thirty years and is only now being resolved in a manner consistant with the needs of both theoretical spectroscopy and analytical chemistry (1).

1. VIBRATIONAL ANALYSIS

In principle the method of normal co-ordinate analysis can provide a very precise description of the dynamic structure of a molecule,including the internal motions of the atoms. It is based on the definition of a set of forces that prescribe the resistance of each bond to stretch and torsion and of all the interbond angles to deformation. Second and higher order interactions can be incorporated but in an elementary treatment only harmonic inter-atomic forces are considered. In its more refined form this approach to molecular vibrational spectroscopy requires that both the infrared and the Raman spectra be known completely,including the polarization of the Raman bands. It is also helpful to know the contours of the infrared bands in the vapor state spectrum and the principal moments of inertia of the molecule which are obtainable from the microwave spectrum. (1 - 3).

Even if this information is all available,formal treatment by normal co-ordinate analysis tells us only about the structure of the individual molecule in free space or about isolated units of structure, such as an internal element of an infinite polymer chain. In real physical systems,especially

Theo M. Theophanides (ed.), Infrared and Raman Spectroscopy of Biological Molecules, 81–93.

biological systems,we are usually dealing with an interactive
group of molecules bound together by hydrogen bonds or other
weak attractions   such as   π bonds. As the molecule or molecular system increases in size,especially if its symmetry is low,
a state is soon reached where the total number of measured input
parameters (force constants,bond lengths,bond angles,atomic
masses, and (particularly) vibrational frequencies) are insuff-
icient to provide a unique mathematical **solution of the**
equations determining the normal vibrations.To cope with this
various approximations can be made or other simplifying
assumptions introduced into the normal co-ordinate treatment.
A powerful aid is to enlarge the system by including isotopically substituted species,notably the selective replacement of
hydrogen by deuterium atoms.Provided one accepts the assumption
that the substitution of deuterium for hydrogen changes only the
atomic mass and not the bond properties this greatly extends
the facility of normal co-ordinate analysis. This however is
only a palliative technique; as the molecular size increases,
more and more assumptions must be made about the force field
and the uniqueness of the solution of the resulting equations
becomes **increasingly suspect.**

        In the development of molecular spectroscopy this
premium on high molecular symmetry and high vapor pressure (to
permit the inclusion of vapor phase infrared and microwave data)
led molecular spectroscopists to concentrate their attention
on a limited range of simple symmetric molecules of low molecular weight. The information gained from such studies is then
only of indirect help in interpreting the vibrational spectra
of the more complex molecules that are basic to most biological
systems.

2. CHARACTERISTIC GROUP FREQUENCES

        To obtain a balanced perspective on our **subject** we must
consider the alternative method of analysing infrared and Raman
spectra.In its simplest form this consists of a purely empirical
comparative study of the bands.The spectra of a wide range of
compounds of known molecular structure are measured and compared
with one another in order to identify bands which,from their
frequency and intensity,can be correlated with the presence of
specific groups of atoms in the molecule. The classic example
of this approach is the pioneer work of W.W.Coblentz who,**in 1905**
published the results of such an analysis **of the infrared**
spectra of 120 compounds ánd first recognized many of these
"characteristic group **frequencies"** (4).

        There was great activity during the period 1940-1960
when analytical spectroscopists systematically recorded the

infrared spectra (and to a lesser extent also Raman spectra)
of the thousands of organic compounds synthesized in the course
of academic,industrial and medical research. This semi-empirical
approach to analytical vibrational spectroscopy permitted
organic and bio-chemists to formulate a network of relation-
ships between the molecular structure and the positions and
intensities of bands in certain regions of the spectrum. These
group frequency correlations are summarized in Charts I - VI
and the supplementary tables I - V.

In this short review it is not feasible to discuss
individual sections of these charts but the subject is well
covered in a series of standard textbooks and monographs (5-8).
It is to be noted, as a generalization,that such characteristic
group frequencies tend to be associated with one or more
of the following conditions :

(a). A light terminal atom attached by a single bond
to a heavy molecular mass e.g. RO-H, Ph-H, N-H.

(b). A heavier terminal atom or radical linked to a
larger group by a double or triple bond.e.g. R-C=O, R-C≡N.

(c). An internal group where the force constants
or structural factors tend to suppress vibrational coupling
with neighboring atoms e.g. X-C=C-Y,

Of particular relevance to bio-chemistry are the
highly characteristic stretching frequencies associated with the
hydroxyl group and primary and secondary amino groups under
varying conditions of hydrogen bonding, also to be noted are
the characteristic frequencies -S-H and X-S-S-Y (these are
stronger in Raman than in infrared spectra) and strong infrared
bands associated with the P-O linkage.

This type of group frequency analysis has also been
aided considerably by observing the displacement or disappearance of the band on selective replacement of hydrogen by
deuterium. An illustration of the application of group frequency analysis to the interpretation of the infrared spectrum of
a molecule of biochemical interest is provided by the spectrum
of oleic acid shown in Fig.1.

3. THE SYMBIOSIS OF NORMAL CO-ORDINATE ANALYSIS AND GROUP
   FREQUENCY ANALYSIS

As more group frequencies were established it was
natural that an urge should develop to rationalize these obser-
vations.Of particular interest are the secondary effects that
appear such as the characteristic small frequency differences
between the C=O stretching bands in five and six membered rings

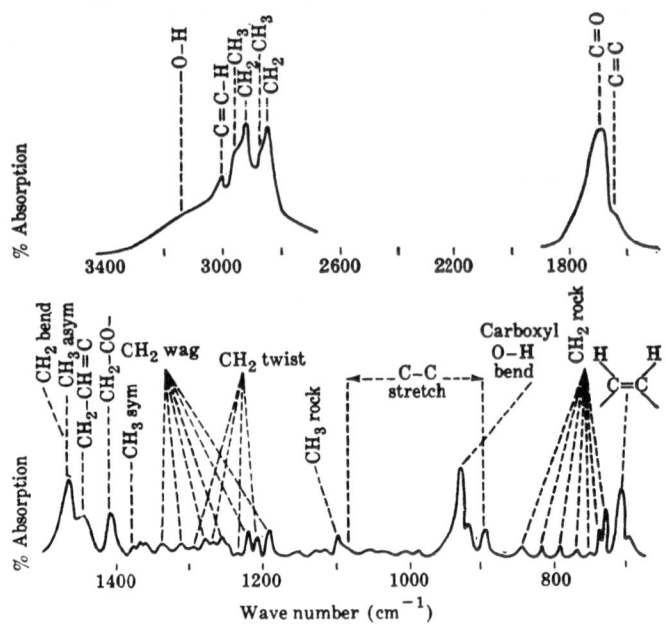

Figure 1 Characteristic group frequencies in the
infrared spectrum of a film of oleic acid at
- 196 °C.

or the characteristic small difference in the C-H
scissoring frequency of a **methylene group α to carbonyl**
as compared with one at a β or more remote position. The
explanations advanced by organic chemists and chemical spectroscopists for such effects tend to be couched in the terminology
of contemporary theories of structural organic chemistry
(e.g. Taft or Hammett constants) rather than within the frame-
work of molecular vibrational theory (5,6,9).

It is a common practice for chemical spectroscopists
to interpret the vibrational spectra of more complex molecules
by a transposition of the results of normal co-ordinate analysis
of simpler molecules, often aided by qualitative comparisons
among the spectra of isotopically substituted species of the
larger molecules and the polarization of their Raman bands.
It has become an accepted practice to include tables of
"vibrational assignments" in publications describing the spectra

of larger molecules.Those who do this often find it difficult to
refrain from making such **"assignments"**for all of the bands of the
spectrum. Many of these assumptions are valid, most are plausible
but they can be highly speculative. Some effort should be made
to restrain this urge for a total interpretation of all the **bands**
This is true particularly in the region of the spectrum between
1400 and 600 cm^{-1} where many of the normal modes of vibration
are associated with extensive coupling of the motions of the
main skeleton of the molecule.

        One of the most important quantities to emerge from
the formal normal co-ordinate treatment is the Potential
Energy Distribution Function (PED). In the classical mechanical
model of the molecular vibrations the total vibrational energy
remains constant but is divided between a kinetic and a potential
energy term. At the maximum displacement of the vibration, when
the atoms are reversing their direction of motion and their
velocities are zero,the vibrational energy is equal to the
potential energy. The basic computer programs for normal
co-ordinate analysis generate a PED matrix which identifies,for
each normal vibration,the distribution of the potential energy
contributed by each internal co-ordinate. These internal
co-ordinates very commonly bear a close relationship to motions
identified with a charactersitic group frequency.From the PED
matrix it is often possible to establish a conceptual link
between the group frequency and the normal vibrations with which
it is primarily involved. This is illustrated in Fig.2.which
illustrates the five principle motions that contribute to the
"C=O stretching band" at 1715 cm^{-1} in the infrared spectrum of
cyclohexanone. The potential energy distribution is computed
to be distributed among these five vibrational motions in the
ratios of 75,13,6,5,2 % respectively. Thus normal co-ordinate
calculations indicate that the empirical assignment of this band
wholly to the C=O stretch motion is justified by calculation to
the extent of 75%.

        Figure 2. Vibrational motions participating in

               the "carbonyl stretch band" of

               cyclohexanone expressed as potential

               energy coefficients.

4. QUANTIFICATION OF THE GROUP FREQUENCY CONCEPT

On the basis of the above considerations one can regard the group frequency assignments as a qualitative formulation of the hypothesis that where such a group can be recognized the total vibrational energy is localized in the specific motion of a few bonds. As such it does not interact with motions occuring in the more remote parts of the molecule and is therefore transferable in good approximation from one molecular structure to another. This provides the means to classify the normal vibrations of molecules in the following terms (10):-

(1). Group Frequencies. Vibrations in which at least 66% of the energy of the vibration, **as defined by the PED,** is localized in one type of force constant (more specifically one element of the $\emptyset$ vector of the $\underset{=}{Z}$ matrix).

(2). Zone Frequencies. These are a sub-category of group frequencies in which two of the **diagonal elements of** PED matrix account for at least 66% of the vibrational energy of the mode.

(3). Delocalized Frequencies- These are vibrations in which not more than 33% of the energy is associated with any one type of force constant.

Reverting to the terminology of group frequency analysis, the zone frequency is a group frequency but one in which the internal motion is not localized in one specific bond. **An example would be the bio-chemically important amide II band** which characteristically identifies the -CO-NHR-unit of a peptide chain but which cannot be classified as either a C=O stretch mode or an N-H bending mode but involves the whole amide group. The delocalized frequencies define bands that depend on motions extending over large parts of the molecule. It is such bands that are sensitive to minor changes in molecular structure and conformation and give the infrared and Raman spectra their high degree of "fingerprint" specificity.

5. CHARTS AND TABLES OF CHARACTERISTIC GROUP FREQUENCIES

CHART I

CHARACTERISTIC GROUP FREQUENCIES OF HYDROCARBONS – I

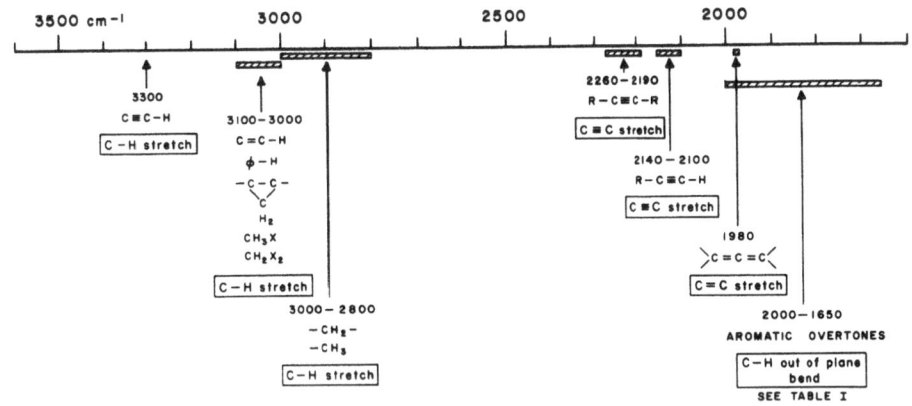

CHART II

CHARACTERISTIC GROUP FREQUENCIES OF HYDROCARBONS – II

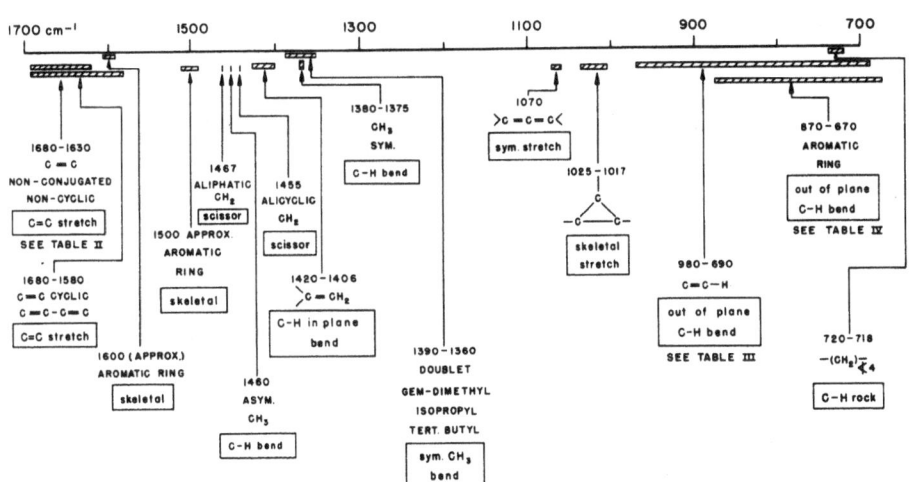

CHART III

CHARACTERISTIC GROUP FREQUENCIES OF OXYGENATED COMPOUNDS – I

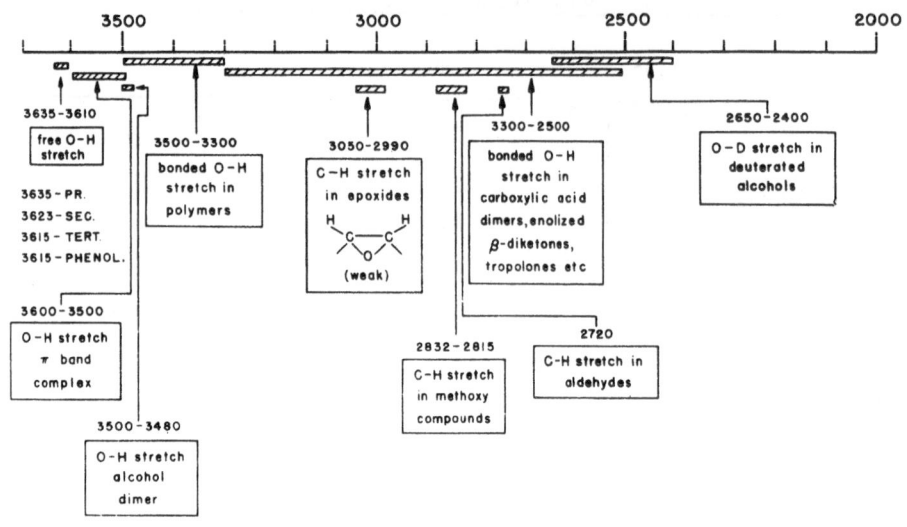

CHART IV

CHARACTERISTIC GROUP FREQUENCIES OF OXYGENATED COMPOUNDS – II

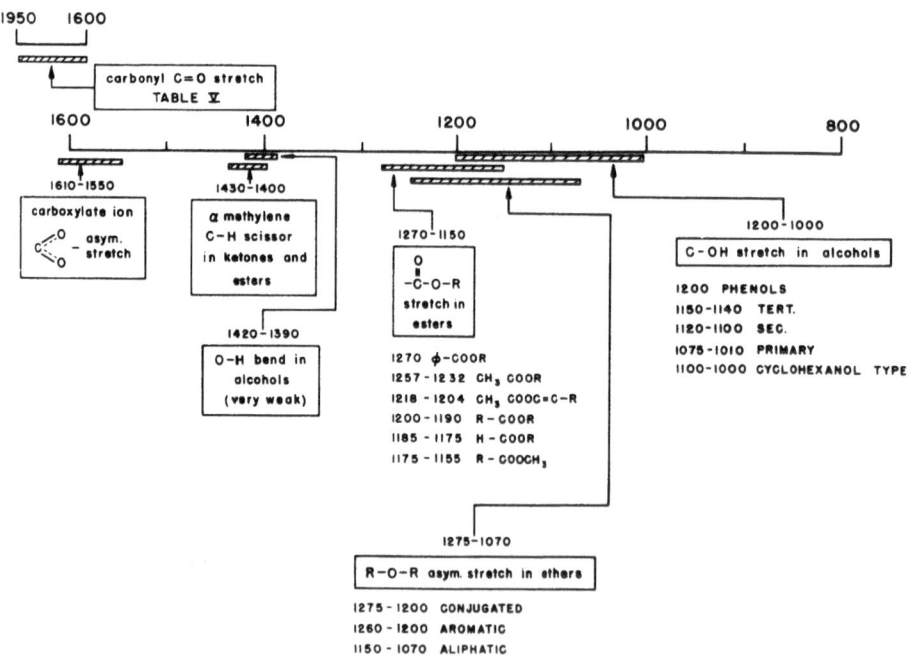

CHART V

CHARACTERISTIC GROUP FREQUENCIES OF NITROGEN, PHOSPHORUS, SULFUR COMPOUNDS — I

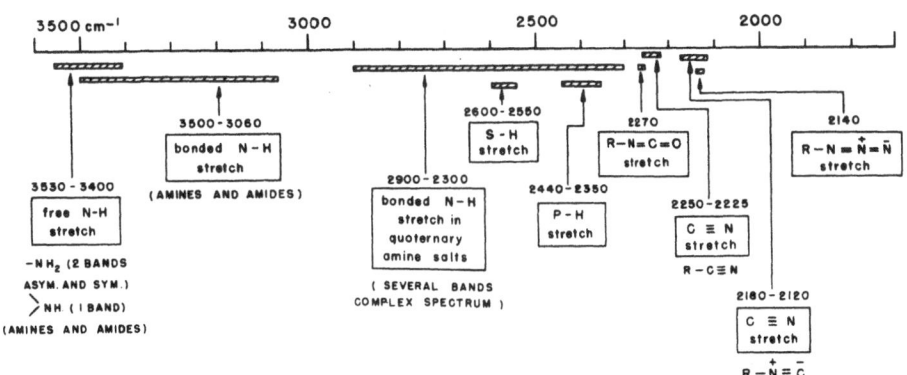

CHART VI

CHARACTERISTIC GROUP FREQUENCIES OF

NITROGEN, PHOSPHORUS, SULFUR COMPOUNDS — II

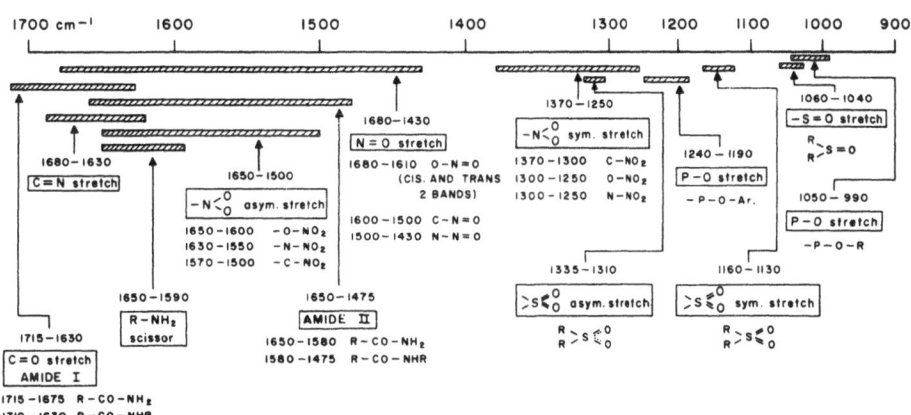

TABLE I

OVERTONE BANDS OF SUBSTITUTED BENZENES

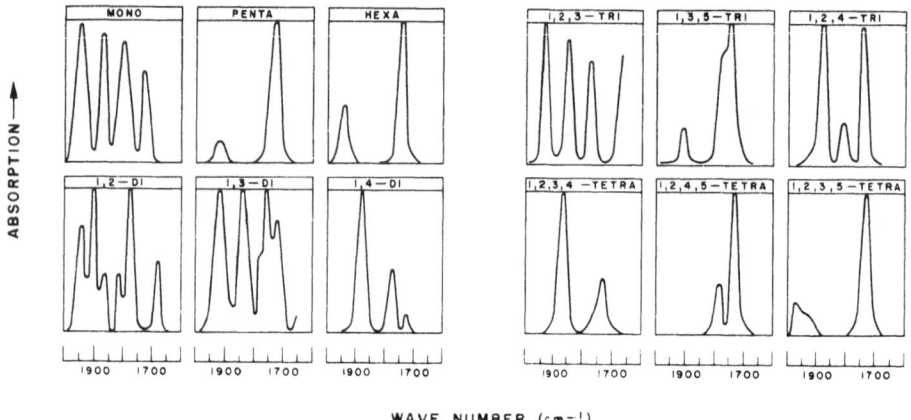

WAVE NUMBER (cm⁻¹)

TABLE II

C=C STRETCHING BANDS OF LINEAR OLEFINS

$$\underset{H}{\overset{R_1}{\diagdown}} C = C \underset{R_2}{\overset{H}{\diagup}} \qquad 1678-1668 \text{ cm}^{-1}$$

$$\underset{R_2}{\overset{R_1}{\diagdown}} C = C \underset{R_3}{\overset{H}{\diagup}} \qquad 1675-1665$$

$$\underset{R_2}{\overset{R_1}{\diagdown}} C = C \underset{R_4}{\overset{R_3}{\diagup}} \qquad 1675-1665$$

$$\underset{H}{\overset{R_1}{\diagdown}} C = C \underset{H}{\overset{R_2}{\diagup}} \qquad 1662-1652$$

$$\underset{R_2}{\overset{R_1}{\diagdown}} C = C \underset{H}{\overset{H}{\diagup}} \qquad 1658-1648$$

$$\underset{H}{\overset{R_1}{\diagdown}} C = C \underset{H}{\overset{H}{\diagup}} \qquad 1648-1638$$

TABLE III

C-H OUT-OF-PLANE BENDING BANDS OF LINEAR OLEFINS

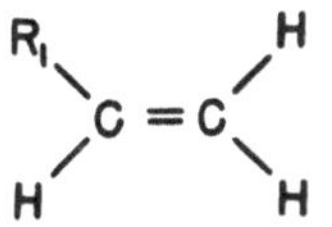

$995-985$ cm^{-1}
$910-905$

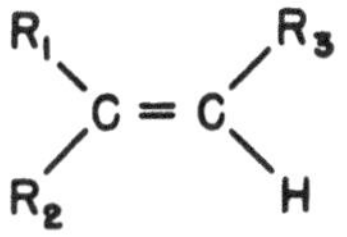

$840-790$

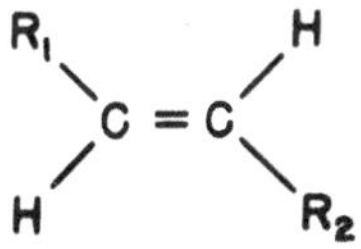

$980-965$

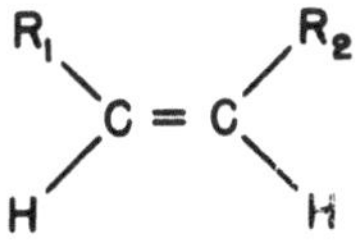

≈ 690

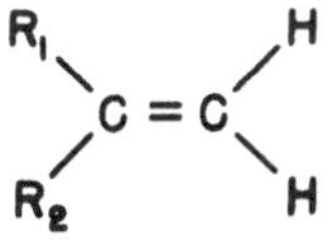

$895-885$

TABLE IV

C - H OUT-OF-PLANE BENDING BANDS OF BENZENE DERIVATIVES

Benzene	671 cm.$^{-1}$	-
Monosubstituted benzene	770-730	710-690
1, 2-Disubstituted	770-735	-
1, 3-Disubstituted	810-750	710-690
1, 4-Disubstituted	833-810	
1, 2, 3-Trisubstituted	780-760	745-705
1, 2, 4-Trisubstituted	825-805	885-870
1, 3, 5-Trisubstituted	865-810	730-675
1, 2, 3, 4-Tetrasubstituted	810-800	-
1, 2, 3, 5-Tetrasubstituted	850-840	-
1, 2, 4, 5-Tetrasubstituted	870-855	-
Pentasubstituted	870	-

. . . .

Five adjacent free hydrogen atoms	770-730	710-690
Four adjacent free hydrogen atoms	770-735	
Three adjacent free hydrogen atoms	810-750	
Two adjacent free hydrogen atoms	860-800	
One free hydrogen atom	900-860	

. . . .

TABLE V

PRINCIPAL CARBONYL C = O STRETCHING BANDS (cm.$^{-1}$)

KETONES

R - CO - R	1725 - 1705	
R - C = C - CO - R	1685 - 1660	
R - CO - Ar.	1700 - 1680	
Ar. - CO - Ar.	1670 - 1660	
R - CO - CO - R	1730 - 1710	
R - CO - CH$_2$ - CO - R (enolized)	1640 - 1540	
Sat. cyclic (six membered or larger ring)		1725 - 1705
Sat. cyclic (five membered ring)		1750 - 1740
Sat. cyclic (four membered ring)		1775
Quinones (2 carbonyl groups in same ring)		1690 - 1660
Quinones (2 carbonyl groups in different rings)		1655 - 1635
Tropolones	1600	

ALDEHYDES

Usually about 15 cm.$^{-1}$ above corresponding ketones.

ACIDS

R - COOH (dimer)	1725 - 1700
R - C = C - COOH (dimer)	1715 - 1690
Ar. - COOH	1700 - 1680

ESTERS

R - COOR	1750 - 1735
R - C = C - COOR	1730 - 1715
Ar. - COOR	1730 - 1715
R - COOAr.	1800 - 1770
R - CO.OC = C - R	1800 - 1770
Sat. γ-Lactone	1780 - 1760
Sat. δ-Lactone	1750 - 1735

OTHER CARBONYL COMPOUNDS

R - CO.Cl	1815 - 1770	
R - CO - O - CO - R	1850 - 1800	and 1790 - 1740
Cyclic anhydride (5-membered ring)	1870 - 1820	and 1800 - 1750
R - CO - O - O - CO - R	1820 - 1810	and 1800 - 1780
Ar. - CO - O - O - CO - Ar.	1805 - 1780	and 1785 - 1755
R - CO - NH$_2$	1690 - 1650	
R - CO - NHR	1680 - 1630	

6.BIBLIOGRAPHY

(1).   R.N.Jones. Forward to "Normal Co-ordinate Analysis"
       by H.Fuhrer,V.B.Kartha,K.G.Kidd,P.J.Krueger and
       H.H.Mantsch. In Computer Programs for Infrared
       Spectrophotometry, Volume 5. National Research
       Council of Canada Bulletin No.15 (1976).

(2). Ian M.Mills. "Force Constant Calculations for Small Molecules". In Infrared Spectroscopy and Molecular Structure.Ed.Mansel Davies. Elsevier Publishing Co. Amsterdam,London,New York. Chapt.V. (1963).

(3). T.Shimanouchi and I.Nakagawa. "Force Fields in Poly-Atomic Molecules". Annual Reviews of Physical Chemistry., 23, 217 (1972).

(4). William W.Coblentz."Investigations of Infra-red Spectra". Carnegie Institution of Washington Publication No. 35 (1905). Reprinted by the Coblentz Society (1962).

(5). R.Norman Jones and Camille Sandorfy. "The Application of Infrared and Raman Spectroscopy to the Elucidation of Molecular Structure." In Technique of Organic Chemistry. Volume 9. Ed.A.Weissberger. Interscience Publishers Inc. New York,London. First Edition, Chapt.IV (1956).

(6). L.J.Bellamy. "Advances in Infrared Group Frequencies". Methuen and Co.Ltd. London (1968).

(7). Margareta Avram and Gh.D.Mateescu. "Infrared Spectroscopy -- Applications in Organic Chemistry". Wiley Interscience.New York,London,Sydney,Toronto.(1972).

(8). R.R.Hill and D.A.E.Rendell. The Interpretation of Infrared Spectra -- A Programmed Introduction". Heyden and Sons Ltd. London,Bellmawr,Rheine (1975).

(9) A.R.Katritzky and R.D.Topsom."Infrared Intensities: A Guide to Intramolecular Interactions in Conjugated Systems. Chemical Reviews, 77, 639 (1977).

(10). H.Fuhrer,V.B.Kartha,P.J.Krueger,H.H.Mantsch and R.N.Jones. "Normal Modes and Group Frequencies -- Conflict or Compromise ? An In-depth Vibrational Analysis of Cyclohexanone. Chemical Reviews, 72, 439 (1972).

APPLICATIONS

VIBRATIONS OF CHAIN MOLECULES

Jack L. Koenig

Department of Macromolecular Science
Case Western Reserve University, Cleveland, Ohio 44106

When chemical units are repeated on a chain in a regular fashion, all of the units have the same energy so they are potentially capable of resonating or coupling their vibrational motions. This intramolecular vibrational coupling can lead to development of a series of resolvable vibrational modes characteristic of the length of the coupled units.

For paraffins, these observations have been confirmed as shown in Figure 1, where the changes in frequencies for the different CH_2 modes

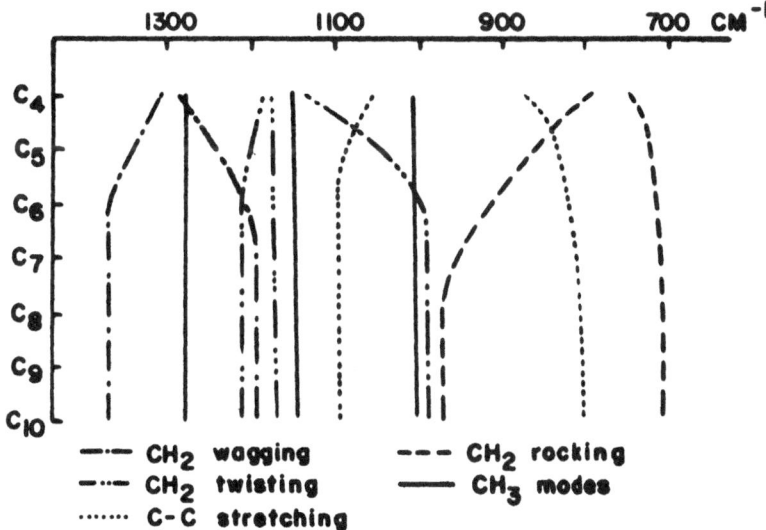

Figure 1. The zone boundaries for the different vibration types.

Theo M. Theophanides (ed.), Infrared and Raman Spectroscopy of Biological Molecules, 97–107.

are plotted as a function of the chain length. The methyl modes have a
constant frequency since the coupling to the end methyl groups is
constant. This plot shows only the boundaries of the vibrational modes
with no indication of the number of modes within each boundary. For
each type of vibrational mode, it is expected that an additional mode of
the same type will be generated with the addition of each methylene unit.
Unfortunately, due to resolution limitations, selection rules and weak
intensities, one seldom observes all of the additional modes. In fact,
as will be shown later, when the chain becomes infinite, the selection
rules are highly restrictive and only a single optical mode for each
vibrational type of the repeating unit is expected.

So the vibrational pattern for a regular polymer chain depends on the
number of coupled oscillators or units, the normal modes of an isolated
repeating unit and the extent of coupling of the vibrational modes with
other repeat units.

These salient features of the spectra of ordered polymers may be
demonstrated by analyzing a uniform one-dimensional lattice of point
masses. The N frequencies for a linear chain of N atoms acting
as parallel dipoles with fixed ends (including only nearest neighbor
interactions) are given by the following equation

$$W_s^2 = W_o^2 + W_1^2 (1 + \cos \theta)$$

$$\theta = \frac{s\pi}{N+1} = 1, 2, \ldots N$$

where W_o is the frequency of the uncoupled mode
    W_1 is the interaction parameter
    s   is an integer going from 1 to N.

The N frequencies (W_s) are a function of θ) which physically corresponds to the phase difference of the vibrational modes of adjacent
atoms or cells. When $\theta = 0$ (not possible with finite chains), all atoms
have the same motion and as θ increases, the difference in phase
increases to a limiting value of 180° corresponding to the atoms on one
unit reaching its maximum displacement at the same time, its adjacent
unit is reaching its minimum position. Physically, each value of K
corresponds to a standing wave of different wavelength in the molecule.
If the interaction parameter, W_1, is small or if the normal modes
fall in the neighborhood of W_o a single unresolved mode will be
observed. Thus, the vibrational mode with W_1 small has no dependence
on chain length. This results reminds one of the concept of group

frequencies where the "internal" modes were relatively independent of their environment. This is the case, for example, with the carbon-hydrogen stretching modes which occur at the same frequencies for all chain lengths of paraffin and in fact for the polymer.

When the interaction parameter is sufficiently large, the energy levels of the N coupled harmonic oscillator are resolvable. The vibrational modes generated as the chain length increases are shown diagramatically for chains with N up to 8 in Figure 2. An additional mode is expected An additional mode is expected for each atom added to the chain.

Not all of these coupled modes will be observed because of selection role restrictions, but sufficient data does exist to verify the theoretical predictions. In Figure 3, the square of the observed frequency for some of the S-modes are plotted versus 1+cos θ for $C_{24}H_{50}$. The limiting frequency extrapolates to 716 cm^{-1} which corresponds to the proper frequency while the interaction constant is 423 cm^{-1} giving 1107 cm^{-1} as the other extreme value of the rocking frequency.

The relative intensities of the modes as s increases are substantially different making detection of the modes difficult.

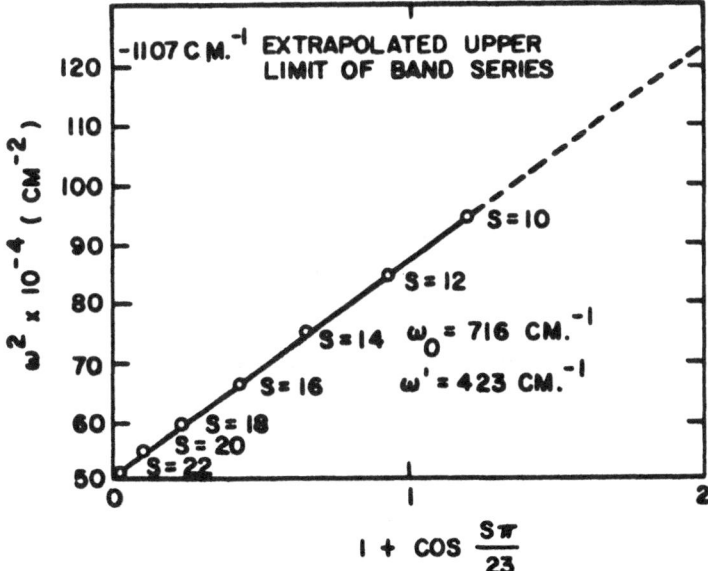

Figure 2.   Frequencies of the modes plotted vs. the phase factor.

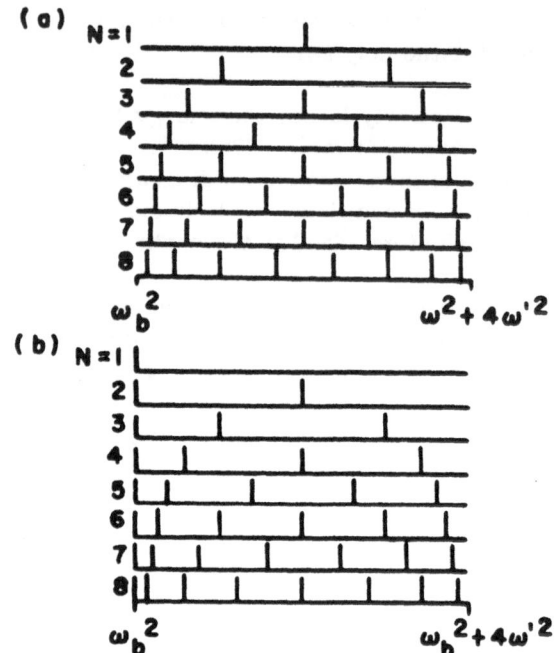

Figure 3. Normal vibrational frequencies for a chain of N–coupled
oscillators N=1, 2, . . . , 8 (a) fixed ends; (b) free ends.

For example consider the case of two coupled oscillators with dipoles
pointing in the same direction. For one of the two modes, the dipoles
vibrate in phase while for the other mode out of phase by 180°. The
in phase mode is infrared active since there will be a total dipole
moment change during the vibration. In the out of phase mode, the
dipole moment changes of the two dipoles cancel each other and the
vibration is infrared inactive. Both modes are Raman active.

Similar considerations apply to a longer chain of coupled units and the
relative intensity depends on the phasing (θ) of the vibrations of the
units. The relative infrared intensities for a coupled chain of N=8 are
shown in Figure 4. Often only the limiting mode will be observed, but
this mode will be dependent on the ordered chain length up to a limiting
value of N which will be different for each mode. The lowest frequency
is always the strongest band in the series while the second strongest is
calculated to be from 25% of the highest to 3% of the strongest.

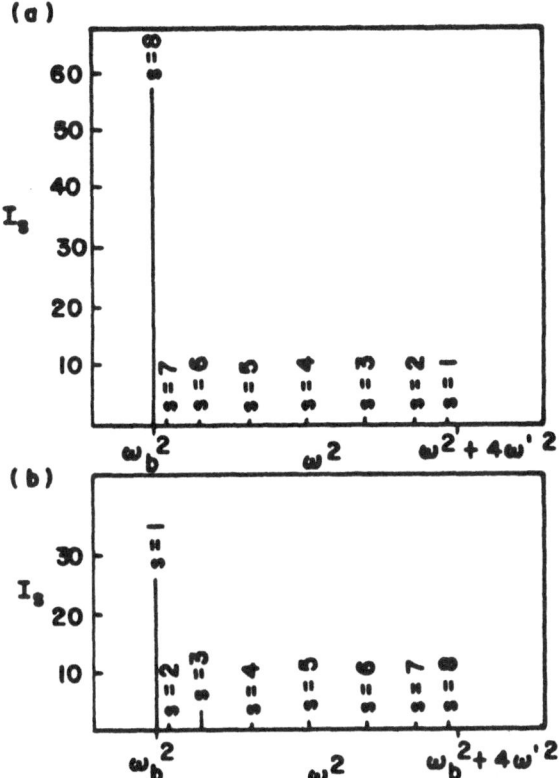

Figure 4. Expected infrared absorption spectrum of 8 coupled dipoles
with free ends. (a) Parallel dipoles; (b) antiparallel dipoles.

This analysis and comparison with experimental data leads to useful
conclusions. First, some vibrational modes show a chain length
dependence particularly for short chains. Secondly, some vibrational
modes are independent of chain length and are useful for structure-
frequency analysis in the same manner as other organic molecules.

The final question meriting our attention is the question of what
happens when the regular chain structure is disrupted by a structural
defect such as a different geometric or steric isomer or chemical
defects. In general these defects decouple the vibrational modes
causing a dramatic change in the spectrum. An excellent example is
the infrared spectra of n-butane as an ordered solid and as a liquid.
The introduction of the rotational isomer completely decouples the

two halves of the oligomer as well as added the spectrum of the gauche isomer. The increase in band width arises from changes due to thermal effects. For longer chains, the effect is not as obvious but is real and detectable. Hence, one can conclude that the coupling only occurs through repeat units having the same chemical and geometric structure. Thus structural defects in the chain will produce vibrational spectra of shorter sequences. In fact, if these sequences are sufficiently short, vibrational frequencies may be observed due to these particular frequencies. In the case of copolymers of ethylene with other monomers infrared bands arising from ethylene sequences have been observed as follows:

Frequency of rocking mode

$$\begin{array}{ccc} X & & X \\ | & & | \\ C - CH_2 - C - \end{array}$$ 815 cm^{-1}

$$\begin{array}{ccc} X & & X \\ | & & | \\ C - CH_2 - CH_2 - C \end{array}$$ 752 cm^{-1}

$$\begin{array}{ccc} X & & X \\ | & & | \\ C - CH_2 - CH_2 - CH_2 - C \end{array}$$ 733 cm^{-1}

$$\begin{array}{ccc} X & & X \\ | & & | \\ C - CH_2 - CH_2 - CH_2 - CH_2 - C \end{array}$$ 726 cm^{-1}

$$\begin{array}{ccc} X & & X \\ | & & | \\ C - CH_2 - CH_2 - CH_2 - CH_2 - CH_2 - C \end{array}$$ 722 cm^{-1}

These vibrational modes have been used to characterize the sequence distribution of ethylene propylene copolymers.

Let us consider the spectra of an infinite chain of dipoles. Physically we will never encounter such an infinite chain but, as we have just seen, above a certain chain length the coupling effects can be considered as arising from an infinite chain as addition of more units does not measurably influence the results. In other words, when the chain length reaches a critical value, the repeat unit can be considered as a part of an infinite chain. For an infinite chain of atoms of mass, M, spaced a distance apart held together by a force K between atoms the frequency function is given by

$$\nu = \frac{1}{\pi c} \quad \frac{K}{2m} \quad (1-\cos 2\pi \text{ kd})^{1/2}$$

$$= \frac{1}{\pi} \quad \frac{K}{m} \quad (\sin \pi \text{ kd}).$$

Thus the frequency is a periodic function of the wave number k, the period being equal to 1/d. The results are plotted in Figure 6. The frequency goes through a maximum and has a period equal to 1/d. The first period is called the first Brillouin zone. The maxima occur at

$$\nu_m = \frac{1}{c\pi} \quad \frac{k}{m}$$

which is the limiting frequency. A plot of frequency ν versus the wave number k is known as a dispersion relation. The dispersion relation for a monoatomic chain is shown in **Figure 5.** The motion exhibited by

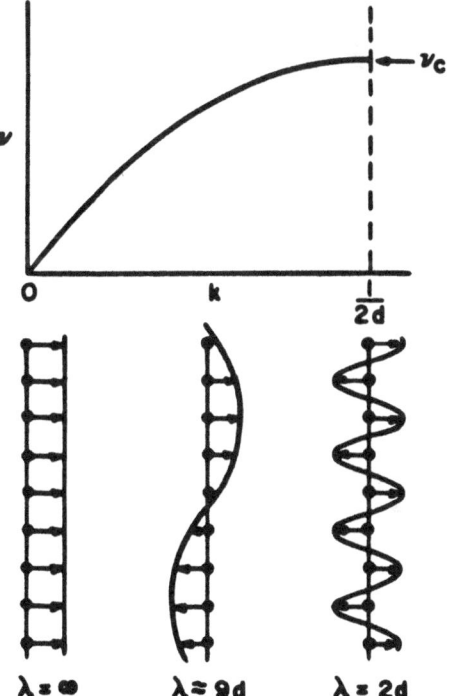

Figure 5. Dispersion relation for the monatomic chain showing forms of the vibrational modes for certain values of k.

the infinite chain can be visualized as standing waves of wavelength $\lambda = 1/k$. When $k=0$ ($\lambda=\infty$) all of the atoms are in phase. As shown in Figure 7, at intermediate values of K, a standing wave motion occurs and when $k=0$ ($X=2d$) all of the particles have the same amplitude of oscillation but alternate ones are out of phase. For a monatomic chain, no spectral activity is expected.

Let us consider a linear diatomic in one dimension chain. For the special case in which all bands are identical and the atoms are equally spaced but have different masses, the result is (2)

$$\nu^2 = \frac{k}{4\pi^2 c} \left\{ \left(\frac{1}{m_1} + \frac{1}{m_2} \right) \pm \left[\left(\frac{1}{m_1} + \frac{1}{m_2} \right)^2 - \frac{4}{m_1 m_2} \sin^2 \pi kd \right]^{1/2} \right\}$$

The dispersion curve is shown in **Figure** 6. Two curves result for the longitudinal vibrations. The curve which passes through the origin is

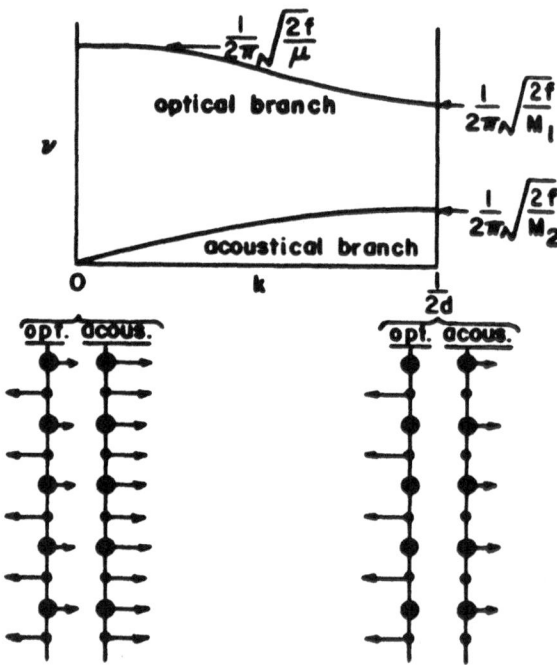

Figure 6. Dispersion curves for the longitudinal vibrations of the diatomic chain.

called the acoustical branch, because the frequencies fall in the region of sonic or ultrasonic waves. The upper curve is the optical branch and falls in the infrared spectral region. There is a frequency gap where no frequencies occur. The forms of the vibrations for each branch are shown in Figure 6 for limiting values $k=0$ and $k=1/2d$. For the optical branch at $k=0$, the optical branch represents a simple stretching of the bond between two given atoms in which their center of mass remains fixed. This mode is obviously optically active (infrared and Raman) since the two atoms are assumed to be different. The acoustical vibration at $k=0$ does not result in a change in dipole moment and is optically inactive. Frequencies corresponding to the intermediate values of k for the optical branch are spectrally inactive since for every atom exhibiting a positive dipole displacement in the infinite chain there will be a corresponding atom moving in the opposite direction so the total will be zero.

When the results are extended to three dimensions for the motion of diatomic chain, there is one pair of dispersion curves (one optical and one optical branch) for each direction in space. For the isolated chain, the two transverse directions are equivalent and each branch is doubly degenerate.

A generalization of these results to a molecular chain is easily comprehended. When the repeating unit contains m atoms there will be 3m separate branches. Three of these modes correspond to acoustical modes so 3m-3 optical branches exist. Of course some of these branches can be degenerate. The general result is shown in **Figure 7**. Many of the modes of the isolated repeat unit will not be severely modified by the insertion in a chain since they are internal frequencies. Others will be coupled in the chain. Solid state physicists prefer to call these modes "lattice" modes.

Normal coordinate analysis of polymer chains have been made and complete dispersion curves obtained as shown in Figure 8 for an infinite polymethylene chain. There are nine branches since the methylene repeating unit has 3 atoms. The three translations and one allowable rotation correspond to the zero modes on the acoustical branches. The infrared and Raman active mode for polyethylene correspond to the frequencies at the intercepts when $\theta=0$ and π. The C-H stretching modes and angle bending modes are independent of the phase as expected while the twisting, wagging, rocking and C-C stretching branches show effects due to the chain.

Interestingly, from the complete dispersion curve of a polyethylene

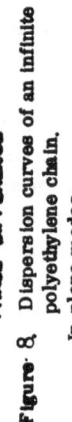

Figure 8 Dispersion curves of an infinite polyethylene chain.

In-plane modes:

ν_1, CH$_2$ symmetric stretching
ν_2, CH$_2$ scissors
ν_3, CH$_2$ wagging
ν_4, CC stretching — CCC bending
ν_5, CCC bending — CC stretching

Out-of-plane modes:

ν_6, CH$_2$ antisymmetric stretching
ν_7, CH$_2$ rocking — CH$_2$ twisting
ν_8, CH$_2$ twisting — CH$_2$ rocking
ν_9, Torsion

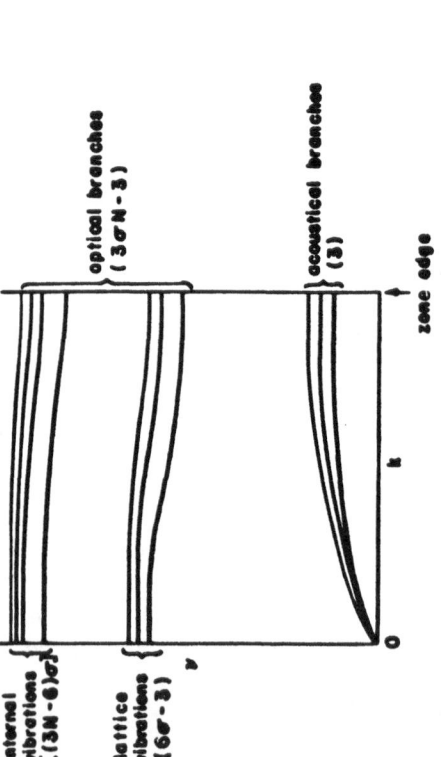

Figure 7. General form of dispersion curves for a molecular crystal.

the spectra of the methylene sequences for all of the oligomers can be obtained simply by dropping perpendicular lines at the appropriate values of the phase angles. The frequencies where these lines intersect the dispersion curves will be the predicted frequencies for the oligomer. Thus, the dispersion curve gives not only the modes for the infinite chain but for finite chains as well.

General References

1. R. Zbinden, Infrared Spectra of High Polymers, Academic Press, N.Y. 1964.
2. G. Turrell, Infrared and Raman Spectra of Crystals, Academic Press, N.Y. 1972.

VIBRATIONAL SPECTROSCOPY OF POLYPEPTIDES AND PROTEINS

Jack L. Koenig

Department of Macromolecular Science
Case Western Reserve University, Cleveland, Ohio 44106

INTRODUCTION

Raman and infrared spectroscopy provide complementary information concerning the structure and conformation of biopolymers. The vibrational modes are infrared and/or Raman active depending on the symmetry of the particular molecule under consideration through the selection rules.

The vibrational selection rules for a polymer chain can be applied rigorously only to chains having at least one dimensional crystallinity. An isolated chain in an ordered conformation such as the alpha helix satisfies this requirement. A chain containing N unit cells with M atoms per unit cell has $3 \times N \times M$ degrees of freedom and after subtracting 3 translational degrees of freedom and the single rotation about the chain axis there remains $(3 \times N \times M)-4$ vibrational motions (1). Obviously this represents a large number of vibrations but in order to have infrared or Raman activity the phase angle between identical vibrations in neighboring unit cells must be zero. This selection rule was derived by Born and von Karmen (2) and is discussed in detail by Krimm (3). Therefore we need only consider the vibrations with a single unit cell. One should not identify the chemical repeat with the unit cell; only in special cases are they identical and generally the number of chemical repeats per unit cell is that number required to complete an integral number of turns of the helix axis. The number of chemical repeats per one dimensional unit cell of polypeptide chain in a planar zig-zag, 3_1 helix and alpha helix (18_5) is 2, 3, and 18, respectively.

Theo M. Theophanides (ed.), Infrared and Raman Spectroscopy of Biological Molecules, 109–124.
All Rights Reserved. Copyright © 1979 by D. Reidel Publishing Company, Dordrecht, Holland.

Molecular symmetry governs the infrared and Raman activity of normal vibrations. Zbinden (4) has thoroughly discussed the symmetry aspects of polymer chain vibrations. Infrared activity requires that the vibration produce a change in the dipole moment of the molecule. The dipole moment is a vector quantity and has components in the x, y, z direction, where z is conventionally taken as the chain axis. Any vibration that transforms under the symmetry operations of the polymer in the manner of a unit vector in the x, y, or z direction will be infrared active. Vibrations that transform like the z unit vector are said to have parallel polarization since they absorb radiation strongest when the electric vector of the radiation lies parallel to the z or chain axis. Similarly, vibrations that transform according to the x and y unit vectors have perpendicular polarization. By measuring infrared band intensities of oriented polymer samples with perpendicular and parallel polarized infrared radiation the symmetry species to which the band belongs may be assigned, or in the case of fibrous proteins (5), the conformation of the polypeptide chain may be determined.

POLYPEPTIDES

Proteins consist of a kaleidoscope of possible amino acid sequences and this complicates the analysis of the spectra. Consequently homopolypeptides have been studied extensively and used to establish structure-frequency correlations. The vibrational modes sensitive to backbone conformation are those of the amide group. Studies on the model amide N-methylacetamide were used to establish the nature of the characteristic vibrations. The amide I mode consists of the carbonyl stretching vibration with small contributions from the C-N-H in-plane bending and the C-N stretching vibrations. The potential energy distribution over these three components have been calculated by a number of authors (8, 9). Work by Krimm and coworkers on monomeric amides and nylons (10, 11) and polyglycine I (12) and II (13) represent the most detailed analysis.

The Raman active amide I mode for the alpha helical conformation appears at the same frequency as in the infrared spectra, i.e. from 1650-1657 cm^{-1}. Strong splitting of the amide I mode is observed for the antiparallel beta sheet conformation. A strong amide I line appears in the Raman spectrum near 1670 cm^{-1} while two lines appear near 1630 cm^{-1} and 1685 cm^{-1} in the infrared spectrum for this conformation. Miyazawa (14) developed classical perturbation equations to explain the splitting in the infrared spectrum that were based on the weakly coupled oscillator model. The equations were modified by Krimm and Abe (15) to account for the recent Raman data and the results of their normal coordinate analysis of polyglycine I. There are four peptide units in the two dimensional unit cell of polyglycine I (and other beta sheet polypeptides if the side chain is ignored) therefore one expects four amide I vibrations. The allowable intrachain phase angle, δ, and interchain phase angle, δ', are 0 and π and four combinations are listed below, where the first number in the parenthesis refers to δ and the second to δ':

$\nu(0, 0)$ Raman active, infrared inactive, polarized in Raman

$\nu(\pi, 0)_\perp$ Raman active, infrared active, perpendicular IR dichroism

$\nu(0, \pi)_{||}$ Raman active, infrared active, parallel IR dichroism

$\nu(\pi, \pi)_\perp$ Raman active, infrared active, perpendicular IR dichroism

In practice the $\nu(0, 0)$ mode appears in the Raman spectrum near 1670 cm^{-1}. The $\nu(\pi, 0)$ mode appears in the infrared spectrum near 1630 cm^{-1} with strong intensity and perpendicular polarization and the $\nu(0, \pi)$ mode also appears in the infrared spectrum near 1680 cm^{-1} with weak intensity and parallel polarization. The amide I mode for the disordered conformation appears near 1665 cm^{-1} in the Raman spectrum and near 1655 cm^{-1} in the infrared spectrum. By using the amide I mode in the Raman spectrum one can easily differentiate between the alpha helical and beta sheet or disordered conformations but cannot easily differentiate between the latter two conformations.

The amide III mode consists of the C–N–H in-plane bending and C–H stretching modes but may be coupled to other motions that occur in this frequency region. This vibration is not localized in the amide group to the extent of the amide I mode and therefore may be more directly sensitive to the conformation of the polypeptide chain. The amide I mode derives its sensitivity to conformation primarily through dipole-dipole interactions among the carbonyl groups that are functions of the chain geometry (15). The antiparallel beta sheet can be immediately distinguished by an intense amide III line at 1235 cm^{-1} $\pm$ 5 cm^{-1} in the Raman spectrum. High resolution spectra may occasionally reveal another amide III component near 1270 cm^{-1}. The disordered conformation appears near 1245 cm^{-1} in the amide III region of the Raman spectrum. Only weak scattering in this region can be observed for the alpha helical conformation and lines appearing from 1260 cm^{-1} to 1295 cm^{-1} have been assigned to this mode. This remarkable decrease in intensity in the amide III region upon going from the disordered or beta sheet conformation to the alpha helix has been attributed to a hypochromic effect by Koenig and coworkers (16,17) and is analogous to the Raman hypochromism observed by Peticolas and coworkers (18) and

Thomas (19) during the coil to helix transitions of the polynucleic acids.
Lack of any strong Raman lines from 1200–1300 cm^{-1} in the spectrum
of a polypeptide is strong evidence for the alpha helical conformation.

The alpha helical conformation is also characterized by a strong line
appearing near 900 cm^{-1} in the Raman spectrum that has been assigned
to a skeletal stretching vibration (16, 20). This line is either absent or
weak in the infrared spectrum. No strong Raman lines appear in this
region of the spectrum for the beta sheet polypeptides and in cases
where an alpha helix to beta sheet transition may be induced, this line
becomes very weak. The intensity of this line for the disordered conformation is not well behaved. During the pH induced helix to coil
transition of ionizable polypeptides (16, 20) the frequency shifts from
near 930 cm^{-1} to 950 cm^{-1} but does not decrease in intensity. However,
a large decrease in intensity during the solvent induced helix to coil
transition of poly–β–benzyl–L–aspartate has been observed (21).

Poly–L–lysine can be prepared in the α helical, β–sheet and "disordered"
(extended) conformation and the Raman spectra of these forms is shown
in Figure 1, illustrating the characteristics of the conformationally
sensitive lines discussed above.

The frequencies for the amide I, III and 900 cm^{-1} skeletal modes of the
Raman spectra for all of the polypeptides studied to date are summarized
in Table I. In order for the reader to assimilate this data, a histogram
of these frequencies was prepared and appears in Figure 2. The relative
line intensities in the histogram are only approximate since the data
originated from different laboratories using various types of instrumentation.

PROTEINS

The globular and fibrous proteins differ from the simple polypeptides
in at least two basic regards. First, proteins are complicated copolymers containing approximately 24 different amino residues that
have aromatic, aliphatic and reactive functional groups as side–chains.
Several of these side chain vibrations scatter strongly in the Raman
spectrum and to a much lesser degree in the infrared spectrum.
Fortunately these lines do not appear with strong intensity in the conformationally sensitive regions of the Raman spectrum, i.e. the amide
I and III regions. Without this fortiutous "rule of exclusion", the
prospects of protein studies with Raman spectroscopy would be greatly
diminished. The Raman lines due to side–chains are themselves of

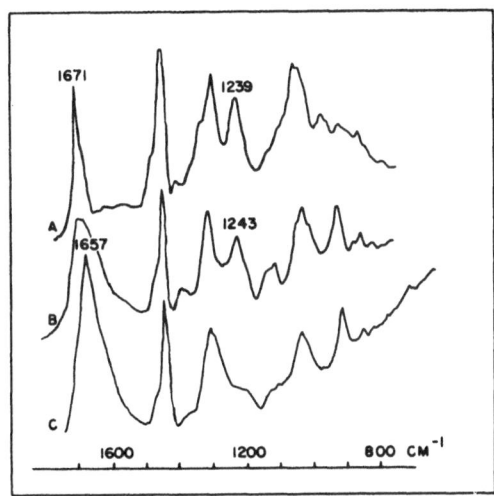

Figure 1. Raman spectrum of $(Lys)_n$
A. β -sheet conformation
B. "Disordered" conformation
C. α helical conformation

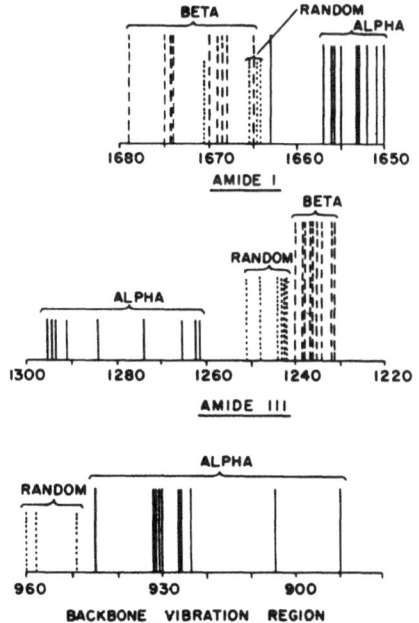

Figure 2. Histogram of conformationally sensitive Raman lines.

Table I

CONFORMATIONALLY SENSITIVE RAMAN LINES OF GLOBULAR PROTEINS

Polypeptide	ALPHA HELIX		Skeletal Vibrations (850–950 cm^{-1})	References
	Amide I	Amide II		
Poly-L-alanine	1655–1659 S[a]	Triplet 1284 VW 1274 VW 1262 VW	905 S	1–5
Poly-γ-benzyl-L-glutamate (PBLG)	1650–1652 S	1294 W	931 S	5,6
Poly-γ-methyl-L-glutamate	1656 S	1250–1350 W	924 S	4
Poly-γ-benzyl-L-aspartate (b)	1663 S		911 M, 890 S	7
Poly-γ-methyl-L-aspartate				8
Poly-L-leucine	1653 S	1261 W 1294 W } C	931 S	3,4,6
Poly-L-methionine	1652 S	1264 W	907 S	9
Poly-L-glutamic acid, solid	1656 S	1246 Wbr	926 S	10
Poly-L-glutamic acid, pH 4.8		1238 M	926 S	10
Poly-L-lysine · HCl, solid at 50% R.H.[a]	1655 M	1295 W	945 M	5
Poly-L-lysine, pH 110	1645 M (H$_2$0) 1632 M (D$_2$0)	1200–1300 W (H$_2$0)	945 S (H$_2$0) 950 S (D$_2$0)	11,9
Random copolymer of L-leucine and L-glutamic acid	1653 S	1200–1300 W	931 S	3
Random copolymer of L and D-leucine	1658 S	1258 W 1294 W } C	931 M	12
Racemic blend of PBLG and PBDG	1650 S	1291 W	931 S	13

Table I (Continued)

Polypeptide	Amide I	Amide II	Skeletal Vibrations (850–905 cm^{-1})	References
ANTI-PARALLEL PLEATED SHEET				
Polyglycine I	1674 S	1234 S, 1221 W	884 W	14,15
Poly-L-alanine	1669 S	1243 S, 1231 M	909 S	3,16
Poly-L-saline	1666–1671 S	1231 S, 1277 W[c]	959 W, 942 W	3-6
Poly-L-lysine gel	1670 S	1240 S	1002 M	12
Poly-L-serine	1674 S	1235 S	894 M	6,9
Poly-β-benzyl-L-aspartate	1679 S	1236 M	911 M	7
Poly-β-methyl-L-aspartate				8
Poly-S-methyl-L-cysteine	1675 S	1238 S	940 M	9
Poly-(Ala-Gly)	1665 S	1238 S, 1271 W	925 M, 890 M	7
Poly-(Ser-Gly)	1668 S	1236 S	983 M	7
RANDOM COIL				
Poly-L-glutamic acid, pH 7.0	1665 S	1248 M	940 S	10,17
Poly-L-lysine, pH 7.0	1665 S (H$_2$O) / 1660 S (D$_2$O)	1243 S	958 M	11
Poly-L-ornithine, pH 7.0	1665 S (H$_2$O)	1242 S	960 M	10
Copolymers of L-glutamic acid and L-tyrosine, pH 10.0		1251 M	930 S	18
Copolymers of L and D-lysine at pH 7.0	1671 S (H$_2$O)	1243 S	959 M	12
3$_1$ HELIX				
Polyglycine II	1654 M	1283 W, 1261 M, 1244 M	884 S	14,19
POLY IMINO ACIDS				
Carbonyl Stretching				
Poly-L-proline I	1650 S (solid)			20,21
Poly-L-proline II	1650 M (solid), 1630 M (in H$_2$O)			20,22,23
Poly-L-hydroxyproline	1628 M (solid), 1630 S (in H$_2$O)			24,25

REFERENCES FOR TABLE I

1. Koenig, J.L. and Sutton, P.L., Biopolymers, 8, 167 (1969).
2. Fanconi, B., Small, E., and Peticolas, W.L., Biopolymers, 10, 1277 (1971).
3. Frushour, B.G., and Koenig, J.L., Biopolymers, 13, 455 (1974).
4. Simons, L., Bergstrom, G., Blomfelt, G., Fores, S., Stenback, H., and Wansen, G., Commentationes Physioco-Mathematicae, 42, 125 (1972).
5. Chen, M.C., and Lord, R.C., J. Am. Chem. Soc. 96, 4750 (1974).
6. Koenig, J.L. and Sutton, P.L., Biopolymers, 10, 89, (1971).
7. Frushour, B.G., and Koenig, J.L., Biopolymers, 14, 363 (1975).
8. Lin, V., and Koenig, J.L., unpublished results.
9. Frushour, B.G., unpublished results.
10. Koenig, J.L., and Frushour, B.G., Biopolymers, 11, 1871 (1972).
11. Yu, T.J., Lippert, J.L., and Peticolas, W.L., Biopolymers, 12, 2161 (1973).
12. Frushour, B.G., and Koenig, J.L., Biopolymers, 11, 1971 (1972).
13. Wilser, W.T., and Fitchen, D.B., Biopolymers, 13, 1435 (1974).
14. Small, E.W., Fanconi, B. and Peticolas, W.L., J. Chem. Phys. 52, 4369 (1970).
15. Abe, Y. and Krimm, S., Biopolymers 11, 1817 (1972).
16. Fanconi, B., et al. J. Chem. Phy. 51, 3993 (1969).
17. Lord, R.C. and Yu, N.T., J. Mol. Biol., 50, 509 (1970).
18. Frushour, B.G. and Koenig, J.L., Biopolymers, 14, 363 (1975).
19. Abe, Y., and Krimm, S., Biopolymers, 11, 1841 (1972).
20. Walton, A.G., Rippon, W.B., and Koenig, J.L., J. Am. Chem. Soc., 92, 7455 (1970).
21. Dwivedi, A.M., and Gupta, V.D., Chem. Phys. Letters, 16, 109, (1972).
22. Smith, M., Walton, A.G., and Koenig, J.L., Biopolymers, 8, 173, (1969).
23. Gupta, V.D., Singh, R.D., and Dwivedi, A.M., Biopolymers, 12, 1377 (1973).
24. Deveney, M.J., Walton, A.G., and Koenig, J.L., Biopolymers, 10, 615 (1971).
25. Srivastava, R.B., and Gupta, V.D., Biopolymers, 13, 1965 (1974).

interest because of their sensitivity to the local environment of the side-chain. Some of the amino acid residues are part of enzyme active sites such as tryptophan and tyrosine and have strong conformationally sensitive Raman lines. One can foresee studying enzyme activation and denaturation via the Raman scattering of these amino acid residues. A second major difference between the polypeptides and proteins is the distribution of secondary structures found in the latter. A typical globular protein like lysozyme will contain a mixture of alpha-helical, beta-sheet and disordered chains. Usually the ordered chain conformations in proteins are not very extensive but are short and often distorted relative to the homopolypeptides (and fibrous proteins). Defects in the ordered structures are the rule rather than the exception. Here the term defect refers to nonlinear hydrogen bonds, distorted helices, breaks in a helical segment to accomodate the cornering or folding of the polypeptide chain and other similar features. The selection rules for the normal vibrations of the ordered polypeptide structures were derived assuming zero defects and infinitely long structures. Relaxation of either assumption implies a loss of symmetry therefore frequency shifts, line broadening and the appearance of new lines are possible consequences. Some of these effects are observed for synthetic polymer chains containing defects and have been treated by Zerbi (22, 23).

Nevertheless, the characteristic frequencies and intensities of the conformationally sensitive lines of proteins correlate reasonably well with the corresponding lines of polypeptides for molecules that have similar conformations. We will now consider three proteins to illustrate this correspondence.

Tropomyosin has approximately 90% α helix and the Raman spectra shown in Figure 3 has the characteristic weak amide III region and a medium intensity skeletal line near 940 cm^{-1}. This spectrum is compared to that obtained from the pH denatured protein and it can be seen that there is a hyperchromic intensity change in the amide III line whereas the skeletal line near 940 cm^{-1} decreases in intensity.

The Raman spectrum of β lactoglobulin, which has extensive regions of β sheet structure, is shown in Figure 4. The strong intensity amide III line near 1240 cm^{-1} is characteristic of β-sheet structures. β-sheet polypeptides with small side chains have strong intensity amide III lines near 1230 cm^{-1}. The frequency difference probably arises from the "twist" in the β-sheet regions of proteins. Upon denaturation (see Figure 4) the amide III line decreases in intensity and shifts to 1246 cm^{-1}. The broad water line near 1645 cm^{-1} obscures the amide I region of the spectrum.

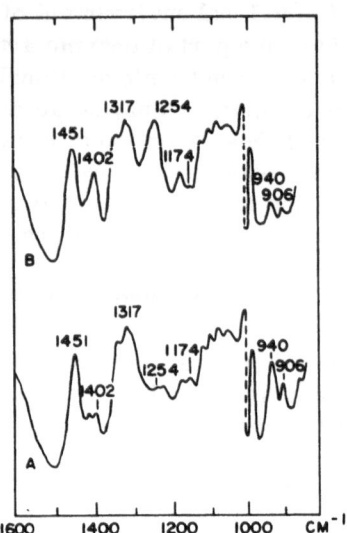

Figure 3. Raman spectrum of native and denatured tropomyosin.
A. Native, 10% concentration at pH 7.5, 0.1 N NaCl.
B. Denatured, 10% concentration at pH 12.0, 0.1 N NaCl.

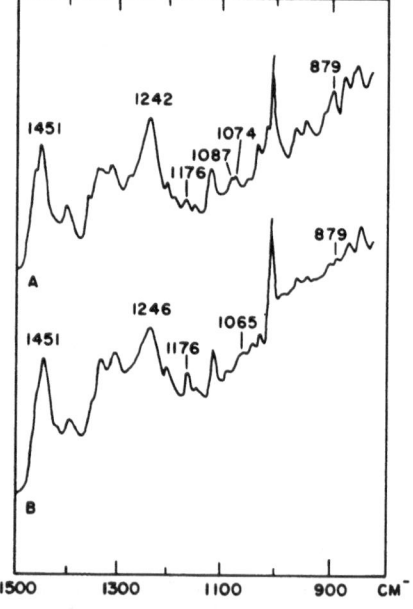

Figure 4. Raman spectrum of native and denatured β-lactoglobulin
A. Native, pH 6.0, 0.1M NaCl, 5% concentration
B. Denatured, pH 11.0.

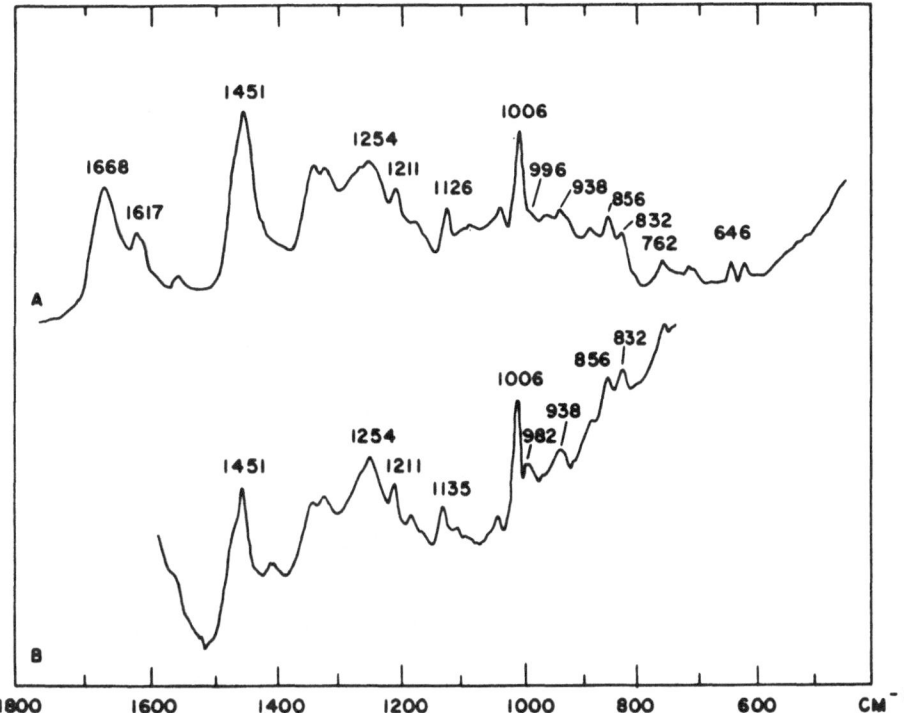

Figure 5. Raman spectrum of α-casein.
         A. Lyophilized. B. 5% solution 0.1 M NaCl.

α-casein is a protein with no ordered elements of secondary structure.
The Raman spectrum of this protein is lyophilized and solution form is
shown in Figure 5. The amide III is a broad, medium intensity line
near 1254 cm^{-1} and the amide I is at 1668 cm^{-1}.

There is an apparent change in intensity of the conformationally
sensitive amide modes and some of the modes of the tyrosine ring upon
dissolution of lyophilized samples. These spectral changes have been
attributed to conformational change; however, results recently obtained
from a study of the model amide N-methylacetamide (24) indicates that
the observed spectral shifts are probably a result of a change in the
electronic properties of the amide group in the presence of a strongly
hydrogen bonding solvent such as water. The frequencies of the conformationally sensitive lines of globular and fibrous proteins are
summarized in Tables II and III respectively. Table IV is a summary
of the frequency/structure correlations made possible by these results.

REFERENCES

1. Boerio, F.J. and Koenig, J.L., J. Macromol. Sci.--Revs.
    Macromol. Chem., C7(2), 209-249 (1972).
2. Bhagavantam, S., and Vankatarayudu, T., Theory of Groups and
    Its Applications to Physical Problems, Academic Press, New
    York, 1969.
3. Krimm, S., Advances in Polymer Science, 1, 103 (1960).
4. Zbinden, R., Infrared Spectroscopy of High Polymers, Academic
    Press, New York, 1964.
5. Elliott, A., Infrared Spectra and Structure of Organic Long-chain
    Polymers, St. Martins Press, New York, p. 83.
6. Snyder, R.G., J. Molecular Spec. 37, 353 (1971).
7. Franconi, B., Tomlinson, B., Nafie, L.A., Small, W., and
    Peticolas, W.L., J. Chem. Phys., 51, 3993 (1969).
8. Miyazawa, T., Shimanouchi, T., and Mizushima, S., J. Chem.
    Phys. 29, 611 (1968).
9. Franconi, B., Small, E., and Peticolas, W.L., Biopolymers, 10
    1277 (1971).
10. Jakes, J., and Krimm, S., Spectrochim. Acta, 27A, 19 (1971).
11. Jakes, J., and Krimm, S., Spectrochim. Acta, 27A, 35 (1971).
12. Abe, Y., and Krimm, S., Biopolymers, 11, 1817 (1972).
13. Abe, Y., and Krimm, S., Biopolymers, 12, 1841 (1972).
14. Miyazawa, T., J. Chem. Phys., 32, 1647 (1960).
15. Krimm, S., and Abe, Y., Proc. Natl. Acad. Sci. USA, 69, 2788
    (1972).

16. Koenig, J.L. and Frushour, B.G., Biopolymers, 11, 1871 (1972).
17. Painter, P., and Koenig, J.L., Biopolymers, 15, 241 (1976).
18. Small, E.W., and Peticolas, W.L., Biopolymers, 10, 69 (1971).
19. Thomas, G.J., Biochim. Biophys. Acta, 213, 417 (1970).
20. Yu, T.J., Lippert, J.L., and Peticolas, W.L., Biopolymers, 12,
    2161 (1973).
21. Frushour, B.G. and Koenig, J.L., Biopolymers, 14, 363 (1975).
22. Rubcic, A., and Zerbi, G., Macromolecules, 7, 754-759 (1974).
23. Rubcic, A., and Zerbi, G., Macromolecules, 7, 759-767 (1974).
24. Painter, P.C., and Koenig, J.L. to be published.

Table II

CONFORMATIONALLY SENSITIVE RAMAN LINES OF GLOBULAR PROTEINS

Predominant Conformation	Amide I	Amide III	Skeletal Modes	References
α-helical				
Tropomyosin	1655 s	(1253) w	940 m	1
Bovine serum albumin	1659 s	weak, broad scattering	934 m	2, 3
Glucagon	1658 s	(1266 w-m)		4
β-form				
β-lactoglobulin	1662 s	1242 s	962 w	5
Concanavalin A	1672 s	1242 s	938 w	6
α-chymotrypsinogen	1669 s	1245 m-s		7
Immunoglobulin G	1673 s	1239 s		6
Proteins with no ordered secondary structure				
α-casein	1668 s	1254 m	962 w	1
Elastin	1668 s	1254 m	938 w	8
Prothrombin	1668 s	1254 m		8
Denatured (solution)				
Tropomyosin	obscured by water	1254 m		1
Bovine serum albumin (70°C)	obscured by water	1246 m		3
β-lactoglobulin (pH 11.0)	obscured by water	1246 m		5
Glucagon		1248 m		4
Lysozyme		1245 m		9
Denatured (aggregated)				
Insulin	1673 s	1227 m		10
Glucagon (gel)	1672 s	1232 s		4
Ovalbumin	1672 s	1239 s		11

REFERENCES FOR TABLE II

1. Frushour, B.G. and Koenig, J.L., Biopolymers, 13, 1809-1819 (1974).

2. Bellocq, A.M., Lord, R.C., and Mendelsohn, R., Biochim, Biophys. Acta, 257, 280-287 (1972).

3. Lin, V. and Koenig, J.L., Biopolymers, 15, 203-218 (1976).

4. Yu, N.T., and Liu, C.S., J. Am. Chem. Soc. 94, 5127-5128 (1972).

5. Frushour, B.G., and Koenig, J.L., Biopolymers, 14, 649-662, (1975).

6. Painter, P.C., and Koenig, J.L., Biopolymers, 14, 457-468, (1975).

7. Koenig, J.L., and Frushour, B.G., Biopolymers, 11, 2505-2520, (1972).

8. Frushour, B.G., and Koenig, J.L., Biopolymers, 14, 379-391, (1975).

9. Lord, R.C., and Mendelsohn, R., J. Am. Chem. Soc. 94, 2133-2135 (1972).

10. Yu, N.T., Liu, C.S., and O'Shea, D.C., J. Mol. Biol. 70, 117-132 (1972).

Table III

CONFORMATION-SENSITIVE RAMAN MODES OF FIBROUS PROTEINS

Predominant Conformation	Amide I	(cm^{-1}) Amide II	Skeletal Modes
α-helical			
Porcupine quill	1653 S (sharp)	~1260 W	935 S
Wool	1658 S (broad)	1261 W (α') 1245 M (disordered)	935 S
β-form			
β-Keratin	1672 S	1237 S	
Cross-β (steam supercontracted wool)	1671 S	1239 S	
Silk	1667 S	1229 S (β) 1259 W (disordered)	880 W (β) 1169 M (β)
Feather	1669 S	1240 S (β) 1274 M (disordered)*	879 W (β) 1162 M (β)
Disordered			
LiBr supercontracted wool	1666 S	1249 S	

Note: S, strong; M, medium; W, weak.

*See text for explanation of discrepancy.

Table IV

SUMMARY OF CONFORMATIONALLY SENSITIVE PROTEIN RAMAN LINES

	Amide I	Amide III	Skeletal Mode
Alpha Helix	$1655-1659 \text{ cm}^{-1}$ (S)*	$1250-1280 \text{ cm}^{-1}$ (W)	$900-960 \text{ cm}^{-1}$ (S)
Antiparallel Beta Sheet	$1667-1672 \text{ cm}^{-1}$ (S)	$1229-1240 \text{ cm}^{-1}$ (S)	$900-960 \text{ cm}^{-1}$ (W)
Random Coil (Disordered)	$1665-1675 \text{ cm}^{-1}$ (S)	$1243-1265 \text{ cm}^{-1}$ (M)	$900-960 \text{ cm}^{-1}$ (W)

*Intensity: (S) strong, (M) medium, (W) weak.

VIBRATIONAL SPECTROSCOPY OF CARBOHYDRATES

Jack L. Koenig

Department of Macromolecular Science
Case Western Reserve University, Cleveland, Ohio 44106

Effective use of infrared and Raman spectroscopy in investigating structure and conformation of carbohydrates requires an understanding of the structural and vibrational origin of the observed frequencies. A complete assignment of the observed vibrations in carbohydrate molecules is quite a formidable task, especially in the region below 1500 cm^{-1}.

The practical difficulties in recording spectra of aqueous solutions of carbohydrates are reflected by the limited amount of spectroscopic work reported for these materials. Goulden (1) has recorded the infrared spectra of a number of carbohydrates in aqueous solution in the region 1000-1500 cm^{-1}. Parker (2) studied the matarotation of α and β-D-glucose and β-D-maltose in aqueous solutions using IR, but only in the region 900-1600 cm^{-1}. Mitchell (3) has also examined mutarotation of several carbohydrates in the 3400 cm^{-1} region.

The intense absorption of water over most of the infrared spectrum restricts the regions where aqueous solutions can be usefully studied. However, water gives only one weak Raman line, in the region 0-2000 cm^{-1}, at approximately 1640 cm^{-1}. A further advantage is that sampling techniques are much simpler for Raman spectroscopy since ordinary glass can be used for sample cells.

The Raman scattering for aqueous solutions of D-glucose, D-maltose, D-cellobiose and dextran have been studied. In particular, we have examined the effect of deuteration on the spectra of the aqueous solutions of these carbohydrates in order to obtain a better identification

125

Theo M. Theophanides (ed.), Infrared and Raman Spectroscopy of Biological Molecules, 125–137.
All Rights Reserved. Copyright © 1979 by D. Reidel Publishing Company, Dordrecht, Holland.

of the COH vibrational frequencies, which occur below 1500 cm^{-1}. We have examined the spectra of these materials in solutions of heavy water (D_2O) with spectra of solutions in normal water (H_2O). In D_2O solution, an exchange of OH to OD occurs. This exchange causes a COH vibrational frequency to decrease in intensity and a new frequency due to a COD vibration to appear in the spectrum.

Frequencies related to CH_2 and CH vibrations have been investigated using several deuterated glucose compounds. The molecules studied were D-glucose-6, 6-d_2 which has a CD_2 in place of CH_2 for the CH_2OH side chain, D-glucose-1-d_1 which has a CD on carbon C_1 rather than CH, and D-glucose-1, 2-d_2 which has a CD on both C_1 and C_2. In the spectra of these materials, the CH_2 and CH related frequencies should decrease in intensity while new bands related to CD_2 and CD are expected to appear. The infrared spectra of these deuterated glucoses were compared to the spectrum of α-D-glucose. Attempts to examine the Raman scattering of these compounds were unsuccessful because of fluorescence problems.

The Raman spectra of solutions of α-D-glucose*, β-D-maltose, D-cellobiose, and dextran in H_2O and D_2O are given in Figures 1 through 4 respectively. There are several features of these spectra which should be noted. The first feature is the similarity of the solution spectra of these four carbohydrates in the region 700-1500 cm^{-1}. The frequencies of the lines observed for solutions in H_2O are listed in Table 1. The similarity in these spectra in the 700-1500 cm^{-1} region is quite important since it may indicate that assignments, in this region, for these carbohydrates will be relevant in studies of cellulose and other polymers of D-glucose.

One feature in this region of the spectrum is the line at 847 cm^{-1} in the D-glucose solution. This line occurs at 846, 841, and 847 cm^{-1} in the D-maltose, D-cellobiose, and dextran solutions, respectively. The intensity of this line is strong in the D-glucose, D-maltose, and dextran spectra while it is weak in the D-cellobiose spectrum. This intensity difference corresponds to the difference in the amount of α-configuration at C_1 for the molecules. The D-glucose exists in a 3:1 ratio of α to β configuration at the anomeric C_1 position in solution. The D-maltose and dextran have the α-configuration at the glycosidic linkage. In

---

*β-D-glucose (in H_2O and D_2O) was also examined, but due to the muta-rotation in solution, its solution spectra were found to be essentially the same as for α-D-glucose.

Table 1

Observed Frequencies for Carbohydrate Solutions

D-Glucose		D-Maltose		D-Cellobiose		Dextran	
1461	S	1458	S	1460	S	1462	M
1405	Sh	1407	Sh	1409	W		
						1386	Sh
1373	S	1372	S	1375	S	1371	M
1349	M(Sh)	1350	S	1350	S	1348	S
1335	M(Sh)	1335	M(Sh)	1342	W(Sh)	1337	S
1328	Sh	1325	Sh	1324	Sh		
1298	W	1296	W	1301	Sh	1300	W
1278	M	1271	M	1264	M	1274	M
		1237	W	1233	Sh	1229	W
1222	Sh						
1206	W			1197	W	1207	W
1152	W	1157	Sh			1153	Sh
		1141	Sh				
1130	VS	1127	VS	1126	VS	1136	S
						1116	W
1071	VS	1079	VS	1080	VS	1084	S
		1070	Sh(S)			1066	W
1041	W	1043	W	1035	W	1041	W
1020	S(Sh)	1022	S	1022	M	1022	M
				990	Sh	955	W
913	S	913	S	915	Sh	917	S
898	S	896	Sh	890	S		
859	Sh	861	Sh	860	M		
847	S	846	S	841	Sh	847	VS
771	M	779	M	771	M	768	M
747	W	740	M				
		722	W	714	W		
705	M	709	W			701	M
635	M	642	W	654	W	656	W
				620	W	633	M
		602	W				
		596	Sh				
		583	Sh	572	S		
				562	Sh		
541	W	543	VS			541	VS
		530	Sh	531	S		
514	VS	515	S	522	S	525	S
498	Sh						
		459	S				
443	S			443	Sh		
				437	VS		

Table 1 (Continued)

D-Glucose		D-Maltose		D-Cellobiose		Dextran	
423	S	426	S	428	VS		
409	Sh	405	Sh			409	S
381	W					379	W
				362	S		
341	M	337	S	343	Sh	343	W
		324	Sh	320	Sh		
294	W	286	W	291	W		
274	W						

S - strong
M - medium
W - weak
Sh - shoulder
Br - broad
V - very

contrast, the D-cellobiose, in which the intensity of the 841 cm^{-1} line is weak, has the β-configuration at C_1. Therefore, the line at ~845 cm^{-1} is assigned to a vibration at C_1 with an α-configuration. This assignment has been previously suggested by Barker, et al (4, 5, 6). Work on carbohydrates has also assigned the band at ~890 cm^{-1} as due to the β configuration at the anomeric C_1 position. Our solution studies do not support this assignment. This band is very strong in cellobiose which has the β-configuration at C_1. However, a strong band also occurs in the spectra of glucose at 898 cm^{-1} and maltose at 896 cm^{-1} in solution. It is possible that the observed line in glucose and maltose does not have the same structural basis as cellobiose, but the results suggest that the line at 890 cm^{-1} in these carbohydrates is not unique to the β configuration at the anomeric C_1 position.

In the region below 700 cm^{-1}, each carbohydrate examined has distinct features in its Raman spectrum, as observed in Figures 1-4. Each compound has a characteristic pattern of observed lines and intensities. Therefore, this region of the Raman spectrum has potential as an identification method for these molecules in solution. This is particularly valuable in view of the difficulties encountered in infrared work in this region, especially in aqueous solution (1, 2, 3).

The solution spectra for each compound were studied in both ordinary water (H_2O) and heavy water (D_2O). The spectra of the different carbohydrate solutions (H_2O and D_2O) are shown in Figures 1 to 4. In D_2O solution, an exchange of OH to OD occurs which decreases the intensity of the COH band. Because of the increased mass of deuterium (versus hydrogen), the COD vibration occurs at a different frequency.

The first Raman line with such an observed intensity decrease upon deuteration occurs at 1349 cm^{-1} in D-glucose, Figure 1. This line occurs in H_2O solution but is absent in D_2O solution. All observed frequencies are from the sample and not the water (H_2O or D_2O). Examination of Figures 2, 3, and 4 indicates that this same intensity decrease of the shoulder at 1349 cm^{-1} occurs in each sample on going from H_2O to D_2O. For D-maltose, this line decreases in intensity but because of the overlap does not apparently disappear. However, the assignment of the 1349 cm^{-1} line to a COH mode is further confirmed by comparison of the spectra of the solid α-D-glucose which had been recrystallized from methanol (MeOH) and deuterated methanol (MeOD), Figure 5. This line (no longer badly overlapped in the solid) is absent for α-D-glucose recrystallized from MeOD in which exchange of OH to OD has occurred.

The next frequency to be considered is at 1071 cm^{-1} for D-glucose. In

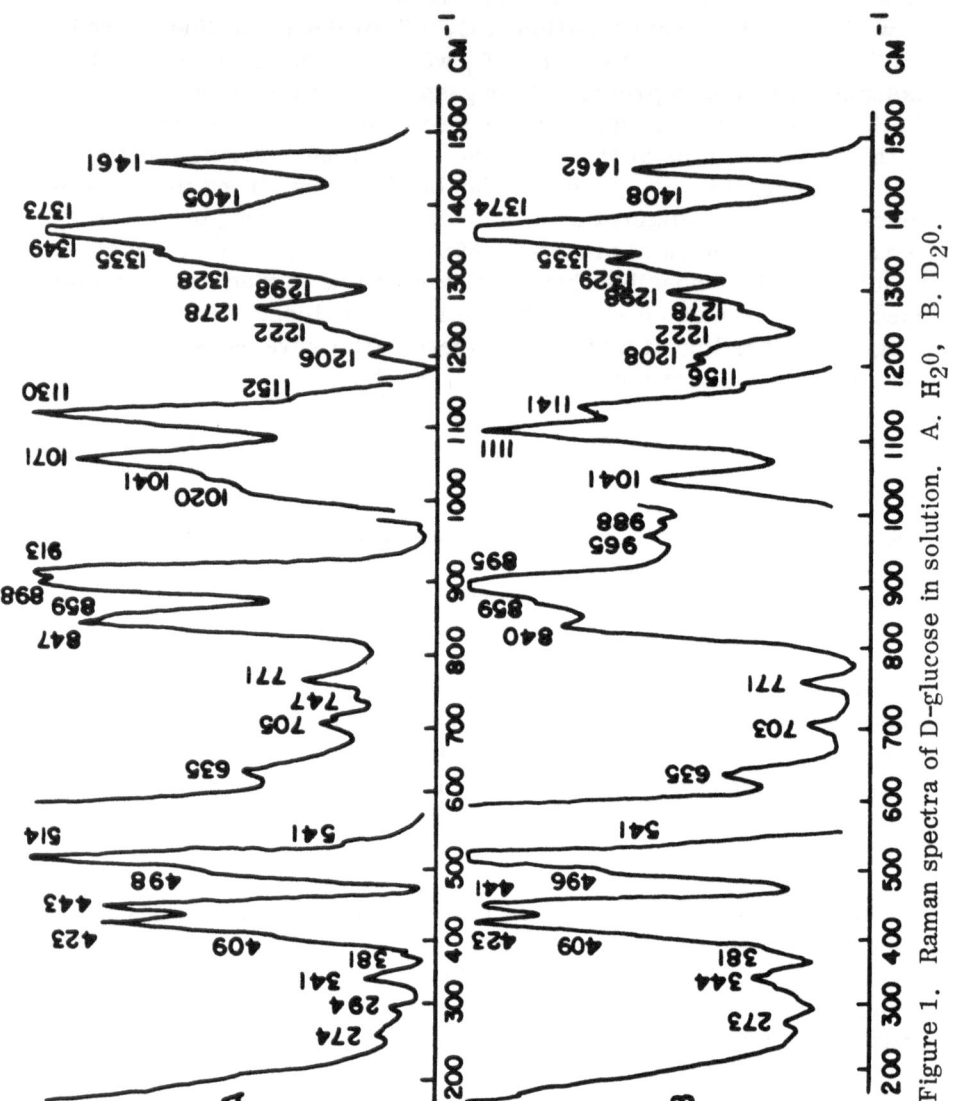

Figure 1.   Raman spectra of D-glucose in solution.   A. H_2O,   B. D_2O.

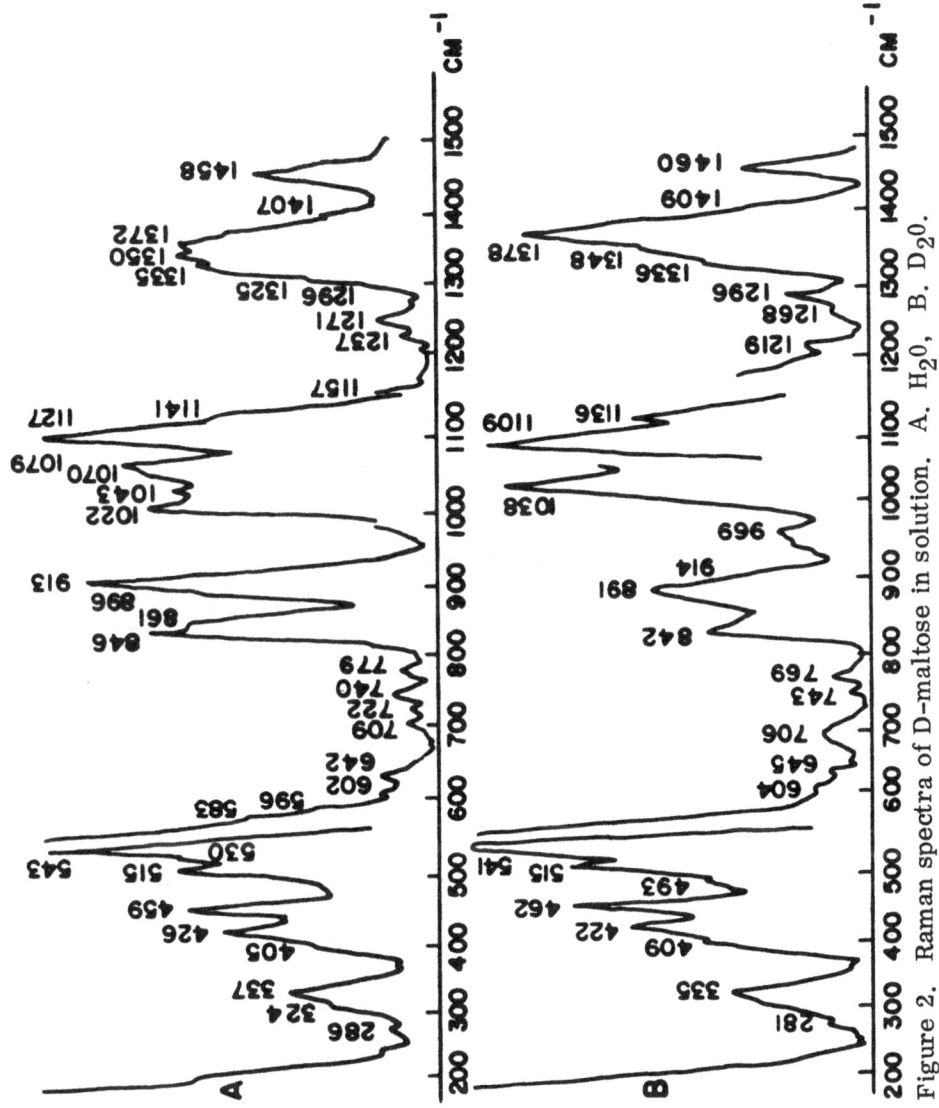

Figure 2.   Raman spectra of D–maltose in solution.   A. H_2O,   B. D_2O.

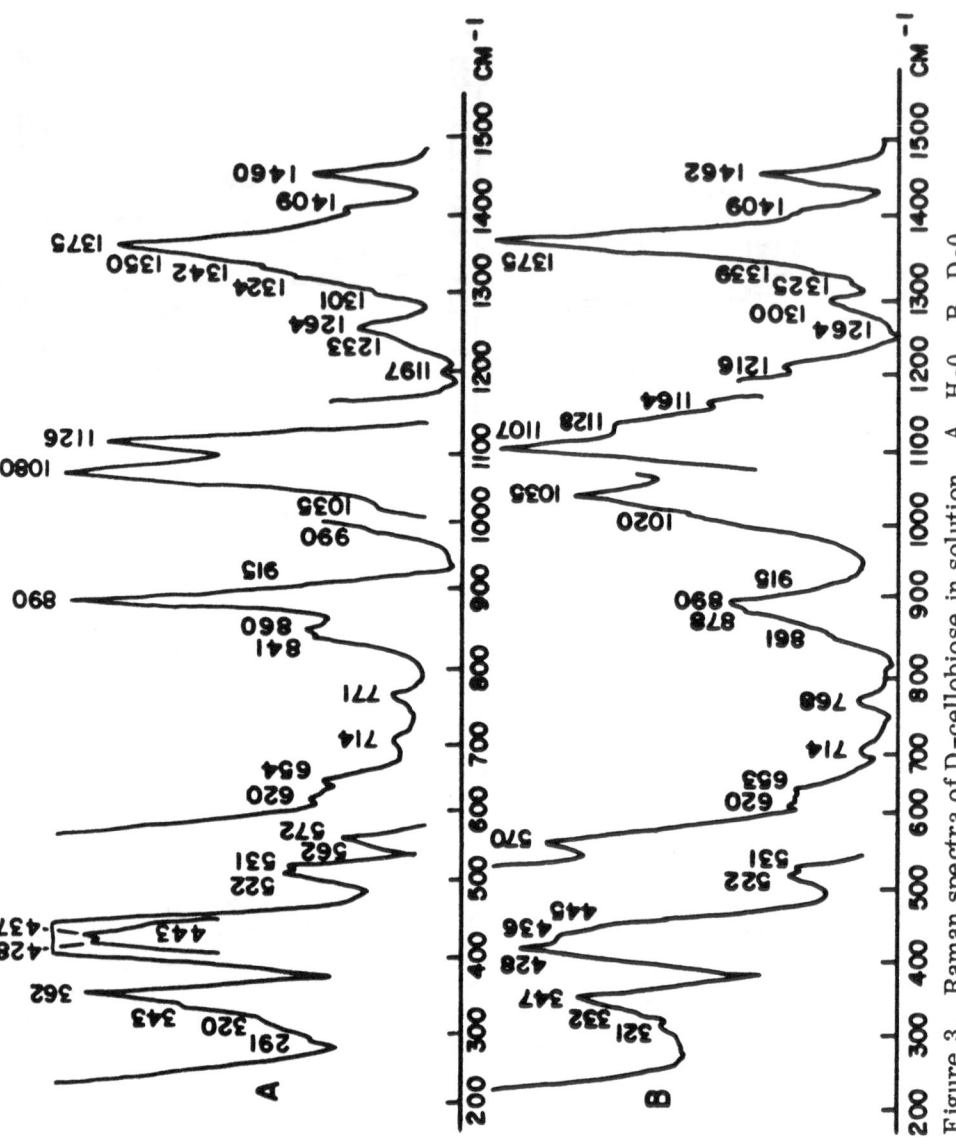

Figure 3.   Raman spectra of D-cellobiose in solution.   A. H_2O,   B. D_2O.

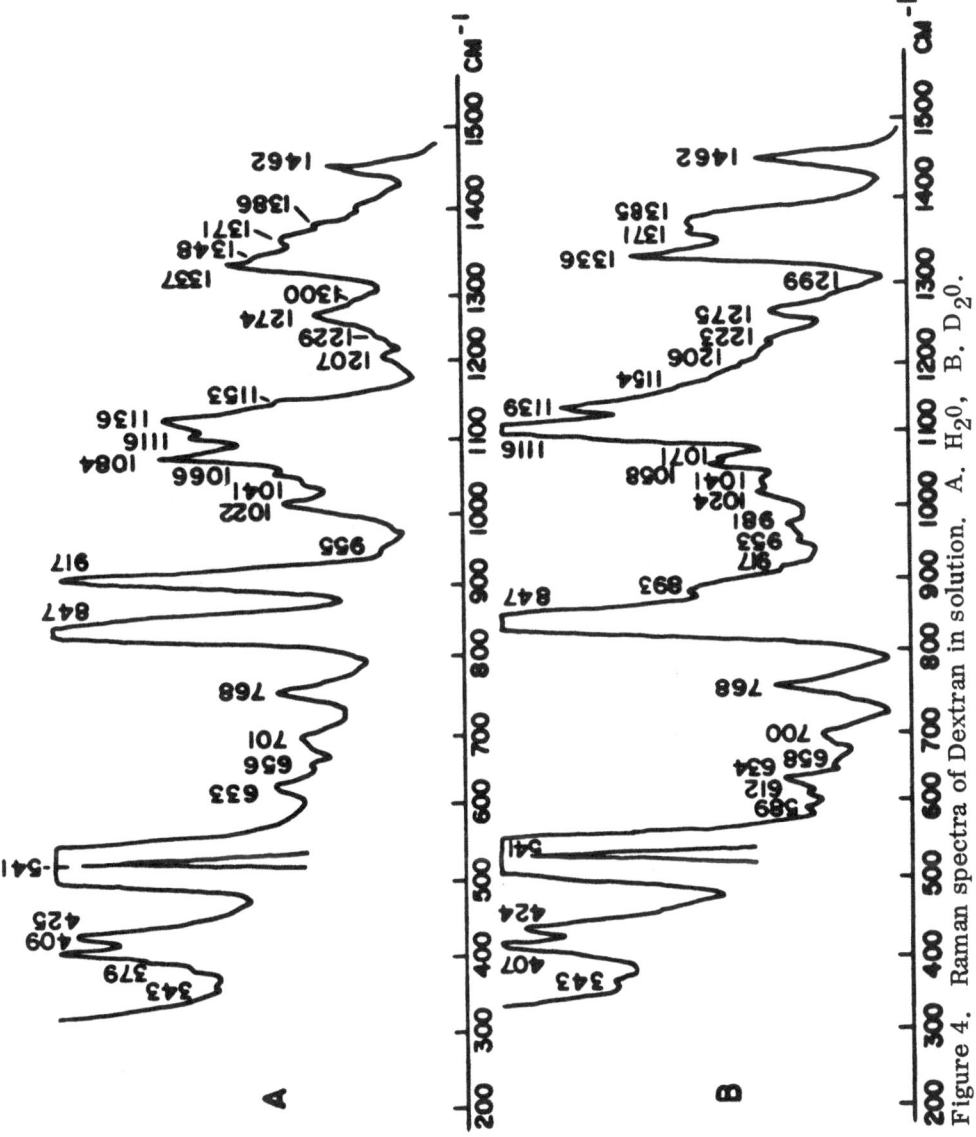

Figure 4. Raman spectra of Dextran in solution. A. H_2O, B. D_2O.

the spectrum of D-maltose, there are two lines in this region at 1079 and 1070 cm^{-1}. Both these frequencies decrease in intensity upon deuteration and appear to be due to COH vibrations. For D-cellobiose, there is a COH line at 1080 cm^{-1}, and for dextran at 1084 cm^{-1}. In each case, the observed line in H$_2$0 solution is not observed in D$_2$0 solution.

The line at 1020 cm^{-1} in D-glucose is also assigned as a COH frequency. This line is observed in the spectra of D-maltose at 1022 cm^{-1}, D-cellobiose at 1020 cm^{-1}, and dextran at 1024 cm^{-1}, and in each case the line is absent from the spectrum for the respective D$_2$0 solution.

The final line assignable to a COH vibration is at 913 cm^{-1} in the spectrum of D-glucose. A similar line occurs for D-cellobiose at 915 cm^{-1}, and for dextran at 917 cm^{-1}. These lines are absent from the spectra of the D$_2$0 solutions.

For D-cellobiose, the line at 890 cm^{-1} decreases considerably upon deuteration and is thought to be due, at leasd partially, to a COH vibration.

When deuteration occurs and a COH frequency decreases in intensity, a COD mode should be observed at a lower frequency in the spectrum. A line appears at 1110 cm^{-1} upon deuteration of all the samples. This line is assigned to a COD vibration, and is the only COD frequency identified by the deuteration. The other COD modes apparently occur in regions of the spectra where they cannot be observed due to line overlap or they are inherently too weak to be observable.

The spectra of D-glucose in solution in mixtures of H$_2$0 and D$_2$0 in the 1300 cm^{-1} region have been studied. As the ratio of H$_2$0 : D$_2$0 goes from pure H$_2$0 to pure D$_2$0, the 1278 cm^{-1} line decreases and the 1298 cm^{-1} line increases. Similar results occur for D-maltose and D-cellobiose. However, no such change occurs for the 1274 cm^{-1} line for dextran.

The exchange reactions for D-glucose and dextran in D$_2$0 solution have been compared. One must conclude that these changes for D-glucose are probably related to deuteration of the CH$_2$OH side group, since this position is not available for exchange in dextran. Thus, at least in part, the 1278 cm^{-1} line of glucose is assigned to a vibrational mode of the CH$_2$OH group. However, an additional contribution occurs in this frequency range from another structural source as evidenced by the presence of a line at 1274 cm^{-1} in dextran and the fact that the line at 1278 cm^{-1} in glucose does not completely disappear with deuteration.

The gradual nature of this intensity change with increasing D_2O concentration, would indicate that perhaps the CH_2OH hydroxyl exchanges more slowly than the other OH's.

The infrared spectra for the C-D deuterated glucoses have been obtained. The observed frequencies are given in Table 3 together with the assignments based on the spectral changes observed. In order to compare the observed frequencies to those for undeuterated glucose, the frequencies for α-D-glucose are also listed in Table 2 as recorded in our laboratory and the literature (7, 8).

There are four bands which can be identified as due to CH_2 motions in α-D-glucose as seen in Table 2. These bands occur at 1460, 1219, 1011, and 836 cm^{-1}. The 1460 and 836 cm^{-1} bands are much decreased in D-glucose-6, 6-d_2 while the bands at 1219 and 1011 cm^{-1} are absent completely from this spectrum. A new band in the spectrum of D-glucose-6, 6-d_2 appears and is assigned as a CD_2 motion at 967 cm^{-1}. It is difficult to identify any other bands as CD_2 related because of weak intensities.

Deuteration of the CH_2 to CD_2 does not affect the 1262 cm^{-1} band. In the previous section, the results indicate that this particular band is affected by deuteration of the CH_2OH (CH_2OH) group. These infrared deuteration results suggest the additional absorption in this region does not arise from the CH_2 portion of the molecule.

Five bands have decreased intensity associated with deuteration in D-glucose-1-d_1, as shown in Table 2. The spectral changes observed for D-glucose-1, 2-d_2 are the same as for D-glucose-1-d_1, Table 2. The bands occurring at 1142, 1076, 1047, 911, and 836 cm^{-1} are present in the undeuterated sample and absent in the deuterated one. These bands are assigned to C_1-H bending vibrational modes. The vibrational mode may be associated with either the CCH group or the OCH group or a combination of the two.

In the spectrum of D-glucose-1-d_1, there are ten bands which can apparently be assigned as C_1-D related vibrations based on the fact that they are present in the deuterated sample and absent in the undeuterated. These bands occur at 1419, 1346 (partially), 1290, 1264, 1228 (partially), 1158, 1090, 960, 874, and 810 cm^{-1}.

Table 2
Frequencies for Deuterated Glucoses and Band Assignments

a-D-Glucose	D-Glucose-6, 6-d$_2$	Assign.	D-Glucose-1, -d$_1$	Assign.	D-Glucose-1, 2-d$_2$	Assign.
1460 S	1458 Sh	CH$_2$	1459 S		1457 S	
1442 Sh	1440 S				1447 M	
1427 VW	1421 Sh		1419 S-Br	C-D	1421 S	C-D
1402 W	1398 W		1404 Sh			
1378 M	1369 M		1369 M		1370 M	
	1355 M					
1337 Sh			1346 S	C-D (Part.)	1341	CD (Part.)
1328 M	1322 M		1323 W		1329 W	
1293 W	1296 M		1290 S	C-D (Part.)	1294 Sh	
1262 M	1270 M		1264 M		1268 S	
	1250 Sh					
	1238 W					
1219 M		CH$_2$	1228 VS	C-D	1220 VW	
1197 M	1205 S		1205 M		1206 Sh	C-H
1189 W					1179 M	C-D
			1158 S	C-D	1163 W	
1142 M	1142 M			C-H	1143 S	
			1131 Sh			
			1115 Sh			
1104 S	1112 S		1106 S(Sh)		1106 S(Sh)	
	1091 Sh		1090 S	C-D	1088 S	C-D
1076 M	1082 W			C-H		C-H
			1063 S(Sh)	C-D	1065	C-D
1047 M	1049 Sh			C-H	1053 Sh	C-D
1026 S(Sh)	1031 S		1035 S		1032 S	
1011 S		CH$_2$	1014 S-Br		1003 M	
988 S	993 S		989 S		987 S(Sh)	
	967 S	CD$_2$	960 M	C-D	961 M	C-D
					954 M	C-D
					924 W	
911 M	908 S			C-H		C-H
890 Sh					896 W	
			874 VS	C-D	876 W	
	866 W					
					849 W	
836 S	837 W	CH$_2$		C-H		C-H
			810 VS	C-D	812 S	C-D
	799 W				800 W	
768 M	768 M		761 S		761 M	
	743 W					
721 W-Br	717 W-Br		716 M-Br		724 M-Br	
645 Sh			640 Sh		640 Sh	
622 S-Br	621 S-Br		627 W		624 M-Br	
603 S-Br	594 S		612 S-Br			
555 W-Br	558 W		553 M		552 M	
	536 Sh		534 W		533 W	
522 W						
	508 Sh		497 VW			
	450 W				451 VW	
			431 W		427 W	
	418 W					
			401 W		403 W	

S - strong
M - medium
W - weak
Sh - shoulder
Br - broad
V - very

Spectral studies have allowed the identification of some of the frequencies in the carbohydrate spectra arising from COH, CH_2, and CH vibrational modes. These assignments aid interpretations of the vibrational spectra of carbohydrate molecules.

The results obtained further indicate that many of the frequencies in the fingerprint region of carbohydrate spectra are due to complicated, coupled modes of vibration.

REFERENCES

1. Goulden, J.D.S., Spectrochimica Acta, 9, 657 (1959).
2. Parker, F.S., Biochim. Biophys. Acta 42, 513 (1960).
3. Michell, A.J., Carbohyd. Res. 5, 229 (1967).
4. Barker, S.A., Bourne, E.J., Stacey, M., and Wiffen, D.H., J. Chem. Soc. 171 (1954).
5. Barker, S.A., Bourne, E.J., Stevens, R., and Wiffen, D.H., J. Chem. Soc. 3468 (1954).
6. Barker, S.A., Bourne, E.J., Stevens, R., and Wiffen, D.H., J. Chem. Soc. 4211 (1954).
7. Casu, B. and Reggiani, M., J. Poly. Sci., C, 171 (1964).
8. Casu, B. and Reggiani, M., Due Starke 7, 218 (1966).

CONFORMATIONAL ANALYSIS OF SNAKE TOXINS BY LASER RAMAN SPECTROSCOPY

Anthony T. Tu

Department of Biochemistry, Colorado State University, Fort Collins, Colorado 80523 USA.

ABSTRACT. Conformations of snake toxins, neurotoxins and hemorrhagic toxin were analyzed by laser Raman spectroscopy. All sea snake neurotoxins consist of antiparallel β-sheet and β-reverse turn structures. Hemorrhagic toxin e isolated from western diamondback rattlesnake is a zinc dependent protease. Judging from the Raman spectra, hemorrhagic toxin e consists of some α-helix with a high degree of random coil or β-reverse turn structure. All these toxins have gauche-gauche-gauche conformation of C-C-S-S-C-C according to S-S stretching vibration frequency. The tyrosine residue of most sea snake neurotoxins is buried as I_{830} is stronger than I_{850}. From the 1361 cm^{-1} band, it is concluded that the tryptophan residue of neurotoxins is exposed to the outside of the molecule.

INTRODUCTION

Snake venom is a complex mixture of proteins with diverse biological activities. The proportions of the different substances in venom and their specific characteristics vary among the species. However, usually the closer the phylogenetic relationship of the snakes, the more similar are the venom properties and composition.

In recent years, laser Raman spectroscopy has been applied successfully to different proteins. In this paper, conformational analysis of snake toxins done in our laboratory is presented.

139

Theo M. Theophanides (ed.), Infrared and Raman Spectroscopy of Biological Molecules, 139–152.
All Rights Reserved. Copyright © 1979 by D. Reidel Publishing Company, Dordrecht, Holland.

1. SEA SNAKE NEUROTOXINS

The venoms of sea snakes contain potent neurotoxins which bind to
the acetylcholine receptor in the neuromuscular junction.
Neurotoxins found in sea snake venoms belong to Type I which are
composed of 60-62 amino acid residues with 4 disulfide bridges
(1,2). A number of neurotoxins have been isolated from the
venoms of different sea snakes such as Lapemis hardwickii,
Pelamis platurus, Laticauda semifasciata, and Enhydrina schistosa.

       For a better understanding of conformational analysis, it
is useful to know the amino acid sequence of a toxin. Recently,
the amino acid sequence of the major toxin of Lapemis hardwickii
was elucidated (3).

1.1 Peptide Backbone Conformation

The frequencies of the Amide I and III are most frequently used
for the determination of peptide backbone conformation. The
Amide I band arises from the coupled C=O stretching vibrations
of the peptide bond. The α-conformation normally appears in the
region of 1650-1658 cm^{-1}; random coil at 1664-1666 cm^{-1}; and
β-sheet conformation in the region of 1665-1680 cm^{-1} (4,5).
The Amide III band originates from "in-plane" vibration of the
peptide bond. The α-conformation usually gives very high Amide
III bands of 1260-1298 cm^{-1}; random coil 1242-1252 cm^{-1}; β-
sheet at 1235-1240 cm^{-1}. The β-reverse turn conformation using
oxytocin as a model compound shows the Amide I band at 1663-
1666 cm^{-1} and the Amide III band at 1260-1266 (6). The β-
reverse turn conformation using [1-penicillamine, 2-leucine]

oxytocin and [1-penicillamine] oxytocin as model compounds shows the Amide I band at 1656-1668 cm^{-1} and the Amide III band at 1255 cm^{-1} (7). Judging from the Amide I band at 1672 cm^{-1} of different sea snake neurotoxins, the possibility of α-helix conformation is excluded (Figs. 1-4). The validity of the Amide III can be readily confirmed by observing the shift of Amide III at 1240 cm^{-1} to 980 cm^{-1} after deuteration (Fig. 5). The Amide III of different toxins lies in the range 1240-1248 cm^{-1}. It is concluded that snake neurotoxins are composed of antiparallel β-sheet and β-reverse turn structures.

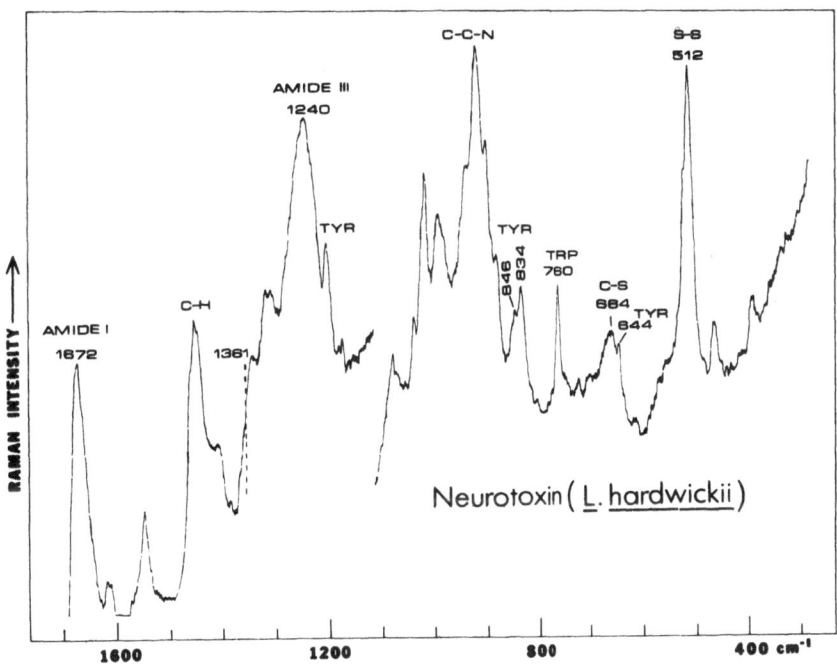

Fig. 1. Raman spectrum of the major toxin isolated from the venom of <u>Lapemis hardwickii</u> (reproduced from reference 13).

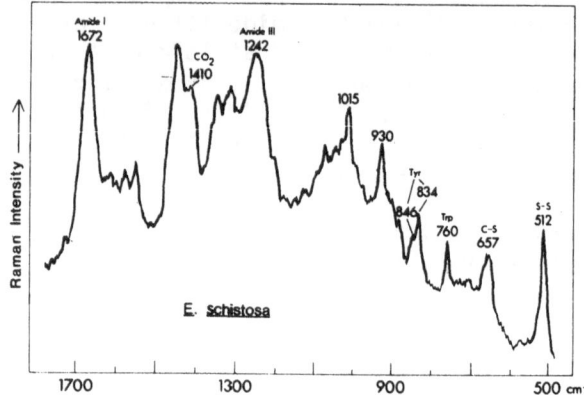

Fig. 2. Raman spec-
trum of the major
toxin isolated from
the venom of
<u>Enhydrina schistosa</u>
(reproduced from
reference 13).

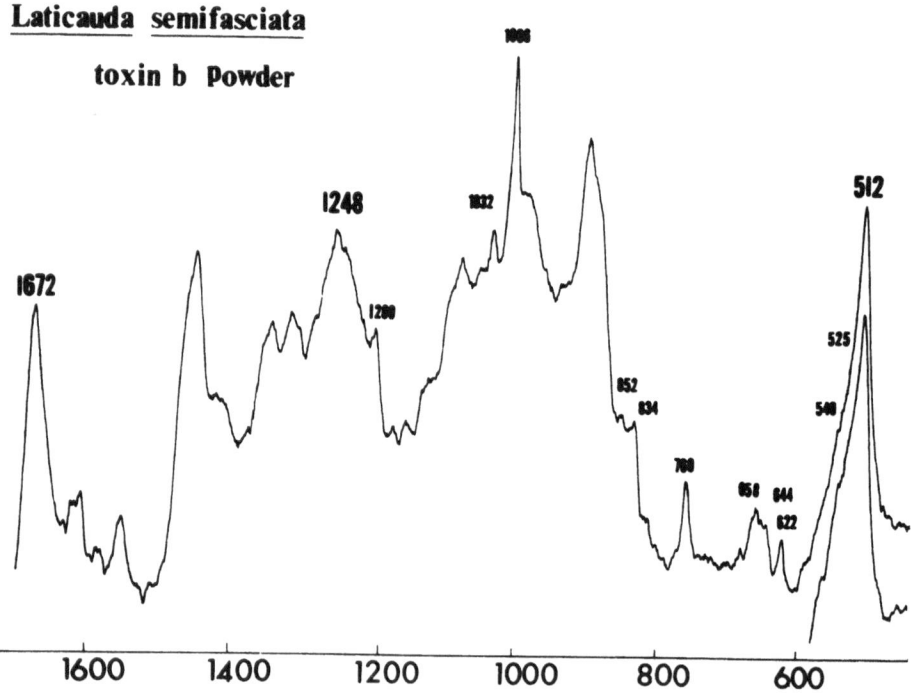

Fig. 3. Raman spectrum of toxin b isolated from the
venom of <u>Laticauda semifasciata</u> (reproduced
from reference 12).

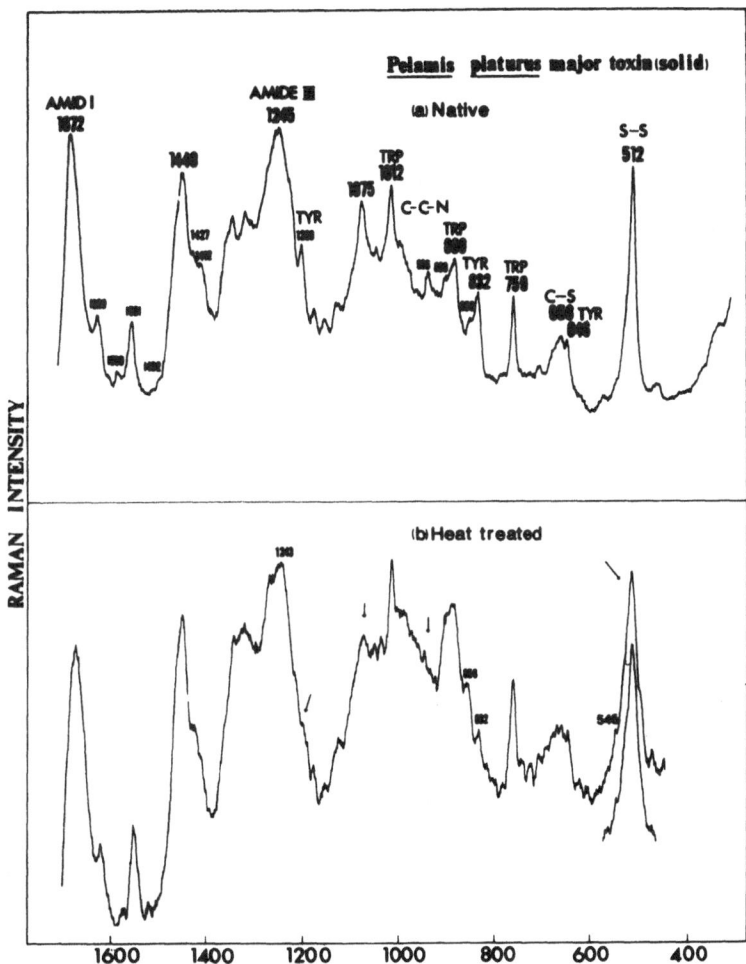

Fig. 4.   Raman spectra of major toxin isolated from
          the venom of Pelamis platurus (reproduced
          from reference 12).

Because of the β-reverse turn and antiparallel conformation
of sea snake toxins, the Amide III bands are found at a frequency between the typical antiparallel band region (1235-
1240 cm⁻¹) and the Amide III frequency of β-reverse turn conformation (1260 cm⁻¹). This conclusion is further confirmed by
other methods. Since the sequence of Lapemis hardwickii is
known, the secondary structure can be predicted by the method

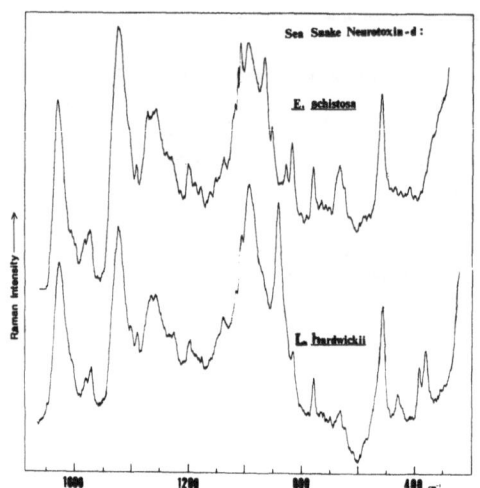

Fig. 5. Raman spectra
of toxins in D$_2$O
(reproduced from
reference 13).

of Chou and Fasman (8). The result indicates that the neurotoxin indeed contains high amounts of β-antiparallel and β-
reverse turn structures (Fig. 6).

In order to get more quantitative information, the
percentage conformation was calculated by the methods
developed by Lippert et al. (9) and Pezolet et al. (10).

Lippert et al. have proposed a technique for determining
the secondary structural content of proteins from their Raman
spectra in H$_2$O and D$_2$O using the relative intensities of the
Amide I and Amide III components. A set of four simultaneous
equations is used:

$$c^{protein} \, I^{protein}_{1240} = f_\alpha \; I^\alpha_{1240} + f_\beta \; I^\beta_{1240} + f_R \; I^R_{1240}$$

$$c^{protein} \, I^{protein}_{1632} = f_\alpha \; I^\alpha_{1632} + f_\beta \; I^\beta_{1632} + f_R \; I^R_{1632}$$

$$c^{protein} \, I^{protein}_{1660} = f_\alpha \; I^\alpha_{1660} + f_\beta \; I^\beta_{1660} + f_R \; I^R_{1660}$$

$$f_\alpha + f_\beta + f_R = 1$$

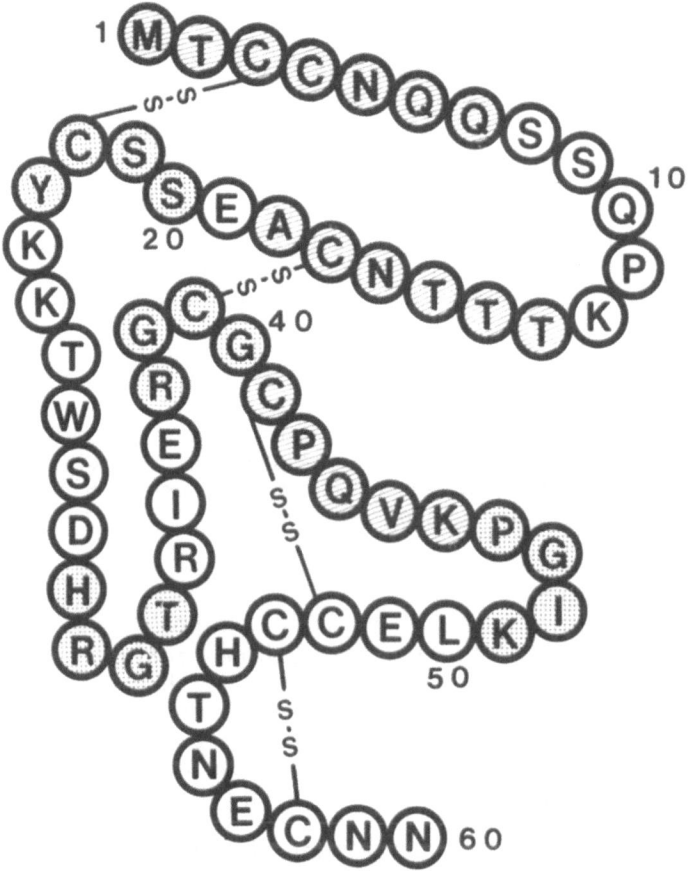

Fig. 6. Two-dimensional structure of <u>Lapemis</u> <u>hardwickii</u> toxin elucidated by the method of Chou and Fasman. Dotted portion indicates the β-reverse turn and lined portion the β-sheet structure.

The experimental values of $I^{protein}$ (intensity) are related to the $I_\nu{}^\alpha$, $I_\nu{}^\alpha$, $I_\nu{}^R$ intensities previously determined for poly (L-lysine) in its various secondary structural forms. $C^{protein}$ is a scaling constant which represents the relative intensity of the methylene band in the protein.

The scaling constant ($C^{protein}$) determined upon solution of the four simultaneous equations was 0.75. The f_β value gave the portion of β-sheet present in the toxin as 35%.

The method of Pezolet et al. (10) was also applied for the determination of β-sheet structure. This method incorporates the intensity of the Amide III band at 1240 +3 cm^{-1} (a requirement of the technique) and the band at 1450 cm^{-1} assigned to the methylene bending modes. The 1450 cm^{-1} band serves as an intensity standard due to its insensitivity to the conformation of the protein. The absolute intensity of the methylene band was calculated using 1.8 as the average number of CH_2 moieties per residue in the major toxin. Using this factor and the intensities of the 1243 cm^{-1} (Amide III) and 1452 cm^{-1} (methylene) peaks the percentage of β-sheet structure in the neurotoxin was determined to be 36 ± 10% in the H_2O neurotoxin sample.

The amino acid sequence method of Chou and Fasman gave the values of 20% β-sheet; 35% β-reverse turn and 47% random coil conformation. The circular cichroism method of Chen et al. (11) gave a negative value which is nonsense value. Presumably this means that for a special protein like snake neurotoxins which have an unusually high content of β-reverse turn, the equation developed by Chen et al. cannot be used.

1.2. Heat Treatment

After vigorous heat treatment, the backbone configuration of the toxin molecule basically remained the same although it was partially denatured (Fig. 4); there were no major changes in the Amide I and III bands. The major peak at 512 cm^{-1} was not altered by the heat treatment but a new shoulder appeared at 546 cm^{-1}. This may suggest that a new type of S-S stretching vibration due to trans-gauche-trans form of C-C-S-S-C-C was produced as a result of heat treatment although the majority of the S-S vibrations remained in the gauche-gauche-gauche orientation (Fig. 7). A substantial change in the interactions between a tyrosine aromatic ring and neighboring residues was apparently caused by the heat treatment (12).

The partial denaturation without major changes in the peptide backbone and disulfide bridge conformations is illustrated in Fig. 8.

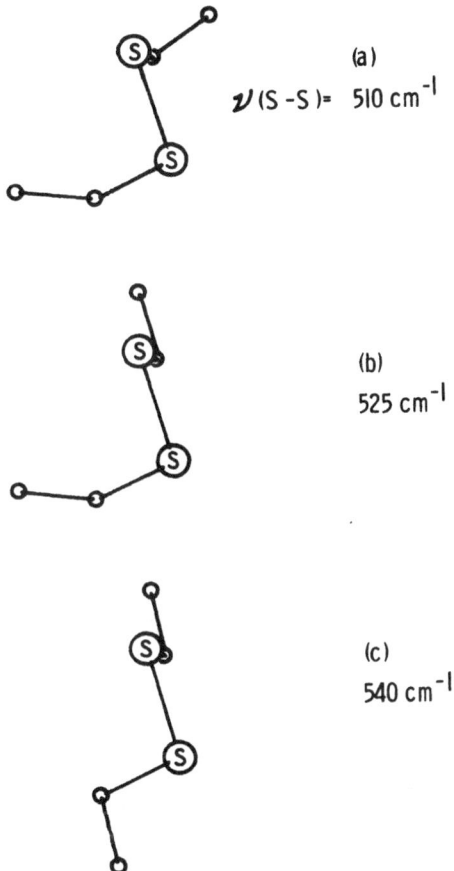

(a)

ν (S - S)= 510 cm^{-1}

(b)

525 cm^{-1}

(c)

540 cm^{-1}

Fig. 7. Dependence of the S-S stretching frequency
on the internal-rotation.
(a) gauche-gauche-gauche form, 510 cm^{-1}
(b) gauche-gauche-trans form, 525 cm^{-1}
(c) trans-gauche-gauche form, 540 cm^{-1}

(reproduced from reference 12).

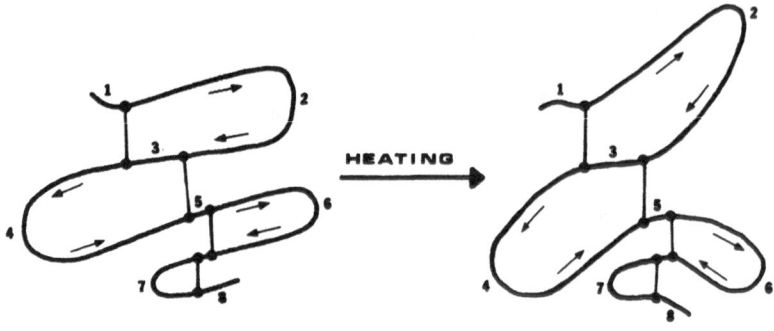

Fig. 8.   Schematic diagram of the neurotoxin backbone
           before and after heat treatment. The numbers
           indicate loop number from the amino terminal to
           the next cysteine residue. For details see the
           review article (ref. 1). The four disulfide
           bonds stabilize the structure. However, anti-
           parallel β and β-turn configuration are
           maintained after heat treatment.

1.3.   Conformation in Aqueous and Solid Phases

There are no changes in the frequencies of the Amide I and
Amide III bands suggesting that they have identical conformations in two phases (Fig. 9). It is well established that
snake neurotoxin Type I consists of 60-62 amino acid residues
with a molecular weight of about 6800. Usually four disulfide
bridges are present, an unusually high content of disulfide
bonds for such a small protein. This property not only makes
the neurotoxin molecule very compact but also is responsible
for its stability. The results of laser Raman spectroscopy
studies suggest a similar stability of sea snake neurotoxins.

1.4.   Disulfide Bonds

Disulfide bonds help maintain the conformation of a protein and
provide extra stability to a protein molecule. There are four
disulfide linkages in Type I neurotoxins. The fact that a very
sharp and symmetrical peak at 512 cm^{-1} was obtained for all
neurotoxins indicates that the four disulfide bonds have gauche-
gauche-gauche form (13). The origin of the S-S and C-S stretching vibrations have been extensively studied by Sugeta et al.
(14) using model compounds. The assignment of different
rotational isomers is shown in Fig. 7.

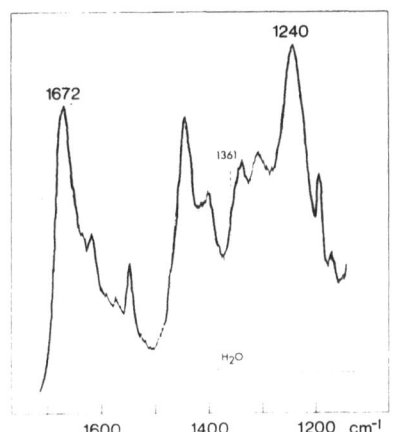

Fig. 9.   Raman spectrum of <u>Lapemis hardwickii</u>
          neurotoxin in aqueous solution at pH 7.

1. .   Side Chain Residue

<u>Tyrosine</u>. It is well known that the relative intensities of
the Raman lines at 850 and 830 cm^{-1} are related to the environment of the tyrosine residue (15). This doublet originates
from the Fermi resonance between the ring-breathing vibration
and the overtone of an out-of-plane ring bending vibration of
the para-substituted benzene (16). From the spectra of neurotoxins, it is apparent that the single tyrosine residue present
in sea snake neurotoxins, with the exception of <u>Laticauda
semifasciata</u> toxin b, is not readily accessible to water
molecules. The Raman spectra are in full agreement with the
fact that only 50% of the tyrosine molecule is nitrated with
tetranitromethan (17).

<u>Tryptophan</u>. Most snake neurotoxins, both of Type I and II,
contain one or two residues of tryptophan, one residue being
predominant. The single tryptophan residue of neurotoxin can
be readily modified by different selective reagents (18,19).
The 1361 cm^{-1} band of tryptophan is very sensitive to the
environment. When the indole ring is buried, the 1361 cm^{-1}
band shows a sharp peak. As the indole ring becomes accessible
to water molecule, the 1361 cm^{-1} line diminishes. The lack of
a distinct peak at 1361 cm^{-1} in Raman spectra of neurotoxins
indicates that the single residue is exposed.

2. HEMORRHAGIC TOXINS FROM RATTLESNAKE VENOMS

Rattlesnake venom causes extensive tissue damages such as
myonecrosis and hemorrhaging. Five hemorrhagic toxins were
isolated from western diamondback rattlesnake (Crotalus
atrox) venom. All of them are zinc proteins possessing pro-
teolytic activities (20). When zinc is removed, both proteo-
lytic and hemorrhagic activities disappear. The change in the
conformation can be detected by the disappearance of 1655
shoulder upon the removal of zinc (Fig. 10).

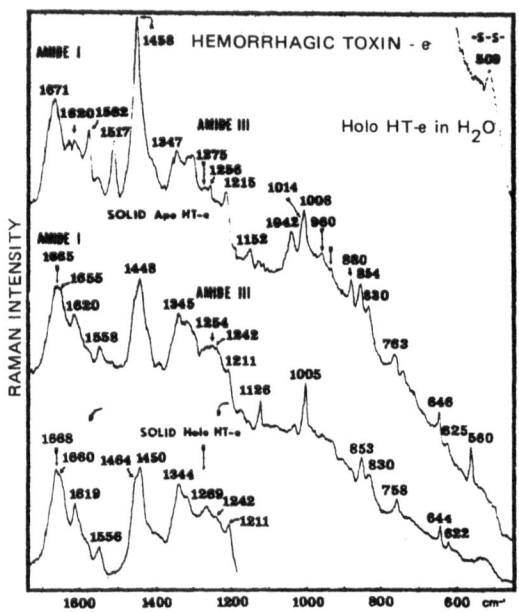

Fig. 10.   Laser Raman spectra of native hemorrhagic
           toxin e (two bottom curves) and hemorrhagic
           apotoxin e (top curve) in the solid state,
           and the disulfide region of aqueous native
           hemorrhagic toxin e (upper right-hand corner).
           (Reproduced from reference 20.)

     This change can also be detected by other methods such as
CD. As can be seen from Figs. 11 and 12, apotoxin has quite
different spectra from the original one. However, by putting
back zinc atom, the CD spectra resembles more the native toxin.
All of this evidence suggests that zinc is essential for
biological as well as well as for conformation of the
toxin molecule.

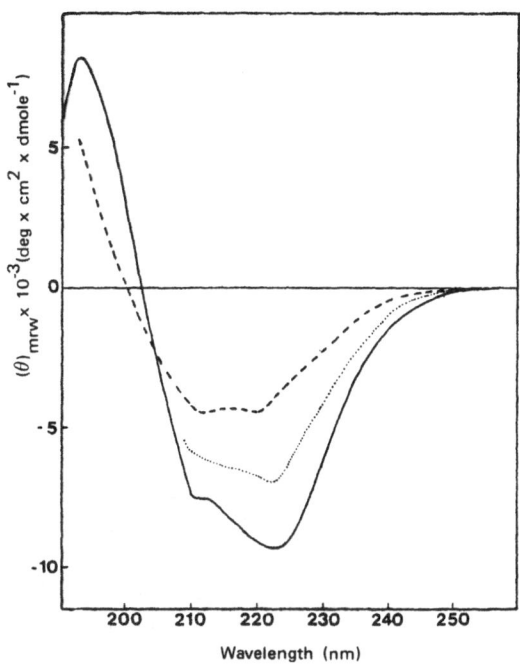

Fig. 11. Circular dichroic spectra of native hemorrhagic toxin e (—), hemorrhagic apotoxin e (---), and zinc-regenerated hemorrhagic apotoxin e (···) in the peptide region. Mean residue ellipticities (θ_{mrw}) based on mean residue weight of 115.

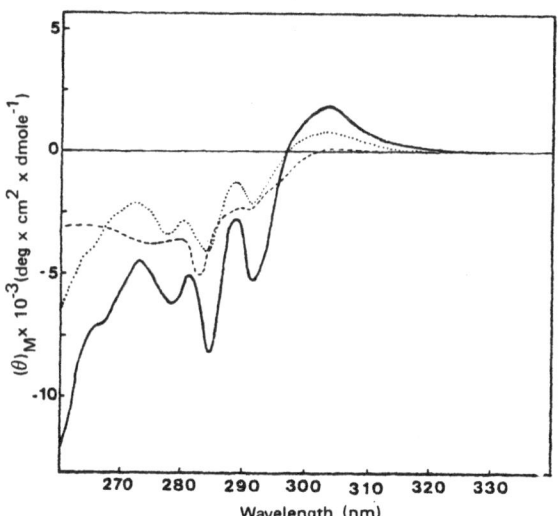

Fig. 12. Circular dichroic spectra of native hemorrhagic toxin e (—), hemorrhagic apotoxin e (---), and zinc-regenerated hemorrhagic toxin e (···) in the aromatic region. Molar ellipticities (θ_M) are based on a molecular weight of 25,000.

ACKNOWLEDGEMENT

The work was supported by NIH grants 5R01 GM19172 and
2R01 GM 15591.

REFERENCES

1. A. T. Tu, Ann. Rev. Biochemistry 42, 235 (1973).
2. A. T. Tu, Venoms: Chemistry and Molecular Biology,
    John Wiley, New York, 560 pages (1977).
3. J. W. Fox, M. Elzinger, and A. T. Tu, FEBS Lett. 80, 217
    (1977).
4. B. G. Frushour and S. L. König, Adv. Infrared and Raman
    Spectroscopy 1, 35 (1975).
5. T. G. Spiro and B. P. Gaber, Ann. Rev. Biochem. 46, 553
    (1977).
6. A. T. Tu, J. B. Bjarnason, and V. J. Hruby, Biochim.
    Biophys. Acta 533, 530 (1978).
7. V. J. Hruby, K. K. Deb, J. Fox, J. Bjarnason, and A. T. Tu,
    J. Biol. Chem., in press (1978).
8. P. Y. Chou and G. D. Fasman, Biochemistry 13, 222 (1974).
9. J. L. Lippert, D. Tyminski, and P. J. Deesmeules, J. Am.
    Chem. Soc. 98, 7075 (1976).
10. M. Pezolet, M. Pigeon-Gosselin, and L. Coulombe, Biochim.
    Biophys. Acta, 453, 502 (1976).
11. Y. H. Chen, J. T. Yang, and K. H. Chan, Biochemistry 13,
    3350 (1974).
12. A. T. Tu, B. H. Jo and N. T. Yu, Int. J. Peptide Prot. Res.
    8, 337 (1976).
13. N. T. Yu, T. S. Lin, and A. T. Tu, J. Biol. Chem. 250,
    1782 (1975).
14. H. Sugeta, A. Go, and T. Miyazawa, Bull. Chem. Soc. 46,
    3407 (1973).
15. N. T. Yu, B. H. Jo, and D. C. O'Shea, Arch. Biochem.
    Biophys. 156, 71 (1973).
16. M. N. Siamwiza, R. C. Lord, and M. C. Chen, Biochemistry
    14, 4870 (1975).
17. M. L. Raymond and A. T. Tu, Biochim. Biophys. Acta 285,
    498 (1972).
18. A. T. Tu, B. S. Hong, and T. N. Solie, Biochemistry 10,
    1295 (1971).
19. A. T. Tu and B. S. Hong, J. Biol. Chem., 246, 2772 (1971).
20. J. B. Bjarnason and A. T. Tu, Biochemistry 17, 3395 (1978).

THE RAMAN SPECTROSCOPY OF NUCLEIC ACIDS

W.L. Peticolas

Department of Chemistry, University of Oregon, Eugene,
Oregon 97403, U.S.A.

and

M. Tsuboi

Faculty of Pharmaceutical Sciences, University of Tokyo,
Bunkyo-ku, Tokyo, Japan

I. Introduction

Nucleic acids are of biological importance because they are the
molecules which are used to store and transmit genetic information in the biological cell.   Deoxyribonucleic acid, DNA, is
known to be the molecule of heredity which forms the genetic
material in the cell.   It is a very long thread-like molecule.
The backbone chain of DNA consists of the sugar molecule,
2'-deoxyribose, linked together by means of esterification with
two of the acid groups of a phosphoric acid molecule.   There
are two general types of nucleic acids: the DNA and ribonucleic
acid, RNA, which differs from DNA in having a ribose instead of
deoxyribose in the chain.   Both DNA and RNA contain the bases
guanine (G), cytosine (C), and adenine (A). They differ in
that RNA contains uracil (U), while DNA contains 5-methy-uracil
(T).

The Raman spectra which are observed from nucleic acids consist
of those Raman bands which come from the backbone chain, and
those which come from each of the nitrogenous bases.   Early
infra-red and Raman measurements gave some initial assignments
for both backbone [1-3] and base vibrations [4].   The backbone
is a fully saturated type molecule which has electronic
transitions which involve sigma-type bonding electrons.   Thus

153

Theo M. Theophanides (ed.), Infrared and Raman Spectroscopy of Biological Molecules, 153–165.
All Rights Reserved. Copyright © 1979 by D. Reidel Publishing Company, Dordrecht, Holland.

the absorption bands which arise from excitation of the electrons
in the backbone chain lie in the vacuum ultraviolet.   On the
other hand, the bases contain a series of rather broad electronic
absorption transitions starting at about 250 nm, and as a result
it is possible to obtain both ordinary Raman spectra and
resonance Raman spectra from the base molecules.   Furthermore,
even the ordinary Raman spectra of these bases show many of the
characteristics of a strong preresonance effect with the
low-lying excited electronic state. This was recognised as
early as 1970 by Peticolas et. al [5, 6], who suggested that
there should be decreases in the intensity of the Raman lines
from certain nucleic acid base vibrations upon base stacking
which would be due to a preresonance Raman effect of these
Raman bands with a hypochromic electronic absorption band.
They suggested that the term Raman hypochromism be used to
describe this phenomenon.   This concept of Raman hypochromism
was put on a much more quantitative and theoretical basis by
Tsuboi and co-workers [8 - 10] and more recently by Pezolet et
al. [11] and Blazej and Peticolas [12].   In this review we
will use the term Raman hypochromism to mean the decrease in
Raman intensity which occurs upon the formation of a stacked
(i.e. ordered) nucleic acid structure.   Actually the effect
is usually measured as the increase in Raman intensity upon
melting.

The vibrations of the backbone are generally discussed in terms
of normal coordinate treatments made for the elements of the
sugar phosphate chain.   Since the first assignments of this
backbone chain were made by Japanese workers [1-3], more
detailed measurements and calculations have been made by others
and will be discussed below.

Reviews on the use of Raman and resonance Raman spectroscopy in
the study for nucleic acids have been given by Hartman et
al. [13] and Tsuboi et al.,[14, 15].   In this brief review we
will discuss the Raman active vibrations of the bases, and the
backbone separately and then discuss the use of this as a tool
for studying the structure of biologically interesting
problems involving, for example, viruses and chromation.
Finally some modern applications of the resonance Raman effect
and Raman kinetic experiments will be given.

2.   Raman Active Vibrations of Nucleic Acid Bases

A.   Normal Modes and Frequencies

A detailed discussion of the vibrational modes of the base
residues of nucleic acid was given by Tsuboi et al. [14].   It
would take too much space to reproduce here all the normal modes

of interest for each of the bases usually found in DNA and RNA.
Reference must be made to this article for detailed pictures of
the atomic displacements involved in the vibrations.   However,
of particular interest are certain strong Raman bands of base
molecules which are listed in Table I [16].

In H_2O, uracil has a strong Raman band at about 1690 cm^{-1} which
is predominantly due to the carbonyl (C=O) stretching mode.
There is also a weaker mode around 1631 cm^{-1}.   In D_2O there is
a splitting and change of frequency of these bands.   In the
double-helical form there is a single band at 1680 cm^{-1}, and
this apparently splits into two bands at 1660 and 1698 cm^{-1} upon
melting of the double-helix. Originally it was assumed that
the splitting was simply due to the breakup of the hydrogen
bonding between the uracil residue and the adenine residue of a
double-helix [6].   However, Morikawa et al. [17] have shown
that it probably is impossible to completely separate the effect
of base-stacking from the effect of hydrogen bond interaction
on the splitting of this band. (For uracil, see also Appendix)

B.   Hypochromism and the Preresonance Raman Effect.   Effect of
Order ⇌ Disorder Transitions on Raman Intensities.

Other Raman vibrations of the RNA bases and their hypochromism
in simple homopolymers are given in Table I. Also, given in
Table I are the ultraviolet absorption bands whose wavelength is
given in nanometers from which the particular base vibration
appears to obtain its intensity through the preresonance
phenomenon [9-12].   Both preresonance and resonance Raman effect
in nucleic acid bases will be discussed below.   One fact of
interest is that the 1480 cm^{-1} band which is a longitudinal
stretching or wave-like mode involving the entire purine ring
along the direction from the C_8 to the N_1 shows little
hypochromism in adenine, but substantial hypochromism in
self-associated 5'-GMP.   In general, the relative intensity of
these bands is measured relative to a backbone chain mode at
1100 cm^{-1} which is due to the O-P-O symmetric stretching
vibration [1-3].   This band will be discussed in more detail
below.

These hypochromic bands are very useful for measuring
independently the amount of base stacking for each type of base
when the RNA or DNA undergoes a transition between ordered
(helical) and disordered states.   Thus it is possible to obtain
separate melting curves for each of the 4 bases.   Lord and
Chen have recently used this effect to study the separate melting
of each type of base pairs in t-RNA [18].

Table I

Some prominent Raman lines (cm^{-1}) due to base vibrations in double helical ribonucleic acid homopolymers.

	Hypochromism			Preresonance
Uracil residue	in Poly (AU)	in Poly A·poly U		
1680(s)(C=O str.)	++			
1634(m)(C=O str.)	no			
1400(m)	++	++		A(265 nm)
1235(s)	++	++		A(265 nm)
785(s)(ring breath.)	++	+		B(210 nm)
Cytosine residue	in Poly C	in GpC		
1657(m)				
1607(m)				
1528(m)	+	+		A(268 nm)
1292(s)	+			A(268 nm)
1240(s)				
782(s)(ring breath.)				B(230 nm)
Adenine residue	in Poly A	in Poly AU	in Poly A·poly U	
1580(s)		no		A(276 nm)
1510(m)	++	no	+	
1484(m)	no	no	−	A(276 nm)
1379(m)	+	++		
1340(s)		no		
1310(s)	++	+	++	
1255(w)	+	no		
729(s)(ring breath.)	+	++	++	C(210 nm)
Guanine residue	in 5' GMP	in GpC		
1582(s)	+			A(276 nm)
1487(s)	++	−		A(276 nm)
1375(m)	+			
1328(m)	++	−		A(276 nm)
670(s)	−−	−		Far

C. Resonance Raman Effect from Nucleic Acids .

Tsuboi et al. [9] and Blazej and Peticolas [12] have shown by
both resonance measurements and preresonance Raman that the
Raman vibrations of adenine at 1580 cm^{-1} and 1484 cm^{-1}
probably obtain their intensity from a weak absorption band at
276 nm.   The former workers also based their conclusions on a
theoretical prediction involving a comparison of the normal modes
of these molecules with the corresponding bond order changes
obtained from molecular orbital calculations on the excited
electronic state.   This suggestion has received rather remarkable
conformation in the work of Blazej and Peticolas [12] who
measured the first resonant Raman excitation profile into the
actual ultraviolet absorption band at 276 nm.   These authors
took as an adjustable parameter the frequency of the excited
electronic state and obtained a least squares fit to an Albrecht
A-type resonant term.

Table II summarizes the least-square fit of the Albrech A-term
to the experimental data. In this table $E_{e,o}$ is the o-th
vibrational level of the excited electronic state of adenine
which is taken as an adjustable parameter since its exact value
cannot be located spectroscopically.   The bottom row of Table II
shows that the vibration at 1484 cm^{-1} of AMP gives a standard
deviation of 0.562 when it is assumed to obtain its intensity
from an absorption band maximum at 269 nm whereas row 4 shows
that a standard deviation of only 0.065 is obtained for an
assumed value of 276 nm for the absorption band.   This was the
smallest standard deviation obtained as the assumed value,
$E_{e,o}$, the absorption energy level was systematically varied
between 260 and 290 nm.   The Raman electronic level for
preresonance of the bands in 1300 cm^{-1} region is not clear.
They obviously are resonance enhanced at 260 nm but the least
squares fit is not so good.

Table II

Calculated vibronic energies and bandwidths of AMP
from U.V. resonance Raman excitation profiles.

$\Omega(cm^{-1})$	$E_{e0}kK(nm)$	$\Gamma_{e0}(kK)$	$E_{e1},kK(nm)$	$\Gamma_{e1}(kK)$	SD
1310	37.2(269)	0.8	38.5(260)	1.2	0.131
1338	37.2(269)	0.9	38.5(259)	1.1	0.131
1380	37.3(268)	0.9	38.7(259)	1.1	0.184
1484	36.2(276)	0.9	37.7(265)	0.6	0.065
1583	36.4(275)	1.0	38.0(263)	0.8	0.091
1338	36.2(276)	0.9	37.5(266)	0.6	0.579
1484	37.2(269)	0.9	38.7(259)	1.1	0.562

Thus, in conclusion we can say that several of the strongly
Raman active bands of the nucleic acid bases derive their
intensity from a preresonance effect with low-lying electronic
absorption bands.   In general, the rule enunciated by Tsuboi
and his co-workers [9, 15, 34] is that if the excited electronic
state geometry resembles that of the normal mode of the ground
state, then the Raman active band will tend to get is intensity
from that low-lying electronic state through a preresonance
effect.   When the bases are placed in juxtaposition, either
longitudinally by means of stacking interactions or laterally
by means of Watson-Crick hydrogen bonding interactions, the
electronic transition moments of the absorption bands may
interact to change the intensity both of the electronic absorption
bands (UV hypochromism), and consequently the Raman intensities
(Raman hypochromism).   Because one can monitor the Raman
hypochromism of each type of base separately the Raman effect
gives a far more sensitive tool for studying specific base-base
interactions than does the intensity of the electronic absorption
bands.   This is particularly true since the electronic absorption
bands of all four of the bases widely overlap and thus are very
difficult to resolve.

3.   Raman Active Vibrations of the Backbone Chains of Nucleic
     Acids.

There are only two Raman active vibrations belonging to the
backbone chain which are clearly identifiable in terms of normal
modes.   These are the Raman band at 1100 cm^{-1} due to the PO_2^-
symmetric stretch and the Raman band at 811 ± 4 cm^{-1}.

This 811 cm^{-1} band may be called the symmetric diester phosphate
stretching mode [3, 6, 7, 19-21] .   Thus if R is the ribose
ring, this mode involves the chain, -R-O-P($\lesssim_O^O$-)-O-R.   This
vibration has been shown to be very conformationally dependent.
It only exists in DNA or RNA when these chains are ordered A-type
confromation [19-22].   Although X-ray measurements on fibers of
RNA homopolymers always showed these materials to be in the
A-type conformation, there was no completely definitive proof
that this A-type conformation existed in the double helical
structures in solution because until laser Raman spectroscopy
there was no technique for structure determination which would
work equally well both on fibers or crystals where the X-ray
structure is known and also in solution.   However by taking
X-ray diffraction and Raman spectra of the same fibers, it was
possible to correlate the Raman spectrum of the A-type, B-type
and C-type forms of DNA [19, 22, 23].   Since, in solution the
Raman spectrum of ordered RNA always shows the 811 cm^{-1} band of
A-type DNA (usually shifted to 814 cm^{-1} because of the difference
in the sugar substituent at the 2' position) the A-type structure

of RNA in solution is definitely established.   Lord and Chen [18] have shown that the crystals and solution of t-RNA show the same Raman spectrum so that the structure of this material in solution is now well-established.

Again the 814 cm^{-1} band in RNA structures may be used as a measure of the A-type conformation.   Following the suggestion of Tsuboi et al. [24] the 1100 cm^{-1} band is used as an internal standard of reference, and the intensity ratio, I_{811}/I_{1100}, is taken as a measure of the A-type conformation.   The value of 1.65±0.05 is generally taken to be 100% A-type configuration both for DNA [25] and RNA structures [25, 24, 27].

In addition to these two main modes, there are also two other modes which are fingerprint bands of unknown origin for the B and C type conformation.   In DNA a weak band at 835 cm^{-1} is always obtained when the DNA is in the B-type conformation [19, 22, 23].   When the DNA shifts to the A-type conformation this band disappears with the appearance of the band at 811 cm^{-1} [19, 22].   There is in DNA a band at 790 cm^{-1} which is due primarily to the cytosine band.   This band increases substantially in intensity when DNA goes from the B-form to the A-form and it has been suggested that in fact there is a band about 800 cm^{-1} which is the diester stretch for the B-type structure [22].

The C-type structure shows a band at 870 cm^{-1} [19, 23].   The original assignment of this mode has been confirmed independently by Brahms [23] who succeeded in obtaining X-ray and Raman data on the same C-type conformation.   Thus we can say rather conclusively the comparison of the X-ray diffraction patterns and the Raman spectrum of the A, B, and C forms of DNA allow the

Table III
Some Raman bands of the nucleic acid backbone.

Frequency (cm^{-1})	Assignment
1100 cm^{-1}	$(R-O-)_2-P\begin{smallmatrix}O\\O\end{smallmatrix}$ symmetric stretch of $-PO_2^-$ group
811 ± 4 cm^{-1}	DNA or RNA chain mode for A-type conformation. Involves ribose-phosphate diester linkage.
835 cm^{-1}	Weak but prominent Raman mode for DNA in B-form
870-880 cm^{-1}	Similar to mode above but for DNA in C-form

establishment of Raman spectra in the region 800 to 1100 cm^{-1} as a determination of the conformation of the nucleic acid. These results are summarized in Table III.

4.   Raman Spectroscopy of Naturally Occurring Nucleic Acids.

Up to now, we have discussed the Raman spectra of nucleic acid components primarily from model compounds, such as the mononucleotides, simple homopolymers, and helical complexes of the polynucleotides.   In this section we'll discuss more naturally occurring materials such as actual RNAs and DNAs.   Tables IV and V list the Raman spectral changes observed on the melting of nucleic acids taken from the papers of several authors which represent a summary of the frequencies observed in ribonucleic acids both in H_2O and D_2O [26-29].   The absolute intensities of these various bands due to the bases will depend on the relative base composition.   The increases in intensity which are observed upon melting are the inverse of the decrease of intensity due to helix formation, i.e. Raman hypochromism.   It will be seen, however, in Tables I, IV and V the 670 cm^{-1} band of guanine actually decreases upon melting.   Consequently, it increases upon helix formation and must be regarded as hyperchromic. Other than this band virtually all of the bands show increases in intensity due to the standard Raman hypochromism pre-resonance condition.   The changes in the Raman bands due to the changes in the backbone vibrations are also given.

Considerable work on the Raman spectroscopy of viruses has been done by Thomas and his cororkers [26, 27, 30]. In general, the viruses show a mixture of bands due to the nucleic acid components and the protein components.   By careful study of the intensity and frequency of these bands Thomas and his coworkers have been able to show the relative conformation of both the nucleic acid and protein portions of the spectrum.   Particularly impressive are the high quality of Raman spectra of the fld virus which shows a high α-helical content in the protein portions [30].

The first Raman work done on chromatin, the complex of histone and non-histone space proteins with DNA which is always found in the nucleus of eukaryotic cells was done by Mansy et al. [31]. This group reported a large fraction of α-helix in the protein part and a predominantly B-type conformation for the DNA in chromatin.   In addition, they found that in whole chromatin samples the Raman band at 1490 cm^{-1}, predominantly due to guanine, is measurably weaker than in normal double-helical DNA. As was noted above, and as may be seen in Table I, this purine vibration shows no hypochromism in the polyA·polyU or poly(A-U) double-helix due to the stacking interactions between the bases. However, a considerable hypochromism is found in the double-helical

Table IV
Raman spectral changes observed on melting of ribonucleic acids

Frequencies (cm^{-1})		Intensity changes observed upon melting (if any)	Assignment
H_2O pH7	D_2O pH7		
670	670	Large decrease	G
725	720	Large decrease	A
785	780	Increase due to shift in 814 cm^{-1} band	U, C OPO symmetric if chain is disordered
814	814	Disappears completely, shifts to 785 cm^{-1}	OPO symmetric stretch
867	860	No change	Ribose
915	915	No change	Ribose
975	990	Decrease	Ribose
1003		No change	A, U, C
1047	1045	Small decrease	Ribose-phosphate
1100 (5)	1100 (4)	No change	PO_2^- symmetric stretch
1182 (2)	1185 (1)	Small increase	A, G, C
1240 (6)		Large increase	U
1248 (5)		Small increase	A, C
	1310 (7)	Small increase	C
1320 (7)	1320 (7)	Small increase	A, G
1340 (7)	1340 (7)	Increase	A
1375 (5B)	1370 (3B)	No change	A, G
1420 (2)		No change	G, A
1460 (sh)	1460 (sh)	No change	Ribose
1484 (10)	1480 (8)	Small increase	G, A
1527 (2)	1526 (3)	Small increase	C, G
1575 (8)	1578 (10)	Small increase	G, A
	1658 (4)		C=O in U, G, C
1692 (4)	1688 (4)		

aggregate, 5'-GMP. Since in helical 5'-GMP there is a hydrogen bond at the N-7 position, the bold assumption was made [31] that H-bonding at the N-7 position of guanine was responsible for the decrease in the intensity of this band rather than base stacking interactions which are normally supposed to be the cause of Raman hypochromism, as discussed above. Additional evidence, which these authors cited for this assumption is the total absence of this 1490 cm^{-1} band in guanine, which is protonated at the N-7 position. Recent measurements by Thomas et al. [32], and Goodwin and Brahms [23] have confirmed the relatively high α-helical content of the protein portions of chromatin and also confirmed the B-form of the DNA. In addition, these workers

Table V

Raman spectral changes observed on melting of calf thymus DNA

cm^{-1} in H_2O pH = 7	cm^{-1} in D_2O pH = 7	Changes observed upon melting (if any)	Assignment
672	656	Small increase	T
683	677	Decreases	G
729	716	Large increase	A
750	734	Shifts to lower cm^{-1}	T
	765	Large increase in intensity	
	785	No change	PO_2 diester symmetric stretch
786		Small increase in intensity	PO_2 diester symmetric stretch
835	828	Shifts to lower cm^{-1}	PO_2 diester anti- symmetric stretch?
879	867	Decreases	Deoxyribose-phosphate
	893	Decreases	Deoxyribose
920		Decreases	Deoxyribose
	966	Decreases	Deoxyribose
	1013	Decreases	C-O stretch
1015		No change	C-O stretch
1051	1047	Decreases and shifts to a higher cm^{-1}	C-O stretch
1094	1091	No change	PO_2^- dioxy symmetric stretch
1144		Disappears	Deoxyribose-phosphate
1186		Moderate increase	T
1214		Moderate increase	T
1225		Moderate increase	A
1240		Large increase	T
1259	1260	Small increase	C, A
1303	1300	Moderate increase	A
1340	1343	No change	A
1378	1375	Moderate increase	T, A
1421	1418	Small increase	A, G
1463		Decreases	Deoxyribose-phosphate
	1484	Shifts to lower cm^{-1} deuteration of C-8 proton on A and G	
1491		Moderate increase	G, A
	1501	No change in in- tensity, shifts to lower cm^{-1}	A

Table V (continued)

cm^{-1} in H_2O pH = 7	cm^{-1} in D_2O pH = 7	Changes observed upon melting (if any)	Assignment
1521	1520	No change in intensity, shifts to lower cm^{-1}	A
1534		Moderate increase in intensity	C
1579	1575	Large increase in intensity due mainly to G	G, A
	1620	No change	C?
1660	1673	Large increase in intensity, shifts to lower cm^{-1}	C=O of T

have shown that the decrease in intensity of the 1490 cm^{-1} band does not occur in purified nucleosomes but only on certain whole chromatin samples [23, 32]. Thus if this decrease in intensity is due in fact to H-bonding at the N-7 position, then it comes from an interaction with guanine and a non-histone component of the chromatin [23].

5. Time Dependent Raman Spectroscopy of Nucleic Acid Disordering

The mechanism of nucleic acid unfolding is still a matter of controversy in spite of many experiments. This is due to the fact that most of the measurements made today involve only ultraviolet hyperchromism which, while it measures nicely the effect of unstacking of the bases, cannot distinguish between unstacking and backbone deterioration. Recently a method for obtaining rapid Raman spectra has been developed and a start has been made on the determination of the kinetics of the base stacking, hydrogen bonding, and backbone conformation change involved in this process [33]. The experimental problems are very difficult, as will be discussed in detail in the lecture.

REFERENCES

1. Tsuboi, M., J. Am. Chem. Soc., $\underline{79}$, 1351 (1957).
2. Sutherland, G.B.B.M. and Tsuboi, M., Proc. Roy. Soc. (London), $\underline{A239}$, 446 (1957).
3. Shimanouchi, T., Tsuboi, M., and Kyogoku, Y., in Advances in Chemical Physics, J. Duchesne, Ed., London, Interscience, Vol. VII, pp. 435-498 (1964).
4. Lord, R.C. and Thomas, G.J., Jr., Spectrochim. Acta, $\underline{23A}$, 969 (1967).
5. Tomlinson, B.L. and Peticolas, W.L., J. Chem. Phys., $\underline{52}$, 2154 (1970).
6. Small, E.W. and Peticolas, W.L., Biopolymers, $\underline{10}$, 69 (1971); $\underline{10}$, 1377 (1971).
7. Peticolas, W.L., Procedures in Nucleic Acid Reseach (Cantoni, G.L. and Davies, D.R., eds.), Harper & Row, Vol. 2, pp. 94-136 (1971).
8. Tsuboi, M., Takahashi, S., Muraishi, S., and Kajiura, T., Bull. Chem. Soc., Japan, $\underline{44}$, 2921 (1971).
9. Tusboi, M., Hirakawa, A.Y., Nishimura, Y., and Harada, I., J. Raman Spectroscopy, $\underline{2}$, 609 (1974).
10. Nishimura, Y., Hirakawa, A. and Tsuboi, M., Chemistry Letters, 907 (1977).
11. Pezolet, M., Yu, T.J. and Peticolas, W.L., J. Raman Spectrosc., $\underline{3}$, 55 (1975).
12. Blazej, D.C., and Peticolas, W.L., Proc. Nat'l. Acad. Sci., $\underline{74}$, 2639 (1977).
13. Hartman, K.A., Lord, R.C., and Thomas, Jr., G.J., Physico-chemical Properties of Nucleic Acids (J. Duchesne, ed.), $\underline{2}$, Chapter 10, pp. 92-143 (1973).
14. Tsuboi, M., Takahashi, S., and Harada, I., in Physico-Chemical Properties of Nucleic Acids, Vol. 2 (J. Duchesne, ed.) Chapter 11, pp. 91-145, Acad. Press, London (1973)
15. Tsuboi, M., et al., Advances in Infrared and Raman Spectroscopy, Clark and Hester, eds., in press.
16. Tsuboi, M., Proc. 5th Inter. Conf. on Raman Spect., Freiburg, et al., ed., Hans Ferdinand Schulz Verlag, D-7800, Freiburg Im Bresigau, pub., 135-143 (1976).
17. Morikawa, K., Tsuboi, M., Takahashi, S., Kyogoku, Y., Mitsui, Y., Iitaka, Y., and Thomas, Jr., G.J., Biopolymers, $\underline{12}$, 790 (1973).
18. Chen, M.C., Geige, R., Lord, R., and Rich, A., Biochemistry, $\underline{14}$, 4385 (1975).
19. Erfurth, S.C., Kiser, E.J., and Peticolas, W.L., Proc. Nat. Acad. Sci. USA, $\underline{69}$, 938 (1972).
20. Thomas, Jr., G.J., Biochim. Biophys. Acta, $\underline{213}$, 471-423 (1970).
21. Lafleur, L., Rice, J., and Thomas, Jr., G.J., Biopolymers, $\underline{11}$, 2423 (1972).
22. Erfurth, S.C., Bond, P.J., and Peticolas, W.L., Biopolymers, $\underline{14}$, 247, 1259 (1975).

23. Goodwin, D.C. and Brahms, J., Nucleic Acid Research (in press).
24. Tsuboi, M., Takahashi, S., Muraishi, S., Kajiura, T., and Nishimura, S., Science, 174, 1142 (1971).
25. Brown, K.B., Kiser, E.J., and Peticolas, W.L., Biopolymers, 11, 1855 (1972).
26. Thomas, Jr., G.J., and Hartman, K.A., Biochim. Biophys. Acta, 312, 311 (1973).
27. Thomas, Jr., G.J., Chen, M.C., and Hartman, K.A., Biochim. Biophys. Acta, 324, 37 (1973).
28. Small, E.W., Brown, K., and Peticolas, W.L., Biopolymers 11, 1209 (1972).
29. Erfurth, S. and Peticolas, W.L., Biopolymers, 14, 247 (1975).
30. Thomas, Jr., G.J. and Murphy, P., Science, 188, 1205 (1975).
31. Mansy, S., Engstrom, S.K., and Peticolas, W.L., Biochem. Biophys. Res. Comm., 68, 1242 (1976).
32. Prescott, B., Thomas, G.J., and Olins, D.E., Biophysical Journal, 17, 114a (1977); Science, 197, 385–388 (1977).
33. Sturm, J., Savoie, R., and Peticolas, W.L., Indian Journal of Pure and Applied Physics, Golden Jubilee Celebration of the Discovery of the Raman Effect (1978), in press.
34. Hirakawa, A.Y. and Tsuboi, M., Science, 188, 359 (1975); Tsuboi, M. and Hirakawa, A.Y., J. Raman Spectroscopy, 5, 75 (1976).

Appendix IV ESTIMATION OF THE DISTORTION OF THE GEOMETRY OF
NUCLEIC-ACID BASES IN THE EXCITED ELECTRONIC
STATE FROM THE ULTRA-VIOLET RESONANCE RAMAN
ENHANCEMENT OF CERTAIN NORMAL MODES.*

Warner L. Peticolas and Dan C. Blazej

Department of Chemistry, University of Oregon,
Eugene, OR 97403

The resonance Raman intensity of the totally symmetric
(i.e. in plane) modes of the nucleic acid bases is derived
from shifts of the excited state potential energy along the
various 3N-6 normal modes, Q_j.[1-6] If the shift is large the
dependence of the Raman intensity on the displacement is rather
complicated.[5-6] However if the shift is small the dependence
is simply proportional to the square of the displacement, Δ_j,
along the various normal modes.[3,6]

From detailed excitation profiles, it is possible in
principle to obtain a good value of Δ_j for each of these modes
using the displaced harmonic oscillator model.[1-6] However, in
this simple treatment we want to show how to obtain an estimate
of the distortion of the geometry in the resonant excited
electronic state from a single measurement of the enhancement of
Raman intensities. By enhancement of Raman intensities, it is
meant the ratio of the resonant Raman intensity to the Raman
intensity far from resonance. This ratio can vary from zero to
10^6. If it is assumed that the Raman intensity far from
resonance is obtained from the sum over the infinite number of
excited electronic states while the resonance Raman intensity is
obtained only from the resonant state, then the ratio gives
direct measure of the resonance enhancement. It is well-known
that the relative intensity of the various Raman active modes of
a molecule can differ markedly in going from far-off resonance
into resonance. Some modes which are not observed at all in the
RRE can be relatively strong far from resonance where all Raman
intensities are weaker. Even when strong pre-resonance effects

* Supported in part by grant no. PCM76-82222 from the National
Institutes of Health (USA) and grant no. GM 15547 from the
National Science Foundation.

Theo M. Theophanides (ed.), Infrared and Raman Spectroscopy of Biological Molecules, 167–173.
All Rights Reserved. Copyright © 1979 by D. Reidel Publishing Company, Dordrecht, Holland.

are present as they are in the nucleic acid bases [7-10] considerable change in the intensity distribution is observed in going
from the RE to the RRE. Consequently in this simple treatment,
it will be assumed that the resonance Raman enhancement I_j of
the jth normal mode can be obtained from the ratio of the Raman
intensities at 36.34 kK (275 nm) to the corresponding intensity
at 19.44 kK (514.5 nm). Furthermore it will be assumed that the
shift in the excited electronic state is sufficiently small
along the normal coordinate, Q_j, that I_j is proportional to
$(\Delta_j)^2$.

Let $\underline{x} = (x_1, x_2 . . . x_{3N})^T$ be the vertical column vector of
the 3N cartesian displacement coordinates between $\underline{x}_g^\circ$, the
equilibrium coordinates of the ground state, and $\underline{x}_g$ the displaced coordinates of the ground state. The following well-
known relations hold between the cartesian displacement
coordinates, $\underline{x}$, the mass reduced coordinates, $\underline{q}$, the internal or
valence coordinates $\underline{R}$, the force constant matrix F, the normal
coordinates, $\underline{Q}$, the potential energy, V, and the diagonal
matrices,

$$\underline{M} = \text{diag} \{M_1, M_2, . . . M_{3N}\} \tag{1a}$$

and

$$\underline{\Lambda} = \text{diag} \{\omega_1^2, \omega_2^2, \omega_3^2, . . . \omega_{3N-6}^2\} \tag{1b}$$

$$\underline{R} = \underline{B} \ \underline{x} = \underline{B} \ \underline{M}^{-1/2}\underline{M}^{1/2} \ \underline{x} = \underline{B} \ \underline{M}^{-1/2}\underline{q} \tag{2}$$

$$2V = \underline{R}^T\underline{FR} = \underline{q}^T\underline{M}^{-1/2}\underline{B}^T\underline{FBM}^{-1/2}\underline{q} = \underline{q}^T\underline{F}^q\underline{q} \tag{3}$$

To diagonalize the potential energy, we solve the eigenvalue
equation,

$$(\underline{M}^{-1/2}\underline{B}^T\underline{FBM}^{-1/2})\underline{A} = \underline{F}^q\underline{A} = \underline{A} \ \underline{\Lambda} \tag{4}$$

Since $\underline{F}^q$ is symmetric (or Hermitian), $\underline{A}$ is orthogonal and

$$\underline{A}^T\underline{F}^q\underline{A} = \underline{A}^T\underline{M}^{-1/2}\underline{B}^T\underline{FBM}^{-1/2}\underline{A} = \underline{L}^T\underline{FL} = \underline{\Lambda} \tag{5}$$

where

$$\underline{L} = \underline{BM}^{-1/2}\underline{A} \tag{6}$$

It is apparent from Eqn. (3) that the potential energy, Eqn. (2),
may be diagonalized by either of the transformations

$$\underline{q} = \underline{A} \ \underline{Q} \ ; \ \underline{Q} = \underline{A}^T\underline{q} \tag{7}$$

or

$$\underline{R} = \underline{L} \ \underline{Q} \ ; \ \underline{Q} = \underline{L}^T\underline{R} \tag{8}$$

to obtain $2V = \underline{q}^T\underline{F}^q\underline{q} = \underline{Q}^T\underline{A}^T\underline{F}^q\underline{A} \ \underline{Q} = \underline{Q}^T\underline{\Lambda Q}$ $\tag{9}$

or $2V = \underline{R}^T\underline{F} \ \underline{R} = \underline{Q}^T\underline{L}^T\underline{FL} \ \underline{Q} = \underline{Q}^T\underline{\Lambda} \ \underline{Q}$ $\tag{10}$

However only Eqns (7) and (9) leave the kinetic energy in diagonal form. For the purposes of making vibrational assignments we neglect the off-diagonal elements of the $\underline{\underline{F}}$ matrix in the following treatment.

From Eqn. (10) we get the important relation,

$$2V = \sum_j \omega_j^2 Q_j^2 = \sum_j \sum_i L_{ij} F_{ii} L_{ij} Q_j^2 \tag{11}$$

From Eqns. (10-11) we see that f_{ij} the fraction potential energy in the jth normal mode, Q_j, which is due to the internal coordinate R_i is given by

$$f_{ij} = L_{ij}^2 F_{ii}/\omega_j^2 \tag{12}$$

where from Eqn. (10)

$$\sum_i f_{ij} = 1 \tag{13}$$

To use the above equations to determine the distortion of the geometry in the excited state we observe that Δ_j measures the displacement of R_j along Q_j in the excited state so that Eqn. (8) may be written

$$\underline{R}_e - \underline{R}_g = \Delta \underline{R}^e = \underline{\underline{L}} \, \underline{\Delta} \tag{14}$$

or
$$\Delta R_i^e = \sum_j L_{ij} \Delta_j \tag{15}$$

Using Eqn. (12) we have

$$\Delta R_i^e = \sum_j \omega_j \left(\frac{f_{ij}}{F_{ii}}\right)^{1/2} \Delta_j \tag{16}$$

When Δ_j has been determined by a careful analysis of an excitation profile, Eqn. (15) can be used to determine ΔR_i^e. A much simpler procedure is to use equation (15) with the assumption that each Δ_j is small and that $I_j \propto \Delta_j^2$,

$$\Delta R_i = \sum_j L_{ij} \Delta_j = K \sum_j L_{ij} I_j^{1/2} \tag{17}$$

where I_j is the resonance Raman enhancement of the jth mode, i.e. in this paper we will take $I_j = I_j(37.6)/I_j(19.44)$.

Thus we can estimate the change in each ΔR_i^e in the excited electronic state from the equation

$$\Delta R_i^e = K \sum_j \omega_j \left(\frac{f_{ij} I_j}{F_{ii}}\right)^{1/2} \tag{18}$$

The fractions, f_{ij}, as well as the force constants, F_{ii}, for nucleic acid bases have been tabulated by Tsuboi, et. al.[4]

Values for the resonance enhancement of each of the normal modes of UMP (uracil monophosphate) and CMP (cytosine monophosphate) is given in Table I which is taken from the more complete excitation profiles to be published elsewhere.[7] In order to show the relative amount of displacement in each of the internal coordinates we have taken the change in the R_9-N_1 bond stretch displacement as unity and we have ratioed all other displacements, ΔR_i, using Eqn. (18) to this N_1-R_9 displacement taken as unity.

The change in the bond lengths which were calculated using the second improved set of force constants[4] for 1-methyluracil and 1-methylcytidine at 265 nm is given in Table II. It is of interest that none of the normal modes of UMP which involve principly the C_5-C_6 stretch go on resonance. Thus there is considerable variation in the various bond stretch displacements varying from 0.0 for the C_5-C_6 bond to 7.84 for the $C_4 - O_{11}$ stretch. This is evidence of considerable distortion of the uracil molecule in the excited state.

In CMP none of the vibrations involving the C_2-O_{10}, C_2-N_3 of C_4-C_5 stretch are resonance Raman Active. Thus these bonds are presumably not displaced in the excited state. The variation in relative bond displacement in the excited state goes from 0 for these three bond stretch displacements to 4.3 for the N_3-C_4. Again we have evidence for considerable distortion in the excited state.

An alternative to the above approach which would be possible even when the intensities can not be measured far below resonance is simply use the resonant Raman intensities themselves. Thus if I (37.58) alone is used in equation (18) one obtains a slightly different set of relative displacements. The only really significant changes for UMP in this procedure are C_4-O_{11} which changes from 7.8 to 5.0 and C_4-C_5 which changes from 6.1 to 3.8. Table 3 gives a list of the relative displacements using I (37.58) instead of the ratio.

From equations 4, 7 and 9 we see that if one has the $\underline{\underline{A}}$ matrix from the diagonalization of $\underline{\underline{F}}^q$, the position of each of the atoms in the excited electronic state, $\underline{\underline{x}}^e$, may be obtained from the relation

$$\underline{\underline{x}}^e = \underline{\underline{M}}^{-1/2} \underline{\underline{A}} \underline{\underline{\Delta}} \tag{19}$$

Where $\underline{\underline{\Delta}}$ replaces $\underline{\underline{Q}}$ in equation (7). Again $\underline{\underline{\Delta}}_j$ may be either obtained from an excitation profile or estimated from $I_j^{1/2}$ as before.

The treatment so far gives only the absolute magnitude of

TABLE I

Ultraviolet Resonant Raman Enhancement of
Some Normal Modes of UMP[7] and CMP[7].

UMP

VIBRATION		$I(37.58)/I(19.44)$	$I(19.44)$	$I(37.58)$
(obs.)	(calc.)[4]			
783	(826)	4300	0.61	2600
1232	(1200)	6600	1.27	8400
1396	(1375)	8800	0.33	2900
1630	(1651)	13000	0.19	2500
1680	(1710)	7000	0.67	4700

CMP

784	(768)	3200	1.00	3200
1243	(1212)	4800	0.82	3900
1294	(1320)	6400	0.74	4700
1529	(1538)	11000	0.33	3800

the relative change but not its sign. However, the displacement along Δ_j can be either positive or negative. To obtain the sign of the change it is necessary to have information on the change in bond order between ground and excited electronic states as emphasized by Hirakawa and Tsuboi.[2] Unfortunately, no such calculations excist in the literature for pyrimidines. Consequently, until such calculations are available, we can only give the relative size of the change but cannot say whether or not it increases or decreases. Further work is in progress to obtain better estimates of the excited state geometries.

Thus we see that with a reliable normal coordinate calculation for the ground state and a preresonant and a resonant Raman spectral intensity measurement one can obtain an estimate of the relative absolute magnitude of the displacements of the nuclear geometry of the excited resonant electronic state if the displacements along the normal coordinates is not too large.

TABLE II

Calculated Displacements of Bond Stretching Coordinates
in the Excited State (260 nm) of UMP and CMP

R_i (valence coordinate)	F_{ii}[4] $(md/\text{A}°)$	Raman Frequencies and Fractions[4]		Relative Displacement in Excited State
UMP				
N_1-R_9	3.5	783 (0.10)		1.0
N_1-C_2	5.3	783 (0.14)		0.96
C_2-O_{10}	8.0	1680 (0.47)		3.9
C_2-N_3	5.3	1680 (0.24),	1232 (0.15)	5.4
N_3-C_4	5.3	783 (0.08)		0.7
C_4-O_{11}	6.7	1630 (0.34),	1396 (0.27)	7.8
C_4-C_5	5.6	1630 (0.34),	783 (0.10)	6.1
C_5-C_6	5.6	1562 (0.20),	1498 (0.22),	0.0
		1302 (0.21) (calculated)[4]		
N_1-C_6	5.3	783 (0.09)		0.8
CMP				
N_1-R_9	3.5	784 (0.18)		1.0
N_1-C_2	5.3	1529 (0.14)		2.6
C_2-O_{10}	7.8	1652 (0.67) (calculated)[4]		0.0
C_2-N_3	5.3	1521 (0.27) (calculated)[4]		0.0
N_3-C_4	5.3	1529 (0.38)		4.3
C_4-N_{11}	5.3	1243 (0.19),	784 (0.14)	2.3
C_4-C_5	5.6	1588 (0.38) (calculated)[4]		0.0
C_5-C_6	5.6	1294 (0.19)		1.9
N_1-C_6	5.3	1294 (0.28)		2.4

TABLE III

Relative Displacement in Excited State Using I_j at 37.58
instead of $I_j = I_{(37.58)} / I_{(19.44)}$

UMP		CMD	
R_i	ΔR_i^e	R_i	ΔR_i^e
N_1-R_9	1.00	N_1-R_9	1.00
N_1-C_2	0.96	N_1-C_2	1.52
C_2-O_{10}	4.14	C_2-O_{10}	0.00
C_2-N_3	6.45	C_2-N_3	0.00
N_3-C_4	0.73	N_3-C_4	2.51
C_4-O_{11}	4.96	C_4-N_{11}	2.18
C_4-C_5	3.77	C_4-C_5	0.00
C_5-C_6	0.00	C_5-C_6	1.62
N_1-C_6	0.77	N_1-C_6	2.03

References

1. Tang, J. and Albrecht, A.C. (1970) in Raman Spectroscopy,
    Vol. 2, H. Szymanski, Editor, Plenum Press, New York,
    pp. 33-68.

2. Hirakawa, A.Y., and Tsuboi, M. (1975) Science 188
    356-361.

3. Blazej, D.C. and Peticolas, W.L. (1977) Proc. Nat. Acad.
    Sci. USA 74 2639-2643.

4. Tsuboi, M., Takahaski, S. and Harada, I. (1973) Physico
    Chemical Properties of Nucleic Acids, J. Duchesne, Ed.,
    Vol. 2, pp. 92-143.

5. Inagaki, F., Tasumi, M., and Miyazawa, T.J. (1974) Mol.
    Spec. 50 286.

6. Warshel, A. and Dauber, P. (1977) J. Chem. Phys. 66
    5477-5488.

7. Blazej, D. and Peticolas, W.L. (in preparation).

8. Tomlinson, B.L. and Peticolas, W.L. (1970) J. Chem. Phys.
    52 2154.

9. Small, E.W. and Peticolas, W.L. (1971) Biopolymers 10 69.

10. Tsuboi, M. Takahashi, S., Muraishi, S. and Kajiura, T.
    (1971) Bull. Chem. Soc. Japan 44 2921.

APPENDIX I
^{15}N ISOTOPE EFFECTS ON THE RAMAN
SPECTRA OF 5'UMP AND 5'UMP-d$_3$

H. Haruyama, Y. Nishimura, and M. Tsuboi

Faculty of Pharmaceutical Sciences, University of Tokyo
Hongo, Bunkyo-ku, Tokyo, Japan

Preparation of 5'UMP-^{15}N$_2$

Baker's yeast was grown in $(^{15}NH_4)_2SO_4$ as the sole nitrogen
source, and RNA was extracted from it with 1.5% sodium dodecyl
sulfate, 2% ethanol, and phosphate buffer pH 7.2.   After
centrifugation, the supernatant was dialyzed against pH 7.2
phosphate buffer.   The ^{15}N-RNA was digested with nuclease Pl,
and a mixture solution of nucleotides was obtained.   Four
nucleotides, 5'AMP, 5'UMP, 5'GMP, and 5'CMP were separated from
one another by Dowex I-X2 column chromatography.   The final
product, purified 5' UMP-^{15}N$_2$, gave only one spot in a thin-
layer chromatography.   A mass-spectrometric analysis of ammonia
produced by micro-Kjeldahl method from this sample indicated
that 90% of the whole nitrogen atoms in it are ^{15}N.

Raman Spectroscopic Measurements

Raman spectra of 5'UMP-^{14}N$_2$ and 5'UMP-^{15}N$_2$ were examined in
^{1}H$_2$O and in ^{2}H$_2$O.   The measurements were made by the use of a
JEOL JRS-U1 Raman spectrophotometer and the 514.5 nm line of a
Coherent Radiation model 52GA argon ion laser.   The amount of
^{15}N isotope shift was examined for each Raman line by repeating
a short-range scan for a neutral aqueous solution of 5'UMP-^{14}N$_2$
and for that of 5'UMP-^{15}N$_2$ alternately many times by keeping
the condition of the instrument as steady as possible.

Results

Theo M. Theophanides (ed.), Infrared and Raman Spectroscopy of Biological Molecules, 175–178.

In Figs. 1 and 2, are given the observed Raman spectra, observed frequencies (in cm^{-1}), and the amount of ^{15}N-shifts determined.

The amount of ^{15}N-isotope shift is proportional to the mean-square desplacements of the two nitrogen atoms of the 5'UMP molecule in each normal coordinate. It forms therefore a very useful set of data for determining the force constants and normal modes of vibrations of the uracil residue. In addition, an interesting isotope effect on the Raman intensity has been found in our present examination. As may be seen in Fig. 2, the relative intensity of the 1246 cm^{-1} line of 5'UMP-d_3 is markedly raised on ^{15}N substitution. This Raman line (let us call its mode "X") is one of the characteristic lines of pyrimidine bases. This is isolated and strong when it is located at about 1235 cm^{-1} or lower frequency (e.g., in uracil residue, protonated cytosine residue, and uracil residue-d_1, $^{15}N_2$). While, it is weaker and has a few satellite lines when it is located at a higher frequency than 1240 cm^{-1} (e.g., in deprotonated uracil residue, deuterated uracil residue, and cytosine residue). This is the strongest Raman line in the uracil residue and shows the greatest intensity enhancement on bringing the exciting wavelength from 514.5 nm to 351.1 nm. The ^{15}N-isotope effect on the intensity of this line may be ascribed to a change in the vibrational mode "X" caused by the $^{14}N \longrightarrow ^{15}N$ mass-increase (Appendix II). It may be explained also by taking a Duschinsky effect (Appendix III) into account.

It is not improbable that, in the first excited state $\tilde{A}$ (260 nm) of the uracil residue, the $C^5 = C^6$ bond order is very low so that its intrinsic stretching frequency is as low as 1234 cm^{-1}. If so, $C^5 = C^6$ stretching and the "mode-X" have nearly equal frequencies to each other in the upper $\tilde{A}$ state; while, in the ground electronic state, $C^5 = C^6$ stretching (1600 cm^{-1}) and "mode-X" (1234 cm^{-1}) are orthogonal to each other. In such a situation, the Franck-Condon overlap integrals that cause the Raman scattering of the "X-vibration" may have an appreciable value. If the "X-frequency" is exactly 1234 cm^{-1} (as that for 5'UMP-$^{15}N_2,d_3$), then the vibrational coupling in $\tilde{A}$ would be very strong and the "X-Raman line" would be strong. If, on the other hand, the "X-frequency" is slightly higher than 1234 cm^{-1} (as that for 5'UMP-$^{14}N_2$, d_3) the vibrational coupling in $\tilde{A}$ would be weaker and the "X-Raman line" would be weaker.

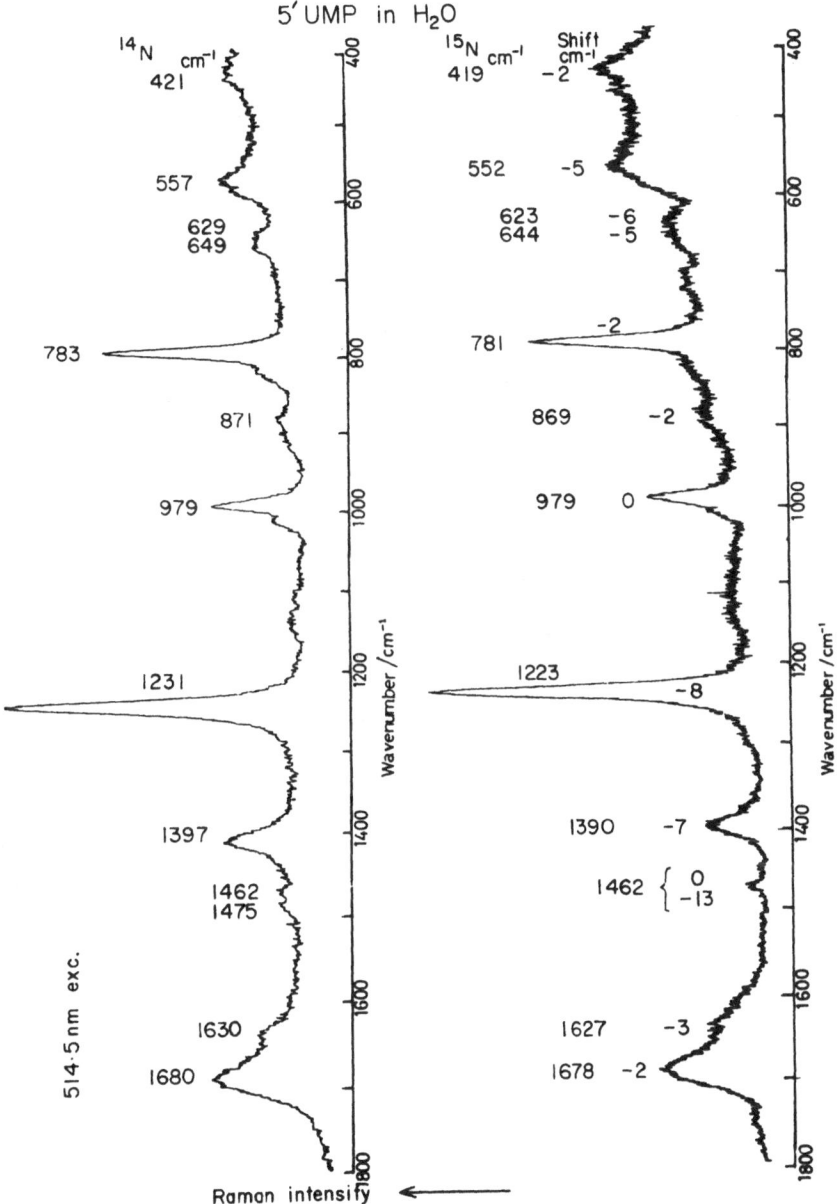

Fig. 1 Raman spectra of β- uridine-5'- phosphoric acid (5'UMP) - ^{14}N$_2$ and 5'UMP - ^{15}N$_2$ in H$_2$O, pH 7.2 at room temperature, concentration 10%.

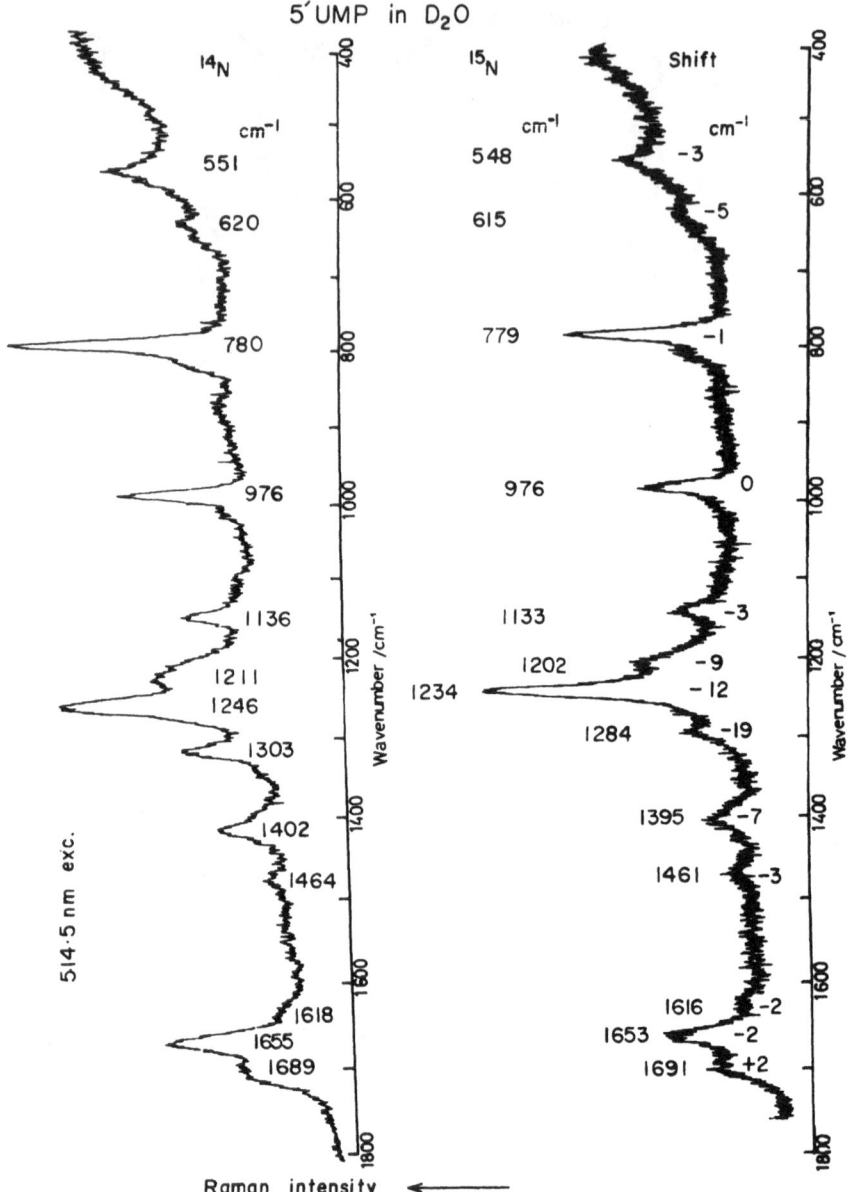

Fig. 2. Raman spectra of β-uridine-5'-phosphoric acid (5'UMP)
-$^{14}N_2$ and 5' UMP-$^{15}N_2$ in H_2O, pH 7.2 at room temperature,
concentration 10%.

APPENDIX II

ISOTOPE EFFECTS ON THE INTENSITIES
OF THE INFRARED AND RAMAN BANDS

M. Tsuboi

Faculty of Pharmaceutical Sciences, University of Tokyo,
Hongo, Bunkyo-ku, Tokyo, Japan

The intensity J of an infrared absorption band assignable
to a normal vibration (a) of a molecule is given by

$$J = (N\pi d/3c^2\nu_a) (\partial M_\sigma/\partial Q_a)^2, \tag{1}$$

where N is the Avogadro number, c the light velocity, d the
degeneracy of the normal coordinate Q_a, and ν_a the vibrational
frequency in cm^{-1} of the normal vibration. M_σ is a component
of the electric dipole moment of the molecule, where

$$\sigma = x, y, \text{ or } z. \tag{2}$$

The intensity I of a Raman band is given as

$$I = \frac{2^7 \pi^5}{3^2 c^4} \nu_{sc}^4 I_0 \left| (\alpha_{\rho\sigma})_{fi} \right|^2, \tag{3}$$

where the Raman scattering tensor $(\alpha_{\rho\sigma})_{fi}$ is expressed as

$$(\alpha_{\rho\sigma})_{10} = \frac{1}{hc} \sum_k \frac{\langle g1| M_\rho |ek\rangle \langle ek| M_\sigma |g0\rangle}{\nu_{ek} - \nu_{g0} - \nu_{exc} + i\Gamma} \tag{4}$$

in a rigorous resonance condition. In other words,[1] we
postulate here that the exciting frequency ν_{exc} (cm^{-1}) is so
close to the frequency $\nu_{ek} - \nu_{go}$ of an electronic transition
e ← g, that only one term shown in Eq. (4) of the Kramers-
Heisenberg expression gives an appreciable contribution. In
Eqs. (3) and (4), ν_{sc} is the frequency of the scattered light,

179

Theo M. Theophanides (ed.), Infrared and Raman Spectroscopy of Biological Molecules, 179–183.
All Rights Reserved. Copyright © 1979 by D. Reidel Publishing Company, Dordrecht, Holland.

I_0 the intensity of the incident laser light, h Planck constant, k vibrational quantum number, M_ρ and M_σ are the dipole moment operators along the ρ and σ directions, respectively, and Γ_e is a damping coefficient. Eq. (4) is rewritten as

$$(\alpha_{\rho\sigma})_{10} = \frac{1}{hc}\, \mu_\rho^\circ \mu_\sigma^\circ \sum_k \frac{\langle 1|\,k\rangle\langle k|0\rangle}{\nu_{ek}-\nu_{go}-\nu_{exc}+i\Gamma_e}$$

$$+ \frac{1}{hc}\left(\frac{\partial\mu_\rho}{\partial Q}\right)^\circ \mu_\sigma^\circ \sum_k \frac{\langle 1|Q|k\rangle\langle k|0\rangle}{\nu_{ek}-\nu_{go}-\nu_{exc}+i\Gamma_e}$$

$$+ \frac{1}{hc}\, \mu_\rho^\circ \left(\frac{\partial\mu_\sigma}{\partial Q}\right)^\circ \sum_k \frac{\langle 1|k\rangle\langle k|Q|0\rangle}{\nu_{ek}-\nu_{go}-\nu_{exc}+i\Gamma_e} \tag{5}$$

where μ_ρ and μ_σ are the electronic e ←g transition moments along ρ and σ directions, respectively, Q is the normal coordinate of the vibration now in question, and superscript $^\circ$ indicates that the value is for Q = O. If Q is a totally-symmetric vibration, the first term of Eq. (5) is usually predominant. If Q is a non-totally-symmetric vibration, however, this is zero; the Raman scattering is attributed to the second and/or third terms.

Let us now examine the isotope effect on $\partial M_\sigma/\partial Q$ or on $\partial\mu_\sigma/\partial Q$; both can be treated in exactly the same way. Let us define that

$$\widetilde{X} = [x_1, y_1, z_1, \ldots\ldots, x_n, y_n, z_n,\ldots.] \tag{6}$$

is the Cartesian displacement vector whose element x_n, for example, is the displacement of the n-th atom along the x-direction from its equilibrium position. Then,

$$\frac{\partial\mu_\sigma}{\partial Q_a} = \sum_n \left[\frac{\partial\mu_\sigma}{\partial x_n}\frac{\partial x_n}{\partial Q_a} + \frac{\partial\mu_\sigma}{\partial y_n}\frac{\partial y_n}{\partial Q_a} + \frac{\partial\mu_\sigma}{\partial z_n}\frac{\partial z_n}{\partial Q_a} \right]$$

$$= \sum_n \left[\frac{\partial\mu_\sigma}{\partial x_n} L^x_{xn,a} + \frac{\partial\mu_\sigma}{\partial y_n} L^x_{yn,a} + \frac{\partial\mu_\sigma}{\partial z_n} L^x_{zn,a} \right] \tag{7}$$

where $L^x_{nx,a}$ is the (nx, a) element of the $\mathbb{L}^x$ matrix which is defined by

$$X = \mathbb{L}^x Q. \tag{8}$$

As was shown by Miyazawa [1], $\mathbb{L}^x$ is conveniently used for estimating an isotope frequency shift. Now suppose that n-th atom is replaced by an isotope, so that $L_{xn,a}$ becomes $L_{xn,a} + \delta L_{xn,a}$. As will be shown below,

$$\delta L^x_{zn,a} = -L^x_{xn,a}\frac{\delta m_n}{m_n} - \frac{1}{2}L^x_{xn,a}\frac{\delta\lambda_a}{\lambda_a} + L^x_{xn,a}\sum_{b}\frac{\{\tilde{L}^x(M\delta\bar{M}^{-1}M)\,\mathcal{L}^x\}_{ab}\,\lambda_b}{\lambda_b - \lambda_a}\,,\quad(9)$$

where m_n is the mass of the n-th atom, δm_n is the difference, m_n(isotope) $- m_n$(normal), $\bar{M}^{-1}$ is a diagonal matrix with $1/m_1$, $1/m_1$, $1/m_1$, $1/m_2$, $1/m_2$, $1/m_2$,...., $1/m_n$, $1/m_n$, $1/m_n$, ...; and λ_b is the frequency parameter of the b-th normal vibration,

$$\lambda_b = 4\pi^2 c^2 \nu_b^2.\qquad(10)$$

Here ν_b is the vibrational frequency (cm^{-1}) of the b-th normal vibration. By the use of such $\delta L^x_{xn,a}$ values,

$$\delta\left(\frac{\partial\mu_\sigma}{\partial Q_a}\right) = \left(\frac{\partial\mu_\sigma}{\partial x_n}\right)\delta L^x_{xn,a} + \left(\frac{\partial\mu_\sigma}{\partial y_n}\right)\delta L^x_{yn,a} + \left(\frac{\partial\mu_\sigma}{\partial z_n}\right)\delta L^x_{zn,a}.\qquad(11)$$

Derivation of Eq. (9) is made as follows:

On the basis of the well-known expression[2],

$$L^x = \bar{M}^{-1}\tilde{B}G^{-1}L = \bar{M}^{-1}\tilde{B}(\tilde{L}^{-1}),\qquad(12)$$

we can write the isotope effect δL^x on L^x as

$$L^x + \delta L^x = [\bar{M}^{-1} + \delta\bar{M}^{-1}]\,\tilde{B}\,[(\tilde{L}^{-1}) + \delta(\tilde{L}^{-1})].\qquad(13)$$

Hence,

$$\delta L^x = \delta(\bar{M}^{-1})\,\tilde{B}\,\tilde{L}^{-1} + \bar{M}^{-1}\tilde{B}\,(\delta\tilde{L}^{-1}).\qquad(14)$$

Here, L matrix is the transformation matrix from the normal coordinates Q to an internal coordinates R, so that

$$R = LQ,\qquad(15)$$

the matrix B is defined [3] by

$$R = BX,\qquad(16)$$

and G is an inverse kinetic energy matrix. Let us now define δP by

$$L + \delta L = L(E + \delta P),\qquad(17)$$

where $\mathbb{E}$ is the unit matrix. Thus, $\mathbb{E} + \delta\mathbb{P}$ is regarded as the transformation matrix from the normal coordinates of the isotopic molecule to the normal coordinates of the nonisotopic molecule.

The first term of Eq. (14) is readily rewritten as

$$\delta(\mathbb{M}^{-1})\,\widetilde{\mathbb{B}}\,\widetilde{\mathbb{L}}^{-1} = \delta(\mathbb{M}^{-1})\,\mathbb{M}\cdot\mathbb{M}^{-1}\,\widetilde{\mathbb{B}}\,\widetilde{\mathbb{L}}^{-1} = \delta(\mathbb{M}^{-1})\,\mathbb{M}\,\mathbb{L}^{x}. \tag{18}$$

This gives the first term of Eq. (9), now in question.

The second term of Eq. (9) comes from the diagonal element of $\delta\mathbb{L}^{-1}$ of Eq. (14). Because

$$\widetilde{\mathbb{L}}\mathbb{F}\mathbb{L} = \Lambda, \tag{19}$$

and because potential energy matrix $\mathbb{F}$ is common for isotopes

$$\left[\widetilde{\mathbb{L}} + \delta(\widetilde{\mathbb{L}})\right]\mathbb{F}\left[\mathbb{L} + \delta\mathbb{L}\right] = \Lambda + \delta\Lambda \tag{20}$$

$$\therefore\ (\mathbb{E} + \widetilde{\delta\mathbb{P}})\,\widetilde{\mathbb{L}}\mathbb{F}\mathbb{L}(\mathbb{E} + \delta\mathbb{P}) = \Lambda + \delta\Lambda \tag{21}$$

$$\therefore\ \widetilde{\delta\mathbb{P}}\Lambda + \Lambda\delta\mathbb{P} = \delta\Lambda \tag{22}$$

$$\therefore\ (\delta\mathbb{P})_{aa}\lambda_a + \lambda_a(\delta\mathbb{P})_{aa} = \delta\lambda_a \tag{23}$$

$$\therefore\ (\delta P)_{aa} = \frac{1}{2}\frac{\delta\lambda_a}{\lambda_a} \tag{24}$$

The third term of Eq. (9) originates from the off-diagonal elements of $\delta\mathbb{L}^{-1}$ of Eq. (14). Because eigenvalues Λ (with λ_a's in diagonal) are given by

$$\Lambda = \mathbb{L}^{-1}\mathbb{G}\mathbb{F}\mathbb{L}, \tag{25}$$

the isotope effect now in question is given by

$$\Lambda + \delta\Lambda = (\mathbb{E} - \delta\mathbb{P})\mathbb{L}^{-1}(\mathbb{G} + \delta\mathbb{G})\mathbb{F}\mathbb{L}(\mathbb{E} + \delta\mathbb{P}). \tag{26}$$

Remembering that

$$\mathbb{F}\mathbb{L} = (\widetilde{\mathbb{L}^{-1}})\Lambda, \tag{27}$$

Eq. (26) is rewritten as

$$\delta\Lambda = (-\delta\mathbb{P})\Lambda + \Lambda(\delta\mathbb{P}) + \mathbb{L}^{-1}(\delta\mathbb{G})\,\widetilde{\mathbb{L}}^{-1}\Lambda. \tag{28}$$

The (a, b) element of this matrix is given as

$$0 = (-\delta P)_{ab} \lambda_b + \lambda_a (\delta P)_{ab} + \left[L^{-1} \delta G \widetilde{L}^{-1}\right]_{ab} \lambda_b. \quad (29)$$

Therefore,

$$(\delta P)_{ab} = \frac{[L^{-1} \delta G \widetilde{L}^{-1}]_{ab} \lambda_b}{\lambda_b - \lambda_a}. \quad (30)$$

Since

$$G = BM^{-1} \widetilde{B}, \quad (31)$$

δG in Eq. (29) is replaced by

$$\delta G = B \delta(M^{-1}) \widetilde{B}$$
$$= BM^{-1} M \, \delta(M^{-1}) MM^{-1} \widetilde{B}. \quad (32)$$

Therefore,

$$L^{-1} \delta G \, \widetilde{L}^{-1} = L^{-1} B M^{-1} (M \delta \bar{M}^{1} M) M^{-1} \widetilde{B} \widetilde{L}^{-1}$$
$$= \widetilde{L}^{x} (M \delta \bar{M}^{1} M) L^{x}. \quad (33)$$

This, in combination with Eq. (30), forms the third term of Eq. (9).

REFERENCES

1. T. Miyazawa, J. Mol. Spectry. 13, 321 (1964).
2. B. L. Crawford, Jr. and W. H. Fletcher, J. Chem. Phys., 19, 141 (1951).
3. E. B. Wilson, Jr., J. Chem. Phys., 9, 76 (1941).

APPENDIX III DUSCHINSKY EFFECT AND RAMAN SCATTERING

M. Tsuboi

Faculty of Pharmaceutical Sciences, University of Tokyo,
Hongo, Bunkyo-ku, Tokyo, Japan.

The discussion in this Appendix starts from the following
expression of the Raman scattering tensor $(\alpha_{\sigma\sigma'})_{10}$ for a totally
symmetric vibration of a molecule:

$$(\alpha_{\sigma\sigma'})_{10} = \frac{1}{\hbar c} \langle e|M_\sigma|g\rangle_0^2 \sum_k \frac{\langle 1|k\rangle\langle k|0\rangle}{\nu_{ek} - \nu_{g0} - \nu_{exc} + i\Gamma_{ek}} \qquad (1)$$

where M_σ is the dipole moment operator along σ direction;
$|g\rangle$ and $|e\rangle$ are the electronic wavefuncions of the ground and
excited states, respectively; $|k\rangle$ is the vibrational wavefunction
in the excited (e) state; ν_{exc} is the Raman exciting frequency;
and Γ_{ek} is a damping coefficient.

Both of the equilibrium geometry and the intramolecular force
field in the excited (e) electronic state are in general
appreciably different from those in the ground (g) electronic
state. Therefore, the normal modes of vibration are also different
in the excited state from those in the ground state. Let us define
the normal coordinates in the excited (e) and ground (g) electronic
states as

$$\tilde{Q}^{(e)} = [\, Q_1^{(e)}, \, Q_2^{(e)}, \, \ldots, \, Q_n^{(e)} \,] \qquad (2)$$

and

$$\tilde{Q}^{(g)} = [\, Q_1^{(g)}, \, Q_2^{(g)}, \, \ldots, \, Q_n^{(g)} \,] \qquad (3)$$

The Franck–Condon overlap integrals which appear in equation (1)
should be considered to be n-fold multiple integrals. To evaluate
such integrals let us use a set of symmetry coordinates in the

Theo M. Theophanides (ed.), Infrared and Raman Spectroscopy of Biological Molecules, 185–186.

ground electronic state,

$$\widetilde{\underset{\sim}{S}}^{(g)} = [\, S_1^{(g)}, \; S_2^{(g)}, \; \cdots \cdots, \; S_n^{(g)} \,]. \tag{4}$$

Let us next assume that symmetry coordinates in the excited electronic state are given as

$$\widetilde{\underset{\sim}{S}}^{(e)} = \widetilde{\underset{\sim}{S}}^{(g)} + \widetilde{\delta \underset{\sim}{S}}^{(e)}$$

$$= [\, S_1^{(g)} + \delta S_1^{(e)}, \; S_2^{(g)} + \delta S_2^{(e)}, \; \cdots \cdots, \; S_n^{(g)} + \delta S_n^{(e)} \,]. \tag{5}$$

Let us further assume that the normal coordinates in the excited and ground electronic states are related with the symmetry coordinates as

$$\underset{\sim}{Q}^{(e)} = (\underset{\sim}{L}^{(e)})^{-1} \underset{\sim}{S}^{(e)} \tag{6}$$

and

$$\underset{\sim}{Q}^{(g)} = (\underset{\sim}{L}^{(g)})^{-1} \underset{\sim}{S}^{(g)} \tag{7}$$

respectively. In general, if there is a change in the composition of the normal motions in terms of symmetry coordinates on going from ground to excited state, the transformation between the normal coordinates for ground and excited state motions includes a rotation in coordinate space. What is caused by such a rotation is called Duschinsky effect [1, 2]. A Franck–Condon overlap integral

$$\langle k | o \rangle = \langle 1\,0\,0\, \cdots\, 0 \,|\, 0\,0\,0\, \cdots\, 0 \rangle,$$

for example, is now given as

$$\langle k | o \rangle = \int \cdots \int \chi_{11}^{(e)}(Q_1^{(e)}) \, \cdots \, \chi_{no}^{(e)}(Q_n^{(e)})$$

$$\times \chi_{10}^{(g)}(Q_1^{(g)}) \, \cdots \, \chi_{no}^{(g)}(Q_n^{(g)}) \, J \, dS_1^{(g)} \cdots \, dS_n^{(g)} \tag{8}$$

where $Q_j^{(e)}$ and $Q_j^{(g)}$ are given by equations (6) and (7), respectively, and

$$J = \{ \, |(\underset{\sim}{L}^{(e)})^{-1}| \, |(\underset{\sim}{L}^{(g)})^{-1}| \, \}^{1/2}. \tag{9}$$

REFERENCES

1.   F. Duschinsky, Acta Physicochim. USSR, 1, 551 (1937).
2.   G. J. Small, J. Chem. Phys., 54, 3300 (1971).

INTERACTION OF METAL IONS WITH NUCLEIC ACIDS

T. Theophanides

National Research Foundation, Physical Chemistry and
Spectroscopy Centre, 48, B. Constantinou Ave., Athens 501-Greece

INTRODUCTION

Every living organism from the simplest one to the more complex contains small amounts of metals. The fundamental importance
of metal ions in biological processes has become clearer in the
last decade. The effects of heavy metal pollution of the environment and the living organisms have started exhaustive studies on
the medical and technical aspects of the toxic metals, mercury,
lead, etc. These are a few of the most dangerous industrial metal contaminants. However, some metals, in small quantities act
as catalysts in very important enzymatic biological reactions.
The ferrodoxins, for example, are the most ancient biological
iron-catalysts. Hemoglobin, the red substance in the blood, contains iron and is very important for life in uptaking oxygen during respiration. Metal ions are involved in a very large number
of reactions and modify either the mechanism of the reaction or
its rate, or both.

The role of inorganic elements in biological systems is extremely important. For example, the alkaline metals play the
role of charge carriers and activators for Na/K-ATPase. The alkaline-

Theo M. Theophanides (ed.), Infrared and Raman Spectroscopy of Biological Molecules, 187–204.
All Rights Reserved. Copyright © 1979 by D. Reidel Publishing Company, Dordrecht, Holland.

earth metals also act as activators of several enzymes. Calcium
is the main component of bones and shells. These metals play an
extremely important structural role in proteins and nucleic acids.
The transition metal ions and post-transition metals are efficient
in blocking active sites of metalloenzymes in redox reactions, and
oxygenations. They act as Lewis acids. The heavy metals, Hg, Pb,
As and Cd play the role of enzyme inhibitors and are very toxic to
the cells. However, some heavy metals depending on the elements
which are attached to them show catatoxic effects and are useful
against several diseases.

Observation of the effects of metal ions on nucleic acids
provides us with knowledge on the active sites and the conformations of the biological macromolecules when they interact with metal ions. The detailed structure and reactions of RNA, DNA and
their constituents, base, sugar and phosphate groups, can be un-
veiled by identifying the potential metal binding sites of complexation and the stabilities of the complexes. For example, the role
of metal ions in DNA replication, the mechanism of the replication
in living cells or in cancer cells are problems which are not well
understood and the metals do play a role in these reactions. The
study of metal-biomolecule interaction may help clarify these biological processes. In some cases the presence of metal ions inhi-
bits these processes, a fact which shows the significance of metal-
nucleic acid chemistry. Different metals attack different sites
and locations in DNA. It would be desirable to have some way of
predicting the effect of metal ions on the course of the above
reactions.

Vibrational spectroscopy and in particular Raman Spectroscopy
has become today a powerful tool for the chemist, biochemist and
biologist, because it provides a wealth of information on the
structural and conformational changes in biological molecules.

The effect of metal ions on nucleic acids can be studied by IR
and Raman Spectroscopy and the spectral changes may be interpreted
to elucidate the mode of action of these molecules. The goal here
is to show some chemical and spectroscopic evidence on the reactivity of metal ions with nucleic acids. We will examine in particular the vibrational spectra of these systems under different
reaction conditions in aqueous solutions and in the solid state,
and show any meaningful patterns that can be revealed.

SOME ASPECTS OF THE METAL ION

Metal ions act as Lewis acids (acceptors) and react with the
non-metals, Lewis bases (donors) to form complexes. The interaction
of metals with nucleic acids is done primarily with the binding
ligand atoms (N, O, P) of the nucleic acid constituents (1-3).
An acid is an electron-pair acceptor and a base an electron-pair
donor. The fundamental metal-molecule reaction in the Lewis
sense is the formation of a coordinate bond between them. All
simple cations therefore, are potential Lewis acids. The acid
strength of the cations is their coordinating ability which increases with an increase in positive charge on the ion and a decrease in ionic radius. For example, the order of increasing
strength of Lewis acids in the following sequence is: $Fe^{++} < Fe^{+++}$
and $K^+ < Na^+ < Li^+$. The transition elements are strong Lewis acids
and form a variety of complexes.

Typical acid-base reactions of cations are shown below:

ACID	BASE	COMPLEX

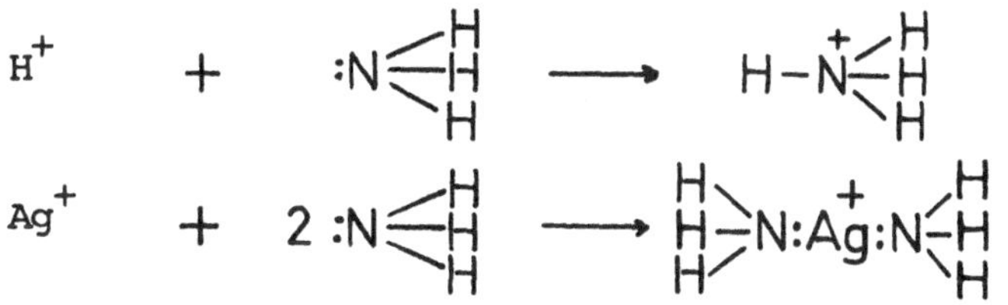

$$M^+ \quad + \quad :L \longrightarrow M^+ \quad : \quad L$$

unfilled	filled	coordinate covalent
orbital in a	orbital in a	bond, M:L, the shairing
valence shell or	valence shell or	of a an electron-pair
electron-pair	electron-pair	
acceptor	donor	

The periodic classification of the chemical elements given in Table I shows the proportion of metals to non-metals from all the chemical elements in nature. The metals constitute about 80 percent of the chemical elements. Some of these metals are essential in living organisms, animals or plants in small quantities (Table I). The inorganic elements Na and K, are the major components of body fluids and of cytoplasm. Sodium is not known to be essential to plants, but potassium is found in plants (3).

A classification of metal ions based on electronegativity was proposed several years ago by Ahrland et al (4). The metal ions were termed as class (a) and class (b). The class (a) or hard acid metals (Table II) are those having the closed shell electronic configuration structure. For example, the alkaline metal ions, Li^+, Na^+, K^+, etc., and the alkaline earth Ca^{++} and Mg^{++} ions are hard acids. The interaction of these metal ions with nucleic acids is electrostatic in nature. The metal ions which have less than half filled d-shells, such as the transition elements V^{3+} and Cr^{3+} are in class (a) whereas those with half-filled ·d-shells, such as Mn^{2+} and Fe^{3+}, begin to become less hard acids and as we go to configurations with more than half filled d-shells, such as, Fe^{2+} and Co^{2+} we move to a borderline region of acidity

TABLE I
PERIODIC CLASSIFICATION OF
THE ELEMENTS

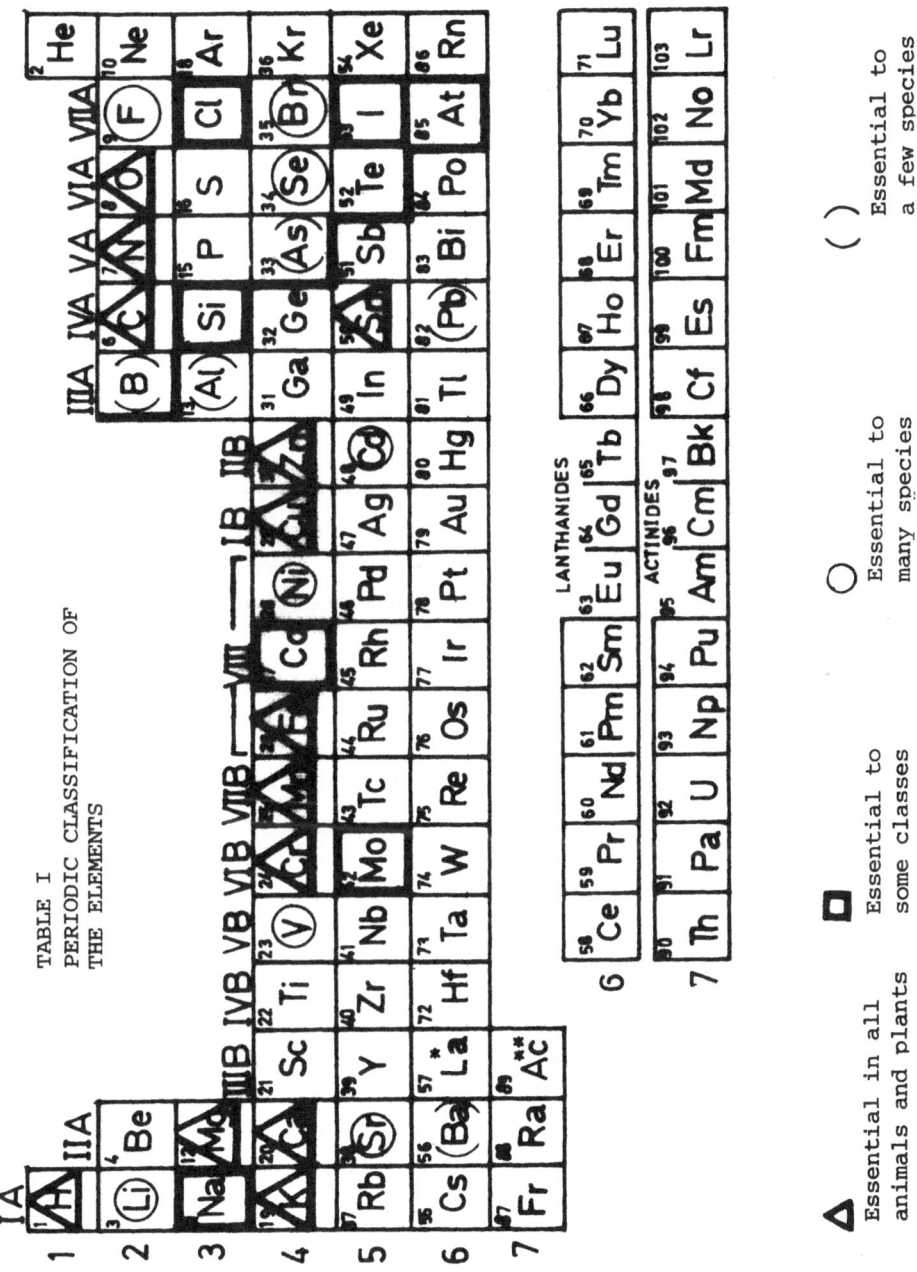

Essential in all animals and plants

Essential to some classes

Essential to many species

Essential to a few species

Table II.
Classification of Metal Ions (hard, soft)*

Hard (s^o, s^2, s^2p^6, d^{3-5})	Borderline (d^{6-7})	Soft (d^{8-10}, $d^{10}s$
H^+, Li^+, Na^+, K^+	Fe^{2+}, Co^{2+}	Cu^+, Ag^+, Tl^+
Mg^{2+}, Ca^{2+}, Sr^{2+}	Cu^{2+}, Zn^{2+}	Pt^{2+}, Rh^{2+}
Al^{3+}, La^{3+}	Pb^{2+}, Sn^{2+}	CH_3Hg^+, Tl^{3+}
Cr^{3+}, Mn^{2+}, Fe^{3+}		

---

*F. Basolo and R.G. Pearson, Mechanisms of Inorganic Reactions 2nd ed., Wiley (1976). In parentheses are given the configurations of the ions.

(see Table II).The transition and post-transition metal ions with a large number of d-electrons are soft acids or class (b) (metals). The most clear-cut cases of (b) class acceptors are the d^{8-10} shell systems with low oxidation numbers. For example, Pt^{2+}, Pd^{2+}, Cu^+, Ag^+, Hg^{2+} CH_3Hg^+ and Au^+.

The polarizability and the charge on the metal are important in determining the class (b) character of the metal. In the post-transition series, Zn^{2+} with a d^{10}-shell is almost an (a) acceptor. However, when we go from first to second and third transition series the polarizability increases and we find Hg^{2+} to be a typical (b) acceptor while Cd^{2+} is in the borderline region (see Table II).   The non-transition metal ions with $d^{10}s^2$-shell electron structure, such as Sn^{2+} and Pb^{2+} are (b) acceptors and the high charge in these ions gives them more (b) type properties. As a result Sn^{4+} and Pb^{4+} with d^8 configurations are better (b) acceptors than Sn^{2+} and Pb^{2+}.

The natural order of the stability of complexes of bivalent transition metals is, $Mn < Fe < Co < Ni < Cu > Zn$. In the metal-nucleic acid interaction the most important affinity sequence for metal ions is the following:

Class (a) metal

$-O^- \gtrsim -OH \gg H_2O \sim Cl^- \sim =C=O \sim -NH_2 \gtrsim -NH- \gtrsim -N=$

Class (b) metal

$=N- \gg -NH_2 > OH^- \gg H_2O$

From the above stability orders and affinity sequence we find that a class (a) acceptor interacts more strongly with oxygen than with nitrogen ($O \gg N$). For example, the hard-metal ions Na^+, K^+, Ca^{++} and Mg^{++} interact mainly with the hydroxide or the keto groups of the nucleic acids, whereas the borderline metal ions, Cu^{++}, Zn^{++} and Co^{++} form complexes simultaneously through the pyrrole, pyridinic and phosphate groups of the nucleotides. The metal ions of (b) character on the other hand form very stable covalent bonds predominantly with the pyridinic nitrogens. The soft metal ions, Ag^+, Hg^{++} and Pt^{++} tend to accummulate in tissues rich in lipids, whereas, the hard-metal ions, Na^+, K^+, Ca^{++}, and Mg^{++} are found in large quantities in blood serum and in tissues, where water is the coordinative ligand.

THE NATURE OF THE LIGAND ATOMS

The nucleic acids have many coordinative sites (electron reach sites) which are the electronegative elements of the purines and the pyrimidines (N,O) and the sugar-phosphate groups (O,P). These are the so called ligand atoms or donors or Lewis bases, with one or more lone pair of electrons to donate and form coordinative bonds. The nitrogen is the most important binding site and from all nitrogens the guanine N-7 is the most reactive with metals of class (b).

3.1 Classification of the ligands

The ligands are classified in monodentates and polydentates. Monodentates are the ligands with one binding atom. There are several types of monodentate ligands:

(i)   with one lone pair of electrons and no valence shell vacant orbitals

(ii) with one lone pair of electrons and with valence shell vacant orbitals,

(iii) with π-electrons and valence shell vacant orbitals;

(iv) with more than one lone pair of electrons and no valence shell vacant orbitals and

(v) with more than one lone pair of electrons and with valence
shell vacant d-orbitals.

The ligand atoms having more than one lone pair of electrons
can react as bridging agents between two or more metal atoms.
Polydentates are the ligands which have more than one binding
atoms. These ligands can be further classified into <u>chelates</u> (from
the Gr. χηλή=claw of a lobster, crab) and <u>non chelate</u> ligands.
The chelates ligands are those that can bind the same metal with
more than one atom. The chelate ligands are further classified
into bidendates, tridentates, tetradentates, etc. Examples:
bidentates, $NH_2(CH_2)_2NH_2$ (en), $NH_2CH_2COO^-$ (gly)

tridentates, $NH_2(CH_2)_2NH(CH_2)_2NH_2$ (trien)

Non chelate polydentate ligands are those that have more than one
binding site but do not form chelates because of steric hidrance
to bind the same metal atom. They bind to two or more metal atoms
depending on the number of binding sites.

The classification of ligands according to Pearson into hard
and soft is due to their reactivity towards the metal atoms, which
are divided into two classes (a) hard acids and (b) soft acids.
(see Table II). According to this classification hard bases prefer to bind with hard acids and soft bases prefer to bind with
soft acids. The more polarizable is a species the softer it is.
Soft reacts with soft and hard with hard. Similar things attract
their own in the world of complexes and similarly in the living
world. The soft bases form chemical bonds of covalent character
as opposed to the hard bases which form ionic bonds, where the
binding forces are governed mainly by electrostatic interactions.

3.2 The Nucleic Acids and their constituents

The cells constitute the simplest microscopic building blocks
of all living organisms (plants and animals). The cell is covered
by a membrane which allows materials to pass in and out and contains the nucleus. The nucleus is also covered by a membrane.
The nucleus contains the genetic information carried by the nucleic acids, DNA and RNA. The cytoplasm contains many compounds
from inorganic ions (Na$^+$, K$^+$, Mg^{++}), etc., to enzymes (5).

The building blocks of nucleic acids are the four letters
A,T,G,C (or U) which form the simplest alphabet through which
information is transmitted. The purine bases are adenine (A) and
guanine(G) in both DNA and RNA and the pyrimidines, cytosine(C)
and thymine(T) in the case of DNA, or cytosine(C) and uracil(U) in
the case of RNA;

The structures of the five bases and the numbering of the
atoms are shown below:

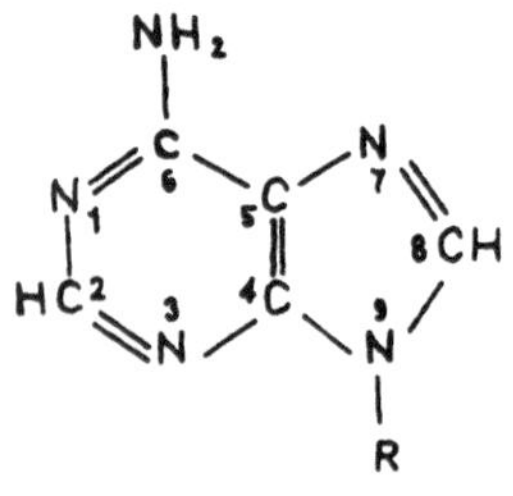

Adenine 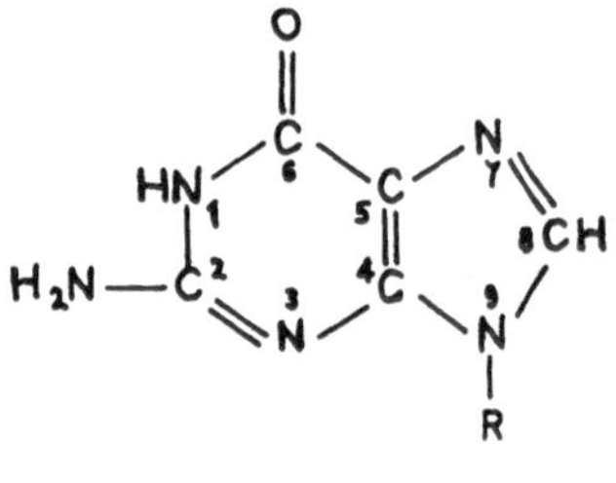   Guanine 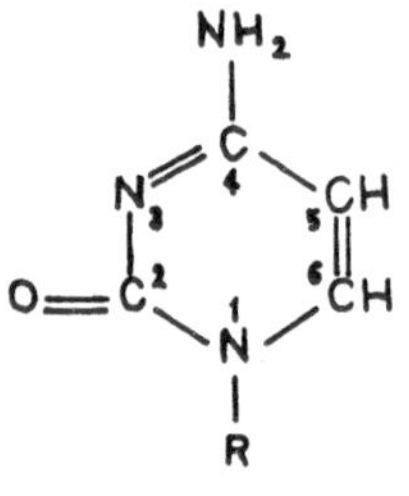   Cytosine

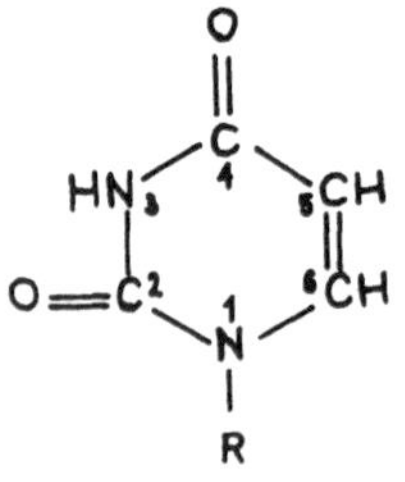

Urasil 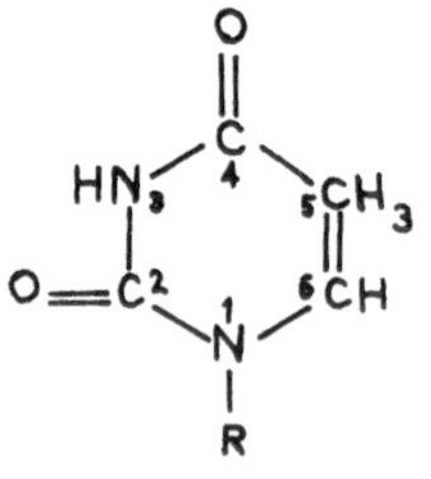   Thymine

if R_r is a ribose we have a nucleoside and when R_{rp}, ribose-phosphate we have a nucleotide. The bases are linked to a sugar and
a phosphate group (orthophosphate, pyrophosphate, or triphosphate)
through the C'5-carbon of the sugar.

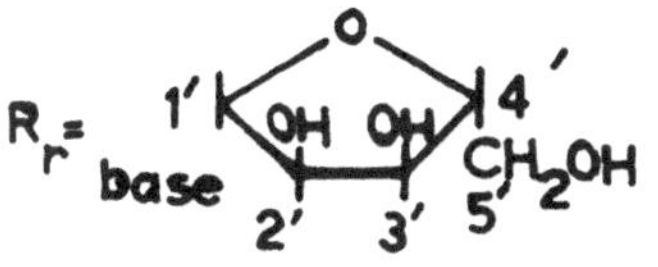

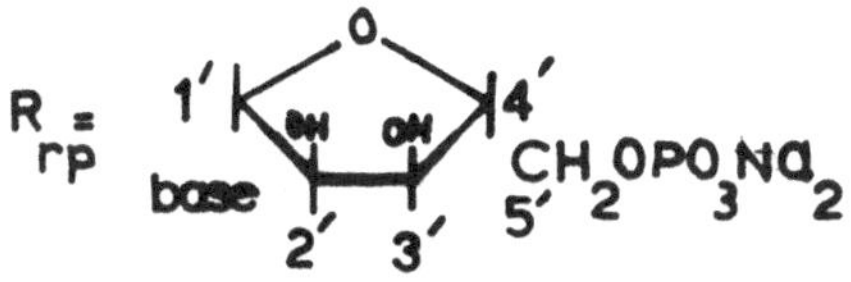

Esterification of the phosphate group with the sugar hydroxyl groups of the nucleoside produces the backbone of the nucleic
acids with the sequence of bases shown in Figure 1.
The nucleotide adenosine triphosphate (ATP) is shown below:

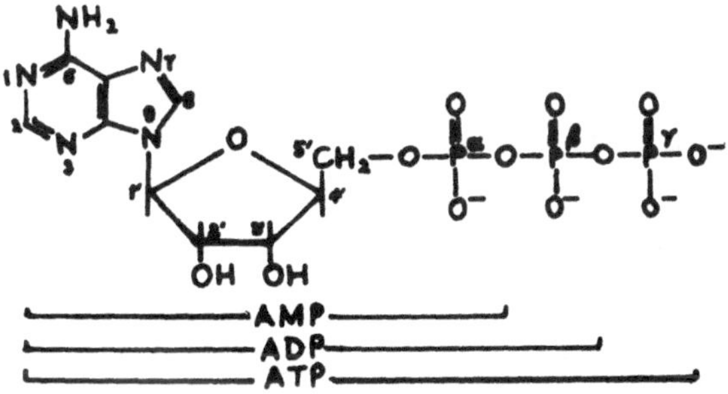

The nucleotides are essential to living organisms as energy carriers, enzymes and coenzymes and their metallic complexes are
biologically important. The nucleic acid constituents can be considered as polydentate ligands. The nucleic acids deoxyribonucleic
acid (DNA) and ribonucleic acid (RNA) are formed by polyester condensation and are normally polymers of nucleotides. In the DNA or
RNA a purine base and a pyrimidine base are paired up in a double
helix through hydrogen bonding forming the combinations A-T or
(A-U) and G-C (Fig. 2).

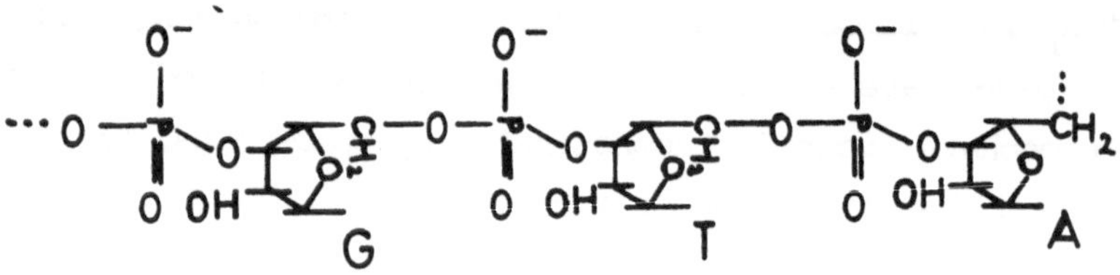

Figure 1.

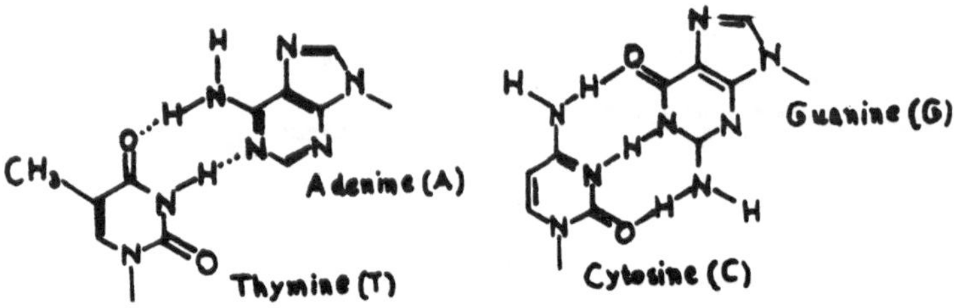

Figure 2.

In the nucleic acid structure there is stored the basic blueprint
of any living being. The information is encoded in a message passed on from generation to generation in the form of a single nucleic acid molecule. Through their specific sequence in the molecular chain the bases represent information which is transfered
in a readable form for purposes of reproduction. The question to
be answered in the metal-molecule interactions is which site is
favored in the presence of a particular metal ion and how many of
these sites are reacting. Another major question to be resolved
is what are the structures of the metal complexes formed.Crystal-
lographic studies are now available on nucleoside and nucleotide
complexes and give us some answers as to the structures of the
most stable complexes that come out of solutions as single crystals (6-10).

THE FORMATION OF METAL COMPLEXES

The reaction of interest on complexation of a metal M with
a ligand L is of the following type:

Metal + Ligand ⇌ Complex

The formation of complex compounds involves successive equilibria of 1:1, 1:2, 1:3,..., 1:n+1:

$$M + L \rightleftharpoons ML \qquad\qquad\qquad 1$$

$$ML + L \rightleftharpoons ML_2 \qquad\qquad\qquad 2$$

$$\cdot \qquad \cdot \qquad \cdot$$

$$\cdot \qquad \cdot \qquad \cdot$$

$$\cdot \qquad \cdot \qquad \cdot$$

$$MLn+ L \rightleftharpoons MLn+1 \qquad\qquad 3$$

The stability of the metal complexes ML, ML_2, $MLn+1$ depends on the enthalpy change or the heat of formation ΔH. However, the stability constant K which is a more convenient parameter is related to the free energy change of the above reaction,

$$\Delta G = -RT\ln K \qquad\qquad\qquad 4$$

The relation of free energy contains the entropy change, ΔS as well as ΔH,

$$\Delta G = \Delta H - T\,\Delta S \qquad\qquad\qquad 5$$

the last relation 5 gives a better picture of the factors involved in the metal ligand interaction and the binding strength. The stability or association constant K of a particular complex is defined by the expressions,

$$K_1 = \frac{[ML]}{[M][L]} \; , \; K_2 = \frac{[ML_2]}{[ML][L]} , \; K_{n+1} = \frac{[ML_{n+1}]}{[ML_n][L]} \qquad 6$$

The constants K_1, K_2... are the stepwise formation constants and the overall formation constant K_T is the product of these:

$$M + (n+1)\,L \rightleftharpoons ML_{n+1} \qquad\qquad 7$$

$$K_T = K_1 K_2 \ldots K_{n+1} = \frac{[ML_{n+1}]}{[M][L^{n+1}]} \qquad\qquad 8$$

K is easier to determine experimentally than ΔH. In the expression
of K_1 when the metal concentration M is equal to the complex concentration, M=ML, i.e., half of the metal ion is complexed then
$K_1 = 1/L$. Therefore, the stability constant, K_1 can be calculated
approximately from the concentration of the free ligand at the
half point of complex formation.

The stabilities of metal complexes with bases, nucleosides
and nucleotides depend on the type of sites (base, ribose and
phosphate) that are involved in the metal-molecule interaction.
Competition between these three types of groups for the metal ion
depends on the nature of the metal. The hard metal ions, i.e.,
the alkaline, alkaline earth and the borderline transition metal
ions (see Table II)prefer the phosphate groups. The ribose group
is the least reactive of the three. It reacts only at high pH values. The soft metal ions react with the nitrogen sites of the
base which are the strongest coordinative sites for these metals.
The order of relative stabilities of the complexes formed with
the soft metals is: G $\gg$ A $>$ C $>$ U,T (11). From crystallographic studies it is found in purines and pyrimidines that the metal binds N-9 in guanine and adenine and N-3 in cytidine (12).

5. METAL COMPLEXES OF NUCLEOSIDES, NUCLEOTIDES AND NUCLEIC ACIDS

In the nucleosides and nucleotides the N-9 binding position
is not available. It is found here that the alkaline earth metal
ions bind to the phosphate groups and the borderline metal ions,
Mn^{2+}, Ni^{2+}, Cu^{2+} and Zn^{2+} bind to both base (N-7) and phosphate
sites (13). These ligands can produce a variety of complexes
with metal ions, depending on the number of phosphorus atoms

present on the sugar, the base and on the nature of the complexing
ion. A chelate structure involving binding first at N-7 and forming a ring with O-6 has also been postulated by several investigators, (14-20) for guanosine and inosine and their phosphate derivatives, mononucleotides, oligonucleotides and polynucleotides.
The N-7, O-6 chelation resembles the stable chelates of 8-hydroxy-
quinoline. There is evidence of O-6 involvement when these molecule bind to metals. Infrared studies show that the carbonyl stretching frequency is perturbed in the presence of metal ions.

The replication process in DNA requires the presence of metal ions, which can even defeat or stop this replication, thus the
significance of metal complexation in the chemistry and biochemistry
of nucleic acids. The metal ions are involved in the genetic code
for the production of proteins and can change the information given by the code. For example, the transcription of DNA into mRNA
is influenced by metal ions.

Some metal ions are capable of distinguishing between RNA
and DNA in enzymatic reactions. In RNA the approximate distance
between the 2' and 3'-hydroxyl groups of the ribose ring is about
2.7Å and if the sugar can form a chelate or bridge between two
metal atoms then it will form a complex. DNA which does not have
2'-hydroxyl groups cannot form complexes of this type and this is
probably how metal ions function in enzymatic reactions, and can
distinguish RNA from DNA.

In the replication of DNA catalysed by DNA polymerase I from
E. coli the reaction requires magnesium(II) ions which involve the
cleavage and the formation of phosphate bonds. However, the metal
participation in this reaction may have some other functions as
well. The msot interesting reaction is when we substitute magne-
sium(II) by manganese(II). In the presence of manganese(II) ions
the synthesis of DNA allows the incorporation of both deoxynucleo-
tides and ribonucleotides into the double helix. Thus magnesium

ions are more selective in that they can screen out the nucleotides with ribose which contains two hydroxyl groups 2'-OH and 3'-OH. This is probably due to the fact that with manganese(II) we may have a dimer containing Mn-Mn bonds which form a bridge between the two hydroxyl groups:

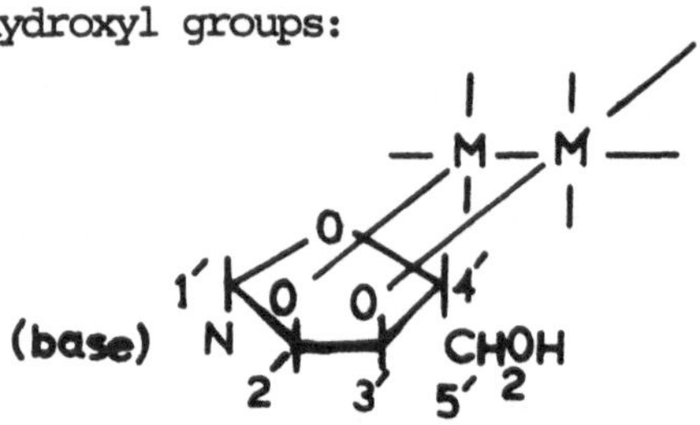

and carry it in the DNA synthesis, whereas with magnesium(II) the dimers Mg-Mg are difficult to stabilize and do not exist. The mechanism of the above selective reaction due to the nature of the metal ion binding to ribose hydroxyls is not known.

Metal ions bind both phosphate and base groups in polynucleotides just as in the monomers. However, little is known about the relative affinity of the various metal ions for phosphate and base sites. Metal ions stabilize the DNA double helix and divalent ions (Mg^{2+} and Co^{2+}) are more effective than univalent ions. The counter ion decreases the repulsion between the negatively charged phosphates. This stabilization phenomenon of metal ions is not universal. It is found that metals binding to bases of DNA compete with the hydrogen-bonding of the double helix and therefore destabilize it. Cu^{2+} ions bind not only to phosphate, but to bases as well. Thus binding to phosphate stabilizes the DNA double helix, whereas binding to bases destabilizes it by H-bond breaking. Reversible unwinding and rewinding of DNA by metals can be achieved in vitro and the effect of temperature and electrolyte on the interaction of metals with DNA is different depending on whether the metal is phosphate or base bound (2). It is interesting to know if such unwinding and rewinding in biological systems may be due to metal ions.

REFERENCES

1. Helmut Sigel (Ed). Metal ions in biological systems,Vol.3
                    Marcel Dekker, inc.New York, 1974,pp 62

2. G.L.Eichhorn, Inorganic Biochemistry. Chapts. 33 and 34, Else-
        vier 1975

3. Ei-Ichiro Ochiai, Bioinorganic Chemistry, An introduction,
   Allyn and Bacon, Inc. Toronto 1977 pp. 18

4. S. Ahrland, J. Chatt, and N.R.Davies, Quart. Rev., 12, 265
   (1968)

5. B.L.Vallee and R.J.P. Williams, Proc.Nat. Acad. Sci. (USA) 59.
   498 (1968)

6. P. de Meester, D.M.L. Goodgame, A.C.Spaski, and B.T.Smith
   Biochem. Biophys. Acta, 340, 113 (1974)

7. D.M.L. Goodgame, I.Jeeves, F.L. Phillips and A.C. Spaski,
   Biochem. Biophys. Acta, 378, 153 (1975)

8. R.W. Gellert and R.Bau, J. Am. Chem.Soc., 97, 7379 (1975)

9. A. Terzis, N. Hadjiliadis, R.Rivest, and T.Theophanides, Inorg.
   Chim. Acta, 12, L5 (1975)

10. C.J.L.Lock and R.A.Speranzini, J. Am. Chem. Soc., 98,7865 (1976).

11. E. Frieden and J. Alles, J.Biol. Chem., 230, 797 (1958)

12. J.A. Carrabine and M.Sandarilangam, J. Am.Chem. Soc., 92,
    369 (1968)

13. H. Sternlich, R.G.Shulman and E.W.Anderson, J.Chem. Phys. 43,
    3123 (1965)

14. A.T. Tu and C.G.Friederich, Biochemistry, 7, 4367 (1968)

15. A.T.Tu and Reinosa, Biochemistry, 5, 3375 (1967)

16. L.E.Minchenkova and V.I. Ivancv, Biopolymers, 5,615 (1967)

17. L.N.Drordov-Tikhomirov and L.I. Kikoin, Biofizika, 12,407 (1967)

18. P.C.Kong and T. Theophanides, Inorg.Chem.13,1167 (1974)

19. (a) J.P. Macquet and T.Theophanides, Biopolymers, 14, 781 (1975)
    (b) J.P. Macquet and T. Theophanides, Bioinorg. Chem., 5,59
    (1975)
      (c) M.M.Millard, J.P. Macquet, and T. Theophanides, Biochim.

Biophys. Acta, 402, 166 (1975)

20. (a) E. Sletten, J. Chem. Soc, Chem. Commun., 558 (1971)
    (b) E. Sletten, Acta Crystallogr., Sect. B, 30, 1961 (1974)

21. N. Hadjiliadis and T.Theophanides, Inorg. Chim. Acta, 16,66, 77 (1976)

22. G. Pneumatikakis, N. Hadjiliadis and T. Theophanides, Inorg. Chem, 17, 915 (1978)

VIBRATIONAL SPECTROSCOPY OF METAL NUCLEIC ACID SYSTEMS

T. THEOPHANIDES

National Research Foundation, Physical Chemistry and
Spectroscopy Centre, 48, B. Constantinou Ave., Athens 501-Greece

INTRODUCTION

The application of IR and Raman Spectroscopy to the study
of the role of metal ions in biological systems has taken promi-
nence during the last decade. Spectral changes caused by metal
interactions, determination of molecular conformation and stacking
interactions are some of the effects that can be studied by Raman
spectroscopy from intensity measurements. The advantage of Raman
Spectroscopy over other techniques is that one can study water
solutions in conditions close to biological systems. Vibrational
spectra of metal-nucleic acid systems provide information on the
backbone chain conformations.

VIBRATIONAL SPECTRA OF THE NUCLEIC CONSTITUENTS

The vibrational spectra of nucleic acids consist of characteristic vibrations of the base, the sugar and the phosphate groups.
These vibrations are observed in narrow regions of the spectrum as
characteristic bands of the vibrations of the atoms. Assignment
of the bands for these groups can be found in the early works (1,2).

Theo M. Theophanides (ed.), Infrared and Raman Spectroscopy of Biological Molecules, 205–223.
All Rights Reserved. Copyright © 1979 by D. Reidel Publishing Company, Dordrecht, Holland.

In the 700–1700 cm^{-1} region there are 9 and 15 in-plane ring
vibrations in the six-membered pyrimidine ring and in the nine-
membered purine ring, respectively. In addition, we have the CH
and NH in-plane and out-of-plane vibrations. The skeletal out-of-
plane deformation vibrations are in the region below 700 cm^{-1} (3).
All the characteristic molecular vibrations of the base residues
are found in the region of 4000 to 300 cm^{-1}. The adenine is represented in Fig. 1., with the possible sites of attack by a metal atom.

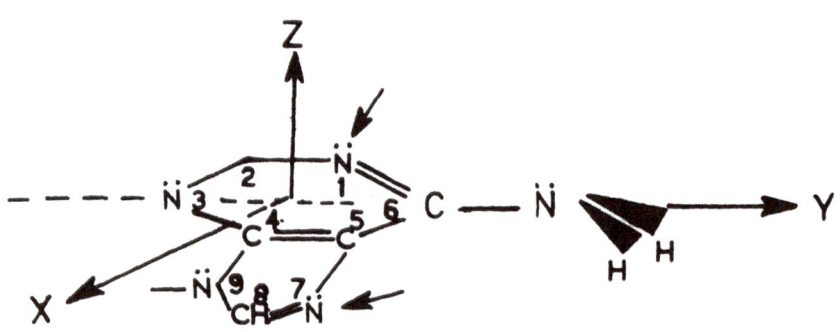

Fig. 1. The adenine residue and the possible metal coordination
        sites (↑).

The amino hydrogens are slightly out-of the plane (XY). However,
the resonance structure (see below) allows the H's to enter in
the plane. The characteristic group frequencies of $-NH_2$ (3) are
shown in Table I, together with the $-ND_2$ frequencies. The carbon-
nitrogen (C=N) carbon-carbon (C=C) double and single bond (C-N,
C-C) stretching vibrations, the nitrogen-hydrogen (N-H) carbon
hydrogen (C-H) and the ring breathing modes are all approximate
modes of vibration due to slight coupling between these vibrations.

Table I

The $-NH_2$ modes of vibration sketched approximately, the experimental group frequencies and the metal effect.

		$-NH_2$ cm^{-1} (3)	$-ND_2$ cm^{-1}	$-MNH_2^+$ (4)
	Asym.stretch	3300	2500	3240
	Sym.stretch	3100	2360	3190
	Def.	1650	1230 (?)	1550
	C-N stretch	1280	1200	1220
	Rock	1100	880	890
	Wag.	600	490	680
	Torsion	200	–	225

The list of the Raman bands and spectral changes given are taken
from the papers of several authors (1-6). The amino group C-N
bond has some double-bond character because of the resonance structure:

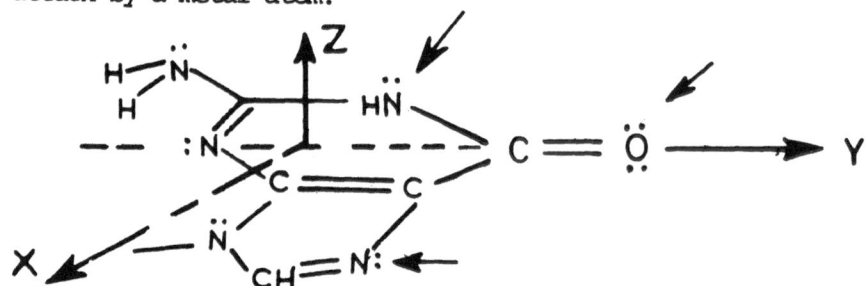

The long pair of electrons on the nitrogen atom is not available
for binding with a metal ion, in the resonance structure. The metals attack preferentially the N-7 and N-1 sites of this molecule
and produce slight local changes in the ring vibrations and their
intensities (5,6).

    The characteristic vibrations of the adenine residue and the
intensity changes are shown in Table II

    The guanine residue is shown in Fig.2 with the possible sites
of attack by a metal atom:

Fig. 2. The guanine residue and the possible metal coordination
        sites (↑).

The carbonyl group (C=O) is in plane (XY) and its characteristic
frequency together with the characteristic carbon-nitrogen, carbon-carbon double and single bond vibrations, the nitrogen-hydrogen, carbon-hydrogen and ring breathing modes are given in Tables
II and III together with the intensity changes of some bands and ten-
tative assignments.

TABLE 2

Guanine residue	Intensity changes				Assignment
	Increasing order	Protonation at N7 & N3	Alkylation at N7	Metallation at N7	
1666(m)					$C_6=O$ stretch
1582(s)	hypo				$C4C5(C=O+C=N)$
1487(s)			hypochramism		wave-like mode involving C_8-H
1375(m)					C8N9, N7C8
1328(m)					NH_2 def. + C-N
1052(m)					
670(s)		hyperchramism + shift			ring breathing mode

Adenine residue	Protonation at N1	Assignment
1623(m)	1668 (hyper)	(C=C + C=N) stret.
1605(m)		
1580(s)	hypo	1584(s) (hypo)	C4C5
1520(m)	hypo	
1510(m)	hypo	hypo
1484(m)	hypo	ring vibration
1379(m)	hypo	
1340(s)		
1255(3)		C8N9, C2N3, C8H ?
729(s)	hypo	ring breathing mode

a hypo: hypochramism; hyper: hyperchramism.

Table 2

continued

Table III

Guanine residue	Intensity changes (3)		Assignment
	Protonation & Deprotonation at N7		
1666 (m)	1690 (m)	1592 (s) (hyper)	$C_6=O$ stret.
1605 (m)	1609 (s) (hyper)		
1577 (s)	1578 (m) (hypo)	1578 (s)	C=N stret.
1565 (m)	1560 (m)		
1539 (m)	1515 (m)	1520 (m)	

Uracil residue neutral	Intensity changes (3)	Assignment
	Deprotonation at N_3 pH 10	
1690 (m)	1640 (m)	$C_2=O$ stret.
1627 (s)	1608 (m) (hypo)	$C_4=O$ stret.
1617 (m)	1584 (m)	
	1508 (m)	

The cytosine residue is shown in Fig. 3 with the possible
sites of attack by a metal atom:

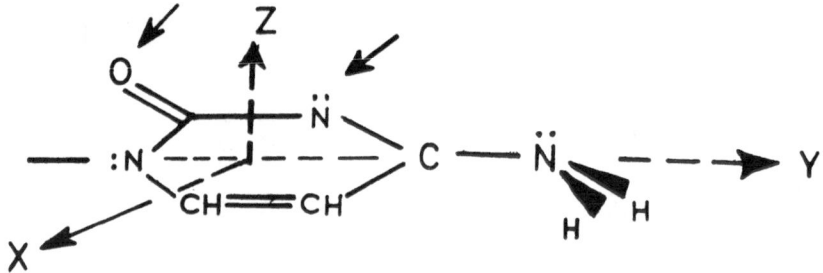

Fig. 3. Cytosine residue and the possible metal coordination
        sites (↑).

In this molecule we have both the amino and carbonyl groups. The
characteristic frequencies of the pyrimidine residue together with
the intensity changes and assignments are given in Tables III and IV

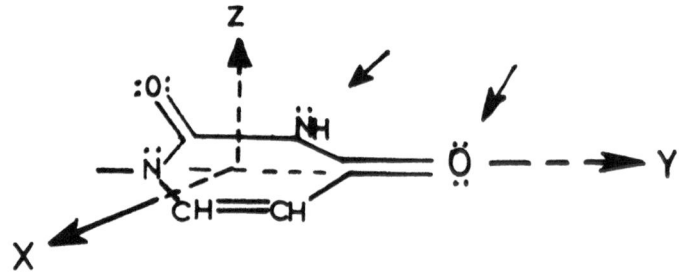

Fig. 4. The uracil residue and the possible metal coordination
        sites (↑).

The characteristic frequencies, the intensity changes of some
bands on perturbation and assignments are given in Tables III and IV

The hard metal M^{n+} attack can be compared to protonation (H^+),
whereas the soft metal attack is similar to methylation because
the metal carries with it other ligands (elements or groups) covalently bound and it forms covalent bonds with the nitrogen sites.
The changes of some frequencies can be explained by redistribution
of the electric charges of the nitrogen and carbon atoms adjacent

Table 4

| Cytosine residue | Intensity changes | | Assignment |
	increasing order	Protonation-Alkylation-Metallation	
1657(m)		1709(m) (shift)	$C_2=0$ stret.
1607(m)		hyper-chramism & shift	C=C
1528(m)	hypo		C=N stret.
1506(s)			C=N stret.
1292(s)			
1240(s)	hypo		
782(s)	hypo		ring breathing mode
Uracil residue			
1680(s)			$C_2=0$ stret.
1630(m)			$C_4=0$ stret.
1400(m)			C4C5
1352(m)	hyper		
1330(s)	hyper		C5C6
785(s)	hypo		ring breathing mode

to the sites of metal attack and the space requirements (steric effects) of the metal ion or metal complex.

THE RIBOSE-PHOSPHATE VIBRATIONS

The addition of a sugar to a base at the position N-9 gives the nucleoside. The sugar residue of a nucleoside is shown below:

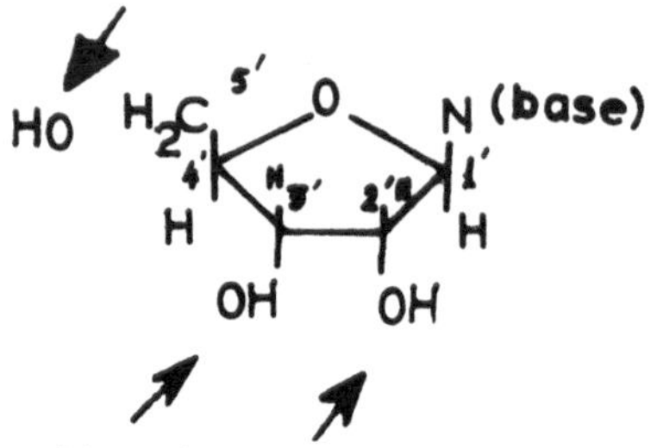

Possible coordination metal sites of a sugar ($\uparrow$).

The hydroxyl groups 2'-OH and 3'-OH show strong and broad characteristic infrared bands near 3400 cm^{-1} and weak Raman bands in the region 1000-1100 cm^{-1} due to the C-O single bond stretching vibrations (3,7,8). The C-H stretching vibrations are shown as strong Raman bands in the 3000-2850 cm^{-1} region, whereas the C-H in-plane bonding vibrations give weak Raman bands in the region 1400-1300 cm^{-1}. In RNA the characteristic bands of the ribose are observed near 1130 cm^{-1} and 810 cm^{-1} whereas in NDA only one band at 1120 cm^{-1} is observed (3). The metal ions react with the hydroxyl groups at pH values above the pK values of these groups by replacing the hydrogen atoms. They can form a variety of complexes, monomers and polymers a few of which are shown below:

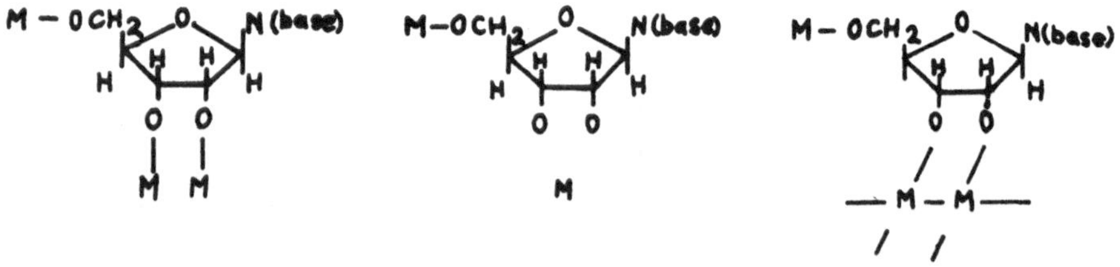

The metal-oxygen bonds are observed in the IR and Raman spectra very often as weak bands in the region 800-400 cm^{-1} depending of the metal and the covalent character of the metal-oxygen bond. Intensity changes of the characteristic frequencies, such as C-H stretching, H-C-O bending and C-O single bond stretching vibrations do take place on metallation of the sugar cycle.

The esterification of the sugar secondary hydroxyl group (5'-OH) with phosphate acid or phosphates gives the nucleotides, whereas the estirification of the tetriary hydroxyl group (3'-OH) gives rise to the polynucleotides and nucleic acids. The phosphodiester linkages take the ionized form for pH values above three:

$$-\overset{|}{C}H-O\diagdown\underset{OH}{\overset{O}{P}} \rightleftharpoons -\overset{|}{C}H-O\diagdown\underset{O}{\overset{O^{-}}{P}} + H^{+}$$
$$-CH_2-O\diagup \qquad\qquad -CH_2-O\diagup$$

Therefore, for physiological solutions (pH=7) of nucleic acids the group PO_2^{-} between the two ribose rings is a repeating unit in the backbone chain of the nucleic acids. The Raman and infrared bands due to the symmetric and antisymmetric stretch of this group are given (3,9) together with other characteristic vibrations in Table V. The binding sites of nucleotides for metal ions are the phosphate oxygens in addition to the base and ribose sites that we have discused. The hard metal ions are fixed on the PO_2^{-} oxygens of the backbone chain of the nucleic acids. The presence of hard metal ions in aqueous solutions of nucleic acids produces changes mostly in the ordering of the single chain helix into the A or B conformations since these metals react with the phosphate groups of the backbone chain. The Raman spectrum of poly U in H_2O at pH 7.0 in the presence of Mg^{2+} ions shows a decrease in intensity of the 1236 cm^{-1} band as the polymer is cooled from 20° to 0° (10a, 11). At low temperatures the band at 814 cm^{-1} is also observed showing an A-form. (See Table V). The intensity changes of the

Table V

Infrared and Raman frequencies of the phosphate groups of the nucleic acid backbone in cm^{-1}.

Infrared (assym.)	Raman (sym.)	Assignment
1230	1100	$(R-O-)_2-PO_2^-$ stretch in DNA
1115	1090	$(R-O-)_2-PO_3^=$ stretch in mononucleotides
	870	O-P-O in DNA in C-form (10b)
	835	O-P-O in DNA B-form
920	815	O-P-O bend in DNA · A-type (10b)

Raman spectra are due primarily to conformational changes of the backbone chain since Mg^{2+} ions are known to bind phosphate groups. There is evidence that Mg^{2+} ions are bound to phosphate sites in DNA as well as in other polynucleotides and stabilize the DNA double helix (12, 13).

The spectra of nucleic acids in the presence of soft metal ions show modifications due to metal-molecule interactions which induce changes in the molecule. The blocking of reactive sites by metals affects the chemistry of the Watson-Crick hydrogen bonded base pairs. The interaction between the highly polarizable nitrogen bases and the metals will cause changes in the intensities of the Raman bands related to base-base interactions. The formation of a coordinative-covalent bond between a soft metal and a nitrogen base can change the local electronic environment of the base and the local geometry. These interactions will produce changes in the Raman bands and intensities.

THE HARD AND SOFT METAL BINDING SITES

The hydrated hard metal ions react with the phosphate groups of DNA and not with the base. The reaction in water is believed to take place according to Scheme I.

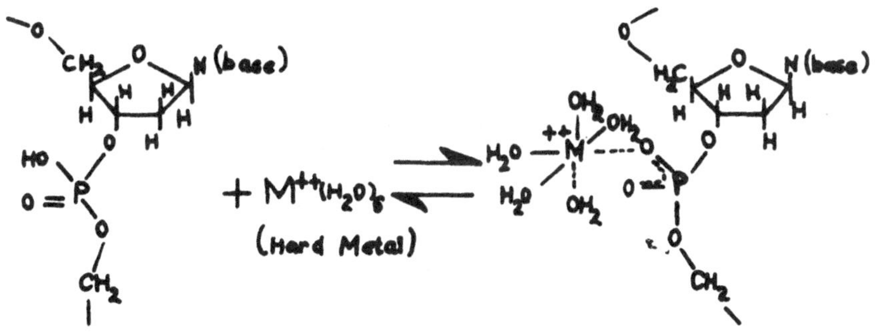

Scheme I

Spectroscopic studies, infrared (14) and Raman (15) agree that the presence of divalent hard metal cations affects only the phosphate groups in mono-di- and triphosphate nucleotides, polynucleotides and nucleic acids.

The infrared vibrations of the phosphate groups, PO_2^-, antisymmetric stretching at 1230 cm^{-1} and $-PO_3^=$ at 1100 cm^{-1} and the P-O-P deformation vibration at 920 cm^{-1} are perturbed on metal complex formation (3). It is found that the uptake of water molecules by NaDNA causes changes in the antisymmetric stretching frequency at 1230 cm^{-1} of PO_3^- (3). The symmetric stretching frequency of PO_2^- at 1115 slightly shifts (10 cm^{-1}) to lower frequencies.

The borderline metal ions (See Table II of Part I) form stable complexes by multiple binding. They may bind to phosphate and base sites, N-7, N-1 or N-3. Both base and phosphate binding has been confirmed by infrared (14, 16) and Raman Spectroscopy (15). The reaction in water is believed to take place according to Scheme II. The border line metal binds to the base at N7 with the monophosphates, whereas with ATP it binds to the base at N7 and to β and γ phosphate groups through the oxygens. For example, the borderline metal ions Ni^{2+} and Mn^{2+} react with 5'-GMP and bind to N-7 of the base and with ATP bind to N-7 and to β and γ phosphate groups (17) (See Scheme II).

Recently, a great deal of attention has been given to the study of complex formation between soft metal ions (See Table II of Part I) and nucleic acids. These heavy metal ions react with DNA, RNA, nucleotides, nucleosides, and other smaller parts of nucleic acids and form very stable complexes. The salts that cause significant changes in the spectra of the bases are CH_3Hg^+, $HgCl_2$, $AgNO_3$, $AuCl_3$, K_2PtCl_4 and other platinum complexes. IR spectra of DNA in solution with $HgCl_2$ show changes in the region 1600-1700 cm^{-1} which suggest DNA-$HgCl_2$ complex formation. Several

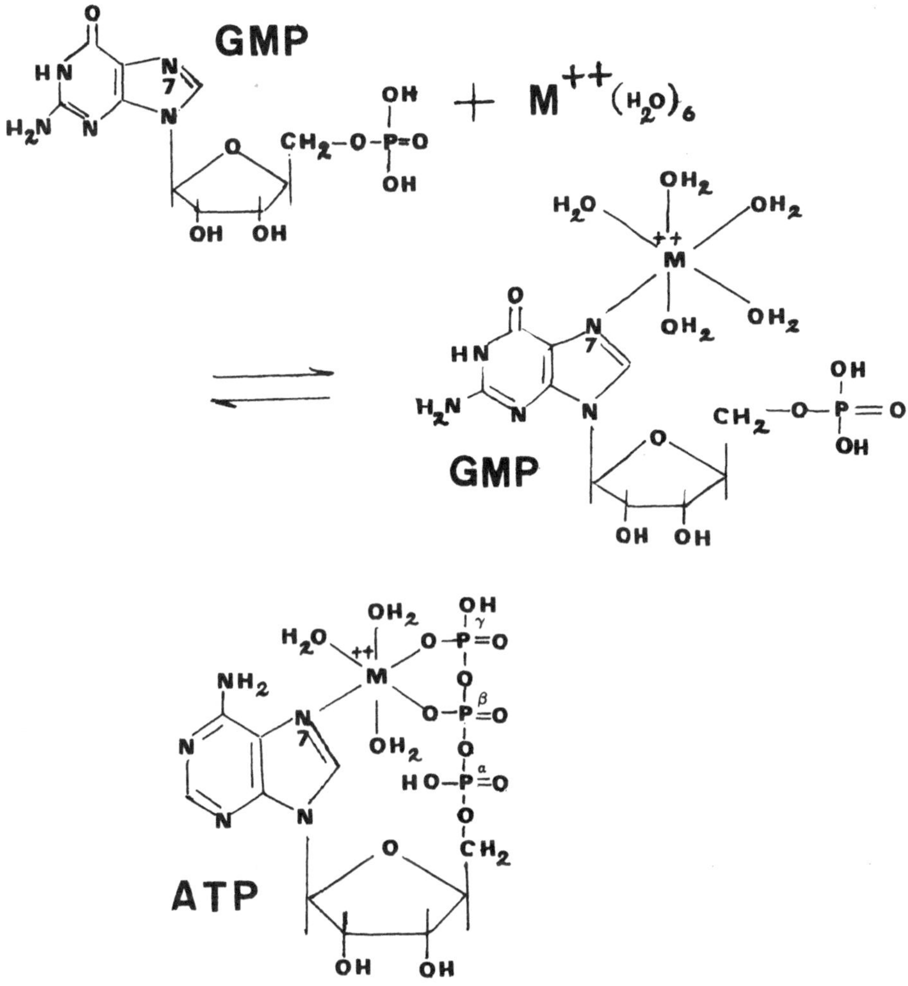

Scheme II

papers have appeared in the literature (18-20) on the interaction of CH_3Hg^+ with various nucleic acid components and DNA. Modifications are observed in the spectra due to interaction with the metal ion. Raman difference spectroscopy was used to study the reactivity of CH_3Hg^+ towards the bases and assign binding sites.

The platinum complex, cis-Pt$(NH_3)_2Cl_2$ shows antitumour activity and seems to react preferentially with DNA. The spectral stu-

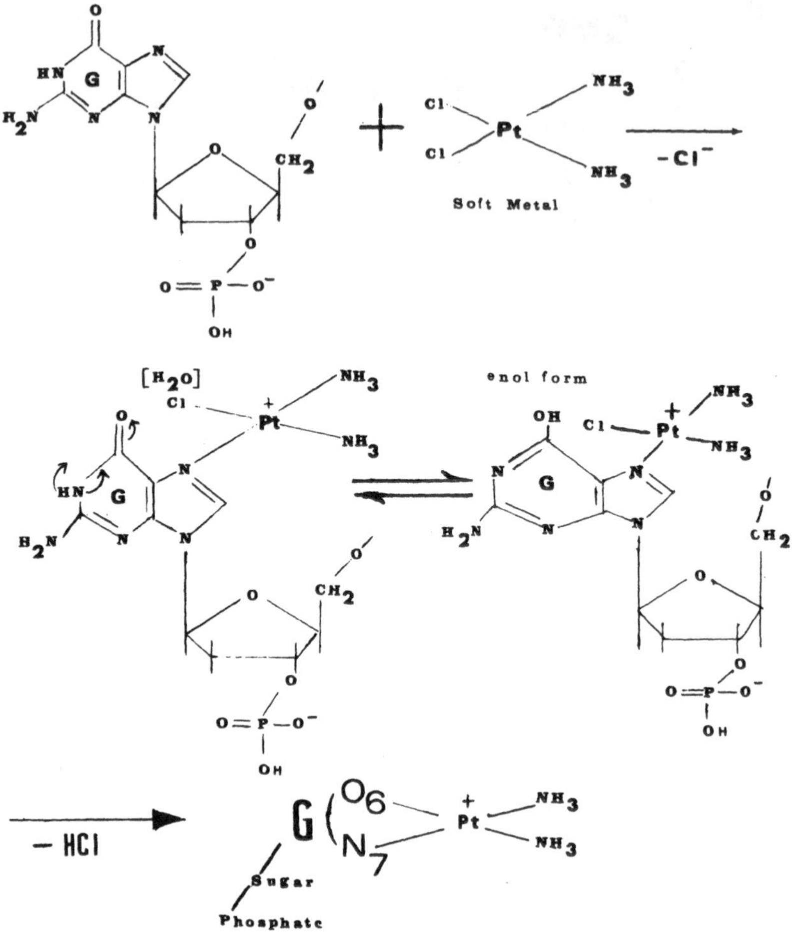

Scheme III

dies with cis-Pt$(NH_3)_2Cl_2$ are consistent with possible binding at
N7, N1, O_6 or O_6-N_7 and O_6-N1 chelation (20). Raman studies show
that the soft metal fixation on the base produces weak shifts in
the bands and important intensity changes (4-6). Reaction of very
small amounts of the antitumour drug, cis-Pt$(NH_3)_2Cl_2$ with DNA or
RNA ($\frac{Pt}{P}$= 0.03) corresponding to three platinum atoms per 100 bases causes significant intensity reduction in the wave-like mode
of G at 1484 cm^{-1} involving mainly the H-C8-N7- portion of the

molecule. A weaker reduction of intensity is also observed for
the band at 1578 cm^{-1} which is assigned to G (C=N + C=O) stretching
vibration of the pyrrole ring (N_7-C_8-N9- (21), for the band at
670 cm^{-1} assigned to G (ring breathing mode) and the band at 1361
assigned to G (C8N9, N7C8). The Raman spectra are very sensitive
and clearly show that the platinum salt reacts preferentially with
G at this very low concentrations (3%). The reaction of this drug
with DNA may take place according to Scheme III.

    In the reaction mechanism of Scheme III the stabilization of
the enol form in G may be induced under special conditions of reaction and the presence of this stable tautomer which has been
established by IR spectra (22, 23) may cause chelation of the soft
metal platinum at N7O6. It is possible that intramolecular hydrogen-bonding between a coordinated water and the carbonyl may occur,
e.g., C_6=O....H_2O-Pt(NH_3)$_2$N7(G) (24). However, the second chloride
atom hydrolizes much slower than the first chloride in platinum
complexes (25). IR and Raman Spectroscopy show that the carbonyl
band of guanosine or inosine (24)disappears or shifts to lower
frequencies under special reaction conditions (26). Cancellation
of the carbonyl band at 1680 cm^{-1} is of course evidence of carbonyl interaction. This takes place only at pH 7 or higher. In
the soft metal-base reactions, the N7 site is the most reactive
site, thus, upon complex formation of cis-Pt(NH_3)$_2$$Cl_2$ with GMP
(20) or DNA in acid or neutral conditions the N7 positions are
attacked preferentially by the platinum metal (19). In this case
the carbonyl is not affected as it has been observed (15, 20).

    Vibrational Spectroscopy is a powerful technique for detec-
ting weak molecule-molecule or metal-molecule interactions or
perturbations even with biomacromolecules such as the nucleic
acids. The intensities of the vibrational frequencies may be the
most sensitive probes to detect such interactions and may give
information about the conformation and other molecular changes in
these biological molecules.

REFERENCES

1. G.B.B.M. Sutherland, and M. Tsuboi, Proc.Roy. Soc. (London) A239, 446 (1975); W.L. Peticolas, Advances in Raman Spectros. 1, 285 (1972)

2. R.C. Lord and G.J. Thomas Jr., Spectrochim. Acta, 23A 969 (1967)

3. M. Tsuboi, Infrared and Raman Spectroscopy in Basic Principles in Nucleic Acid Chemistry Vol. I. Ed. Paul O.P. Ts'o Academic Press, Inc. pp. 407; (1974), M. Tsuboi, S. Takahasi, and I. Harada, In:Physico-Chemical Properties of Nucleic Acids, Vol. 2 (J. Duchesne) Ed, chapter 11, pp. 91-145, Acad. Press, London (1973)

4. N. Hadjiliadis and T. Theophanides, Can. J. Spectros 16, 135 (1971)

5. T. Theophanides, N. Hadjiliadis, M. Berjot, M. Manfait and L. Bernard, J. Raman Spectrosc. 5, 315-323 (1976)

6. T. Theophanides, M. Berjot and L. Bernard, J. Raman Spectros. 6, 1109 (1977)

7. R.C. Lord and G.J. Thomas, Jr., Spectrochim. Acta, Part A23, 2551 (1967)

8. M. Tsuboi and Y. Kyogoku, in "Synthetic Proceedures in Nucleic Acid Chemistry" (W.W. Zorbach and R.S. Tipson, eds.) Vol. 2 (II, p. 215) Wiley, New York (1973)

9. W.L. Peticolas, Proceedures in Nucleic Acids Research (G.L. Cantoni and D. R. Davies, Eds) Harper and Row, Vol. 2 pp. 94-136 (1971)

10(a) S.C. Erfurth, P.J. Bond and W.L. Peticolas, Biopolymers, 14, 247, 1259 (1957); (b) W.L. Peticolas and M. Tsuboi. In this volume.

11. W.L. Peticolas, Biochimie, 57, 417 (1975)

12. J. Shack and B.S. Bynum, Nature, 184, 635 (1959)

13. J. Eisinger, I. Fawaz-Estrup and R.G.Shulman, J. Chem.Phys., 43, 43 (1965)

14. H.Brintzinger, Biochim. Biophys. Acta, 77, 343 (1963); K.A. Hartman, Jr., Biochim. Biophys. Acta, 138, 192 (1967)

15. L. Rimai, M.E.Heyde and E.B. Caren, Biochem. Biophys. Res. Commun., 38, 231 (1970)

16. J. Brigando and D. Colaitis, Bull. Soc. Chim. France, 3445 (1969)

17. M. Cohn and T.R.Hughes, Jr., J. Biol. Chem., 237, 176 (1962)

18. S. Mansy and R.S. Tobias, J. Amer. Chem. Soc., 96, 6874 (1974) id., Inorg. Chem. 14, 287 (1975); id. Biochemistry, 14, 2952 (1975)

19. S. Mansy, T.E. Wood, J.C. Sprowles, and R.S. Tobias, J. Amer. Chem. Soc. 96, 1762 (1974)

20. G.Y.H. Chu, S. Mansy, E.E. Duncan, and R.S. Tobias, J. Amer. Chem. Soc., 100, 593 (1978); 100, 607 (1978)

21. Unpublished results

22. F.B. Howard and H.T.Miles, Biol. Chem., 240, 801 (1965) H.T. Miles and J. Frazier, Biochim. Biophys. Acta, 79, 216 (1964)

23. H.T.Miles, F.B. Howard, and J. Frazier, Science 142, 1458 (1963)

24. D.M.L. Goodgame, I. Jeeves, F.L. Phillips, and A.C.Skapski, Biochim. Biophys. Acta. 378, 153 (1975)

25. M.M. Millard, J.P. Macquet and T. Theophanides, Biochim. Biophys. Acta, 402, 166 (1975)

26. N. Hadjiliadis and T. Theophanides, Inorg. Chim. Acta, 16, 77 (1976)

ABRAHAMS & CROWD VOLUME I, INDEX Author index

14. M. Black, and J.S. B..., G.M. ...a..y, 104, 3..(1959),

13. ... Weidinger, V. Peters-Schauro and C.S. Brothen, J. Chem. Phys.

Vol. 77 C. ...

12. Amithvan, Chemical Society ... ... 77, 213 (1955), ...

INTERACTION OF METAL IONS WITH PEPTIDES IN AQUEOUS MEDIUM AS STUDIED BY INFRARED AND RAMAN SPECTROSCOPY

M. Tasumi

Department of Chemistry, Faculty of Science,
The University of Tokyo, Bunkyo-ku, Tokyo,Japan

ABSTRACT. The infrared spectra of the metal(II) complexes of representative di- and tripeptides in deuterium oxide solutions are discussed with particular emphasis on the variation of complex structure with pD.   The results of recent resonance Raman studies are also mentioned.

1. INTRODUCTION

The importance of metal ions in biological systems is well recognized and, accordingly, an enormous number of studies have been carried out on the metal-protein interactions using various methods.   The interaction of metal ions with polypeptides and smaller peptides has also been studied repeatedly.   Vibrational spectroscopy certainly provides a unique tool for investigating such interactions.

In this chapter recent developments in the infrared and Raman studies of the metal-peptide complexes in aqueous solutions are described, and resonance Raman studies are also included.

2. VIBRATIONAL SPECTRA AND STRUCTURES OF THE METAL(II)-DIPEPTIDE COMPLEXES IN AQUEOUS MEDIUM

2.1 The metal(II)-glycylglycine complexes

The structure of Cu(II) complex of glycylglycine (GlyGly) has been studied by many authors in crystal [1-3] and in aqueous solution [4-9].   The most interesting feature derived from these

225

studies is structural changes of the complex with the increase in
pH (or pD).   All the authors seem to agree that the peptide
group is deprotonated in going from acidic to neutral solution
and that in neutral solution the metal-ligand bonding occurs
through the amino nitrogen, peptide nitrogen (deprotonated), and
carboxyl oxygen.   As for the structure in acidic solution, two
different forms have been proposed with respect to the metal-
ligand bonding.   We discuss this matter through the analysis of
infrared spectra of this complex in deuterium oxide solution.
The structure of the Zn(II) complex is also examined by referring
to the Cu(II) complex-case.

In Fig. 1 are shown the infrared spectra of GlyGly in D_2O at
three pD values.   The corresponding Raman spectra are essentially
the same as the infrared spectra except for considerable
differences in intensities.   Clearly, the pD value has a great
influence on the vibrational spectrum, reflecting the following
dissociation scheme.

$$D_3N^+CH_2CONDCH_2COOD \xrightarrow{\text{pKa 3.1}} D_3N^+CH_2CONDCH_2CO_2^- \xrightarrow{\text{pKa 8.3}}$$

$$D_2NCH_2CONDCH_2CO_2^-$$

From the spectral changes with pD some bands can be assigned
without difficulty.   More detailed assignments can be attained
by observing the spectra of isotopically substituted species (^{13}C,
[15]N, and D).   This was in fact carried out using eight isotopic
species shown in Fig. 2 [10], where the infrared spectra for
neutral solutions are schematically given together with the
assignment of each band.   The band assignments are made most
easily by looking at the frequency shifts and disappearance of
bands due to isotopic substitutions.   The following points should
be noted.
(1) The CO_2^- symmetric and antisymmetric stretching frequencies
are 1397 and 1595 cm^{-1}, respectively, whereas the CO stretching
of COOH is 1727 cm^{-1}.
(2) The band due to the amide I' mode (amide I of COND; mainly CO
stretching) shifts from 1674 to 1635 cm^{-1} on dissociation of the
terminal ND_3^+ to ND_2.   [The amide I' band shifts to a lower
frequency on ^{13}C substitution of the peptide carbonyl carbon atom.]
(3) The band due to the amide II' mode (amide II of COND; mainly
CN stretching) is found at 1487 cm^{-1}.   [The amide II' band shifts
to lower frequencies on substitution of either the carbon or
nitrogen atom of the peptide group by ^{13}C or ^{15}N, respectively.]

The infrared spectra of 1:1 mixture of GlyGly and Cu(II) at
various pD values are shown in Fig. 3.   Comparing the spectra in
this figure with those in Fig. 1, we note the following.
(1) The complex formation between GlyGly and Cu(II) takes place
only when pD is higher than 3.

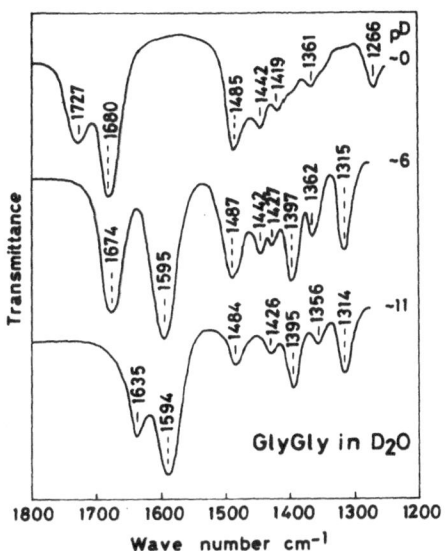

Fig. 1. Infrared spectra of glycylglycine in acidic, neutral, and alkaline D_2O solutions.   Concentration, ca. 0.2 M ; sample layer thickness, 0.05 mm (CaF_2 cell).

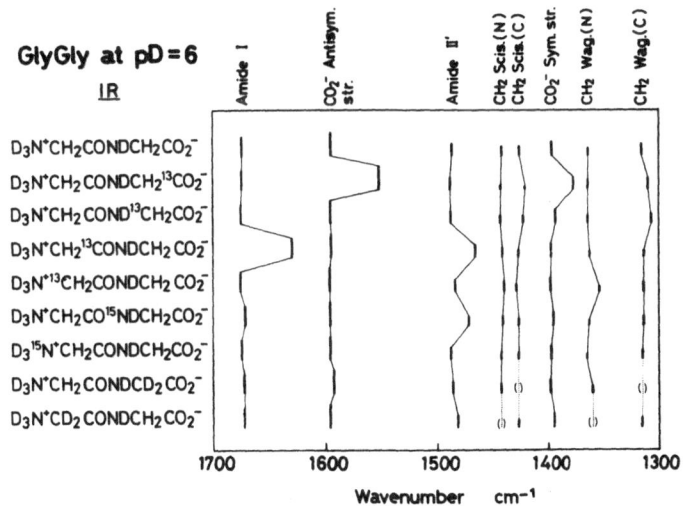

Fig. 2. Schematically shown infrared spectra of glycylglycine and its isotopically substituted species in D_2O solutions of pD 6 [10].

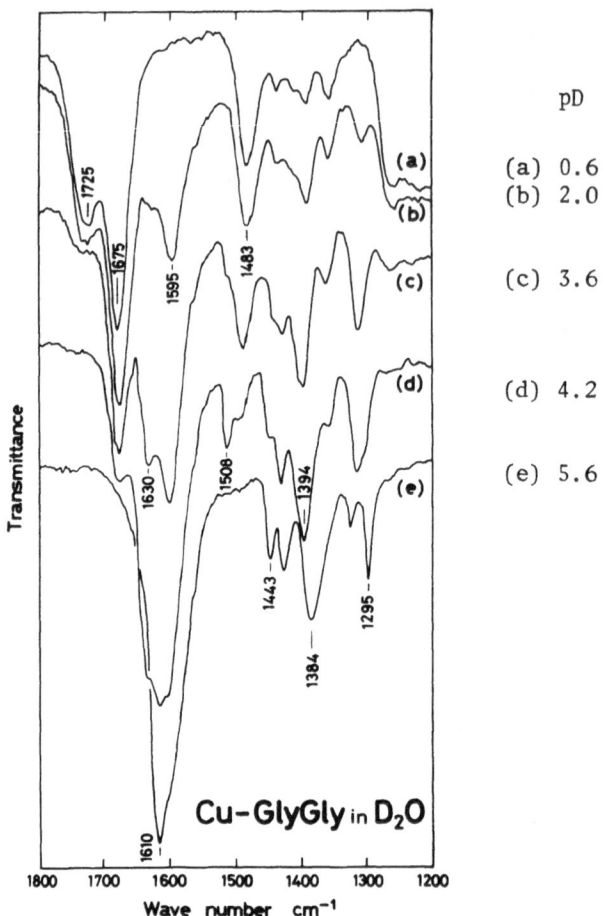

Fig. 3. Infrared spectra of the 1:1 mixture of Cu(II) and glycyl-glycine in D_2O solutions of various pD values [9].

---

(2) In the pD range between 3 and 5 two new bands appear at 1630 and 1508 cm^{-1}, which are undoubtedly due to a form of the Cu(II)-GlyGly complex.

(3) Another step of spectral changes occurs at about pD 5. A new band appears at 1610 cm^{-1} and both the 1508 and 1483 cm^{-1} bands disappear. The CO_2^- symmetric stretching frequency shifts from 1394 to 1384 cm^{-1}.

The spectral changes described above may be interpreted as follows. Three ligand structures (Forms I~III) of Fig. 4 have been proposed for the Cu(II)-GlyGly complex. Here we discuss the ligand structure rather than the structure of whole complex, since

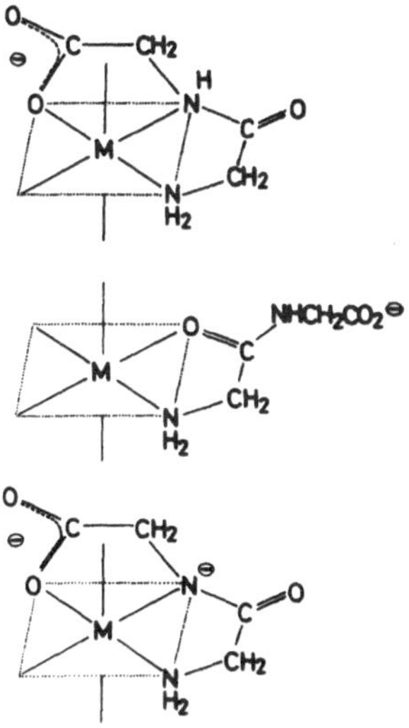

Fig. 4. Structures of the metal-glycylglycine complex.

the vibrational spectra in the region of 1800-1200 cm^{-1} provide direct information on the ligand.

The complex for 3<pD<5.   The two bands at 1630 and 1508 cm^{-1} characteristic of this pD range are assigned to the amide I' and II' modes of the complex, respectively, by referring to the spectra of the Cu(II) complexes of isotopically substituted species shown in Fig. 5.   The shift of amide I' from 1674 to 1630 cm^{-1} on complex formation is partly due to the dissociation of terminal ND$_3^+$ to ND$_2$ [which becomes coordinated to Cu(II)], as was the case for free GlyGly described above.   Another factor giving rise to this frequency shift is the coordination of the peptide carbonyl oxygen to Cu(II) (Form II of Fig. 4).   Some authors have proposed Form I as the ligand structure present in this pD range.   However, the shift of amide II' to 1508 from 1483 cm^{-1} on complex formation seems to be better accounted for with Form II.   The shift of amide II' to a higher frequency is related to the increase in bond order of the peptide CN linkage; in other words, the contribution of the resonance form

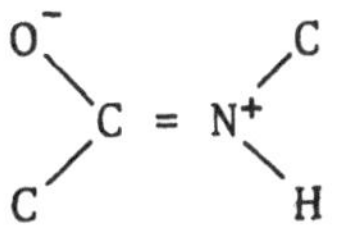

$$\begin{array}{c} O^- \qquad\qquad C \\ \diagdown \qquad\qquad \diagup \\ C = N^+ \\ \diagup \qquad\qquad \diagdown \\ C \qquad\qquad H \end{array}$$

increases with complex formation.   This can be expected of Form

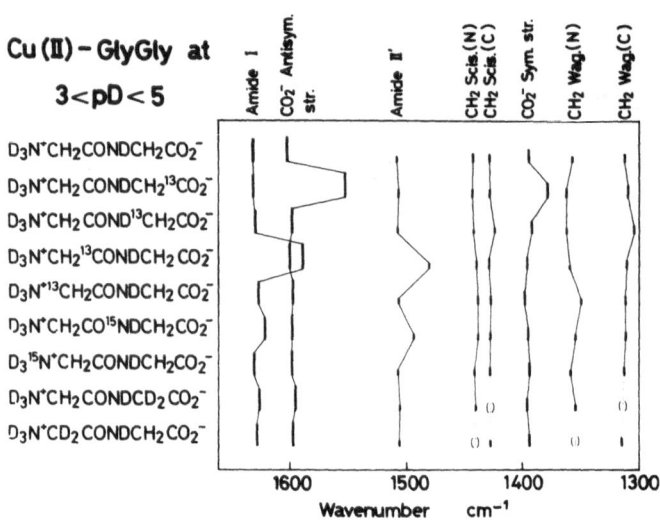

Fig. 5. Schematically shown infrared spectra of the 1:1 mixtures of Cu(II) and isotopically substituted glycylglycines in D_2O solutions (3<pD<5) [10].

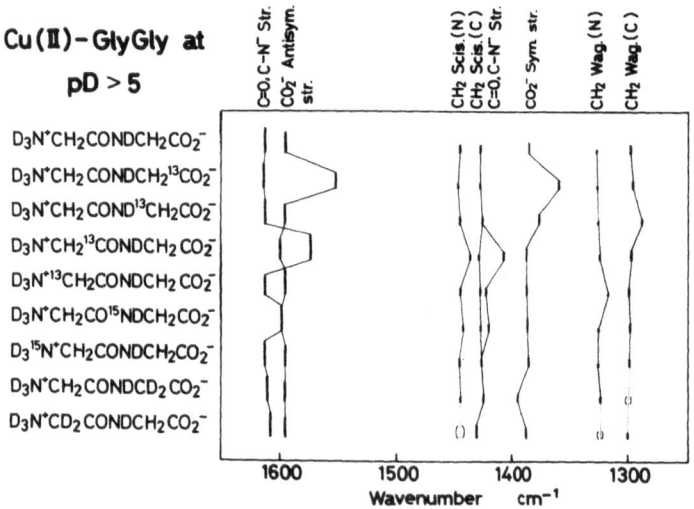

Fig. 6. Schematically shown infrared spectra of the 1:1 mixtures of Cu(II) and isotopically substituted glycylglycines in D_2O solutions (pD>5) [10].

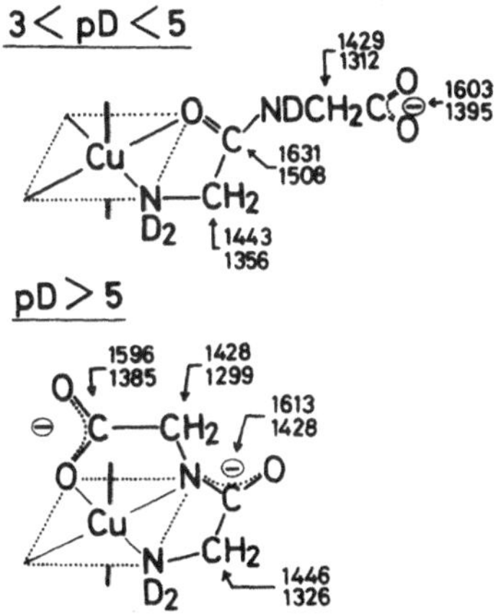

3 < pD < 5

pD > 5

Fig. 7. Assignments of the observed infrared frequencies for two forms of the Cu(II)-glycylglycine complex [10].

II but not of Form I.   In the latter the bond order of the peptide linkage would be reduced through the increased contribution of sp^3 configuration of the nitrogen atom.

The complex for pD>5.   The observed spectra of this pD range are reasonably interpreted on assuming Form III.   The amide I' band appears at 1610 cm^{-1} and no band assignable to amide II' is found around 1500 cm^{-1}.   The spectra of isotopically substituted species shown in Fig. 6 reveal that two bands are overlapped at 1428 cm^{-1}; one is assigned to the $CH_2(C)$ scissoring and the other must be the in-phase (symmetrically coupled) stretching of CO and CN$^-$ of the deprotonated peptide group.   The 1610 cm^{-1} band is assigned to the out-of-phase (antisymmetrically coupled) stretching of CO and CN$^-$.   [Interestingly, these in-phase and out-of-phase stretching frequencies of CO and CN$^-$ are quite close to the symmetric and antisymmetric stretching frequencies of the CO_2^- group at about 1600 and 1400 cm^{-1}, respectively.]   In Form III the terminal CO_2^- group is also coordinated to Cu(II).   In fact, the CO_2^- symmetric stretching band is found at 1384 cm^{-1} in this pD range as compared with 1395 cm^{-1} of free CO_2^-.

The assignments of bands observed in the region between 1750 and 1300 cm^{-1} are collectively given in Fig. 7 for the two types of the Cu(II)-GlyGly complex.

The infrared spectra of the 1:1 mixture of GlyGly and Zn(II) at various pD are shown in Fig. 8, where the following conclusions may be derived.

(1) No complex is formed for pD<4.

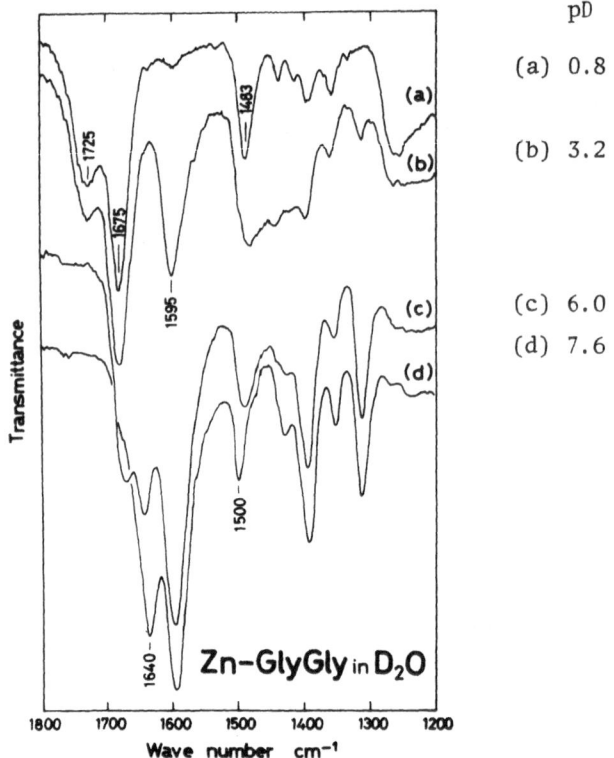

Fig. 8. Infrared spectra of the 1:1 mixture of Zn(II) and glycyl-glycine in D_2O solutions of various pD values [9].

(2) The bands at 1640 and 1500 cm^{-1} begin to appear at pD about 5 and become stronger for higher pD values until the precipitation begins to occur at pD about 9.   At pD 7.6 the amide II' band of free GlyGly (1483 cm^{-1}) is observed as a very weak shoulder. This indicates that Form II is the predominant species in this pD range and that the transition from Form II to Form III does not take place in contrast with the case of the Cu(II) complex.

2.2 The metal(II) complexes of glycyl-L-phenylalanine and
      glycyl-L-tyrosine [11]

The Cu(II) complexes of both glycyl-L-phenylalanine (GlyPhe) and glycyl-L-tyrosine (GlyTyr) are formed like that of GlyGly, though the complex of Type II is recognizable but less stable for GlyPhe and GlyTyr.   For pD>5 the Type III complex is evidently formed for both GlyPhe and GlyTyr.   The Zn(II), Co(II), and Ni(II) complexes of GlyPhe and GlyTyr are quite similar to the Zn(II) complex of GlyGly described above; i.e., the Type II complexes are formed for 5<pD<8.

Table I.   Assignments of the observed infrared bands of the metal(II) complexes of glycyl-L-phenylalanine and glycyl-L-tyrosine in D_2O.

GlyPhe		GlyTyr		Assignments
Cu	Zn,Co,Ni	Cu	Zn,Co,Ni	
1715	1715	1720	1720	CO str.(COOH) [free]
1675	1675	1675	1675	Amide I' [free]
1620	1630	1615	1630-35	Amide I' [Type II]
		1613	1613	Undissociated Tyr ring str.
1600		1600		CO & CN^- out-of-phase str. [Type III]
1600	1595	1600	1598	CO_2^- antisym. str.
		1515	1515	Undissociated Tyr ring str.
1495	1490-95		1492-98	Amide II' [Type II]
		1498		Dissociated Tyr ring str.
1475	1475	1477	1477	Amide II' [free]
1440		1440		CH_2 scis. [Type III]
1420		1420		CO & CN^- in-phase str. [Type III]
1400	1400	1400	1400	CO_2^- sym. str. [free]
1385		1385		CO_2^- sym. str. [Type III]

The observed frequencies for all the metal(II) complexes of GlyPhe and GlyTyr are listed in Table I, together with their assignments.

The results obtained above indicate that Cu(II) is the only metal cation which forms the Type III complex with dipeptides in the neutral pD range.   It has been shown that the rate of hydrolysis of the peptide bond in the presence of a metal(II) ion reflects the type of coordination [12].   The rate of hydrolysis is promoted to a greater extent by the formation of Type II complex, where the peptide carbonyl oxygen is coordinated to the metal(II) ion.   Such coordination occurs also in the active site of carboxypeptidase A when a substrate peptide is bound to Zn(II) [13].   Interestingly, the activity of this enzyme is largely kept by substituting Zn(II) with either Ni(II) or Co(II), whereas it is completely lost with Cu(II).   The present author infers that such singular behavior of Cu(II) in the enzymatic activity is associated with the coordination of the substrate peptide to Cu(II) through N^- of the deprotonated peptide bond as in the Type III complex.   It is certainly desirable to verify this inference by some experimental means such as the x-ray diffraction analysis of the ES complex, where E and S refer, respectively, to Cu(II)-substituted carboxypeptidase A and GlyPhe.

3. VIBRATIONAL SPECTRA AND STRUCTURES OF THE METAL(II)-TRIPEPTIDE COMPLEXES IN AQUEOUS MEDIUM

3.1 The Cu(II)-glycylglycylglycine complex [11]

In Figs. 9 and 10 are shown the pD dependences of the infrared spectrum of glycylglycylglycine (GlyGlyGly) and that of the 1:1 mixture of GlyGlyGly and Cu(II), respectively. The observed features and some conclusions drawn from them are summarized as follows.

(1) The band at 1675 cm^{-1} arises from the amide I' mode of the first peptide group adjacent to the ND_3^+ terminal. Similarly, the band at 1650 cm^{-1} is due to the amide I' mode of the second peptide group adjacent to the carboxyl terminal. The amide I' frequency of the first peptide group shifts to about 1650 cm^{-1} on dissociation of ND_3^+ to ND_2.

(2) The band at 1625 cm^{-1} is assigned to the amide I' mode of the peptide group complexed to Cu(II) through the carbonyl oxygen atom (the Type II complex). At 1595 cm^{-1} two bands are overlapped, one arising from the CO_2^- antisymmetric stretching and the other from the out-of-phase stretching of CO and CN^- of the deprotonated peptide group (the Type III complex.)

(3) The amide II' mode of free peptide group gives rise to the absorption at 1475 cm^{-1}. No appreciable difference is found between the amide II' frequencies of the first and second peptide groups in the free (not coordinated) state. When the first peptide CO becomes coordinated to Cu(II) at pD about 3.5, its amide II' band shifts to 1508 cm^{-1}. This parallels what was observed for the Cu(II)-GlyGly case. As pD is raised to about 5, another band appears at 1490 cm^{-1}. It is most likely that this band is due to the amide II' mode of the second peptide group coordinated to Cu(II) through its carbonyl oxygen.

(4) No coordination of the CO_2^- group to Cu(II) seems to occur over the entire pD range studied, since the CO_2^- symmetric stretching band is always found only at 1390 cm^{-1}.

(5) The spectral features described above lead to the conclusion that the complex structure changes stepwise with the increase in pD as shown in Fig. 11.

3.2 The Cu(II)-glycylglycyl-L-histidine complex [11]

Glycylglycyl-L-histidine (GlyGlyHis) is a model peptide for the Cu(II) binding site of human serum albumin which is considered to transport Cu(II) in blood [14]. Although we have not yet fully analysed the infrared spectra of the Cu(II) complex of this peptide in D_2O, part of the results obtained so far are given here.

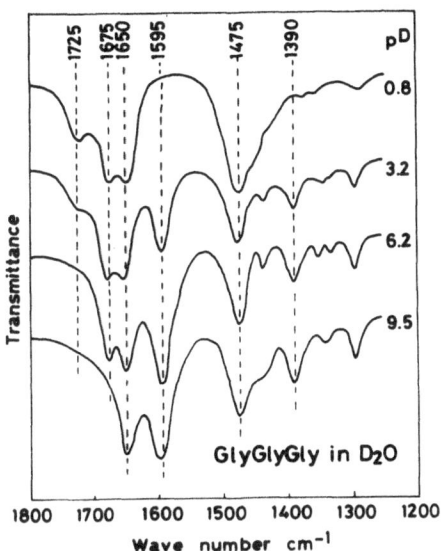

Fig. 9. Infrared spectra of glycylglycylglycine in D$_2$O solutions of various pD values [11].

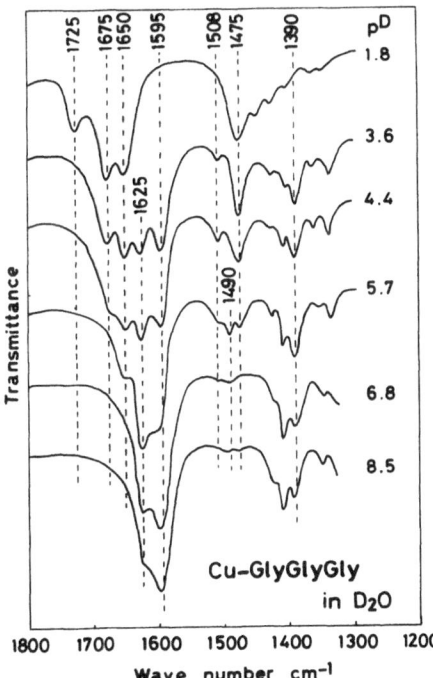

Fig. 10. Infrared spectra of the 1:1 mixture of Cu(II) and glycylglycylglycine in D$_2$O solutions of various pD values [11].

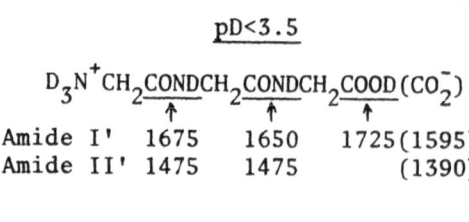

pD<3.5

$D_3N^+CH_2\underline{CONDCH_2}\underline{CONDCH_2}\underline{COOD}(CO_2^-)$

Amide I'	1675	1650	1725 (1595)
Amide II'	1475	1475	(1390)

3.5<pD<5

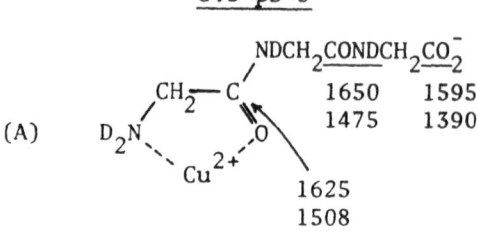

(A)

NDCH$_2$CONDCH$_2$CO$_2^-$

1650   1595
1475   1390

1625
1508

5<pD<6

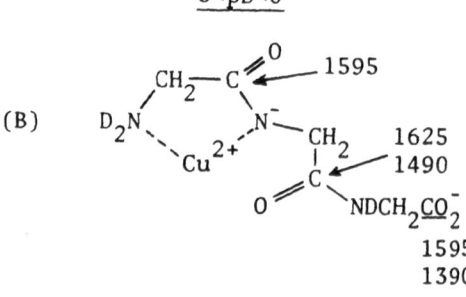

(B)

1595

1625
1490

NDCH$_2$CO$_2^-$
1595
1390

6<pD

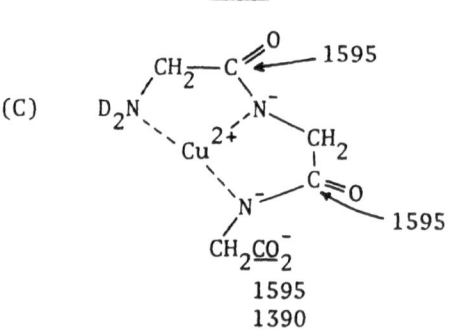

(C)

1595

1595

CH$_2$CO$_2^-$
1595
1390

Fig. 11. Structural variation of the Cu(II)-glycylglycylglycine complex with pD and assignments of the observed infrared frequencies [11].

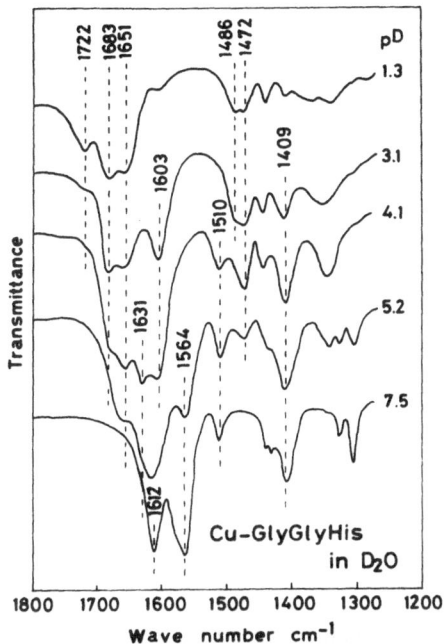

Fig. 12. Infrared spectra of the 1:1 mixture of Cu(II) and glycylglycyl-L-histidine in D_2O solutions of various pD values [11].

The infrared spectra of the 1:1 mixture of GlyGlyHis and Cu(II) at various pD values are shown in Fig. 12. The spectra are similar to those of the Cu(II)-GlyGlyGly system except for a few points to be discussed below. GlyGlyHis does not form a complex with Cu(II) for pD<3.5. Free GlyGlyHis shows two amide I' bands at 1683 and 1651 cm^{-1} in parallel with GlyGlyGly. GlyGlyHis has two bands at 1486 and 1472 cm^{-1} which are assignable to the amide II' modes of the first and second peptide groups, respectively. This is in contrast with the case of GlyGlyGly which has a single amide II' band at 1475 cm^{-1}. As pD is increased above 3.5, the terminal ND_2 group and the first peptide group become coordinated (through CO) to Cu(II) [Type A of Fig. 11]. This is evidenced by appearance of the bands at 1631 and 1510 cm^{-1} and concomitant disappearance of the 1486 cm^{-1} band. In the range of pD>5 the spectra of the Cu(II)-GlyGlyHis system are different from those of the Cu(II)-GlyGlyGly system. A new band at 1564 cm^{-1} increases in intensity as pD is further raised, and the 1510 cm^{-1} band is still present at pD 7.5 and even at higher values of pD. The presence of these two bands as well as the 1612 cm^{-1} band suggests that the complex formed for pD>5 is predominantly of the type shown in Fig. 13. Probably the Type B

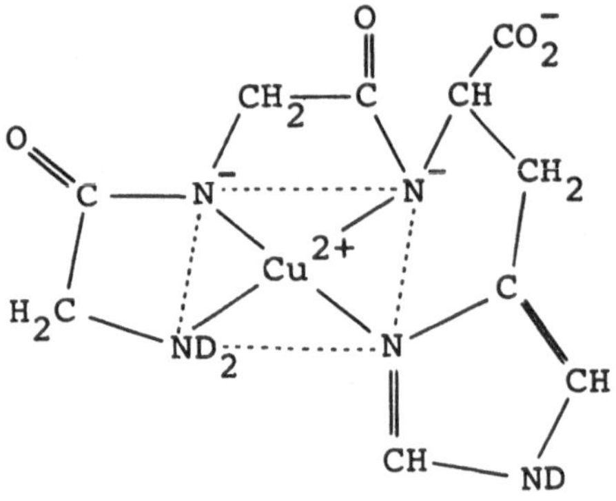

Fig. 13. Structure of the Cu(II)-glycylglycyl-L-histidine complex formed for pD>5 [14].

complex (Fig. 11) is not formed with GlyGlyHis, since the 1490 cm^{-1} band characteristic of this type is not observed. The 1564 and 1510 cm^{-1} bands are associated with the coordinated His moiety. The former arises from the CO_2^- antisymmetric stretching mode of the species with the coordinated imidazole ring and the deprotonated peptide group (Gly-His). [The CO_2^- antisymmetric stretching frequency of L-histidine itself shifts from 1624 to 1617 cm^{-1} on deprotonation of the imidazolium ring (pKa ~6) and again from 1617 to 1578 cm^{-1} on dissociation of ND_3^+ to ND_2 (pKa ~9).] The 1510 cm^{-1} band seems to have two separate origins. One is the amide II' mode of CO-coordinated peptide group (Gly-Gly) and the other is a stretching mode of the coordinated imidazole ring. [A similar band is observed at 1500 cm^{-1} for the Cu(II)-His system. The infrared bands due to the imidazole or imidazolium ring moiety are generally weak in intensity, though the corresponding Raman bands are strong. The coordination to Cu(II) seems to shift the imidazole ring stretching band at 1480 cm^{-1} to 1500 cm^{-1}].

4. RESONANCE RAMAN STUDIES ON THE METAL-PEPTIDE COMPLEXES

The value of resonance Raman measurements in the study of some metalloprotein chromophores is now well established. Resonance Raman studies of heme proteins and related molecules will be treated separately in this book by other authors. As for the works on other metalloproteins, readers may refer to the review by Fawcett and Long [15].

Resonance or "pseudo" resonance Raman spectra of (glycyl-L-histidinato)copper(II) trihydrate, glycylglycinatocopper(II) sesquihydrate, and disodium glycylglycylglycylglycinatocuprate(II) decahydrate as powdered solids were observed by Siiman et al. using the 441.6-nm Cd laser line. For each of above three complexes an intense Raman band was observed around 400~350 cm^{-1} which was assigned to the stretching mode of Cu(II)-N$^-$ (deprotonated peptide group) bands. On Such an experimental basis the coordination around Cu(II) of "blue" copper proteins was discussed [16].

More recently Kincaid et al. succeeded in observing the resonance Raman spectra from very dilute solutions of the Cu(III) complexes of GlyGlyGlyGly and GlyGlyGlycineamide using a rather special sampling technique and a near-UV laser (363.8-nm Ar$^+$) [17]. An intense Raman band at about 420 cm^{-1} was assigned to the in-phase stretching mode of the Cu(III)-N$^-$ bonds in the square-planar complex. [For the corresponding Cu(II) complexes in aqueous solutions the Cu(II)-N$^-$ stretching bands were found at about 388 cm^{-1} (nonresonance Raman) in accord with the observation for powdered solids by Siiman et al.] There is a possibility that resonance Raman measurements may throw a new light upon the biochemical role of copper(III) which is a subject of recent interest.

5. CONCLUDING REMARKS

Infrared spectroscopy provides a useful tool for investigating the metal-peptide interactions in aqueous (D$_2$O) medium. Detailed information can be obtained on the coordination of peptide main chain in particular. The use of Raman spectroscopy for such studies is not as commom yet as it should be. Undoubtedly Raman spectroscopy is more convenient for studying the metal-peptide interactions in H$_2$O solution and will bring more information on the metal-ligand bonding and the states of aromatic side chains.

REFERENCES

1. B. Strandberg, I. Lindgvist, and R. Rosenstein, Z. Krist, 116, 266 (1961).
2. H. C. Freeman, Adv. Protein Chem. 22, 257 (1967).
3. M. Shiro, Y. Nakao, O. Yamauchi, and A. Nakahara, Chem. Lett. 1972, 123.
4. B. R. Rabin, Trans. Faraday Soc. 52, 1130 (1956).
5. M. K. Kim and A. E. Martell, Biochemistry 3, 1169 (1964).
6. A. P. Brunetti, M. C. Lim, and G. H. Nancollas, J. Am. Chem. Soc. 90, 5120 (1968).
7. O. Yamauchi, Y. Hirano, Y. Nakao, and A. Nakahara, Can. J. Chem. 47, 344 (1969).
8. R. F. Pasternack, M. Angwin, and E. Gibbs, J. Am. Chem. Soc. 92, 5878 (1970).
9. M. Tasumi, S. Takahashi, T. Nakata, and T. Miyazawa, Bull.

Chem. Soc. Jpn. <u>48</u>, 1595 (1975).

10. T. Takahashi, T. Miyazawa, and M. Tasumi, <u>Ind. J. Pure Appl. Phys.</u> (The Raman Effect Golden Jubillee Number), March. 1978.

11. M. Tasumi, Unpublished work.

12. T. Nakata, M. Tasumi, and T. Miyazawa, <u>Bull. Chem. Soc. Jpn.</u> <u>48</u>, 1599 (1975).

13. F. A. Quiocho and W. N. Lipscomb, <u>Advan. Protein Chem.</u> <u>25</u>, 1 (1971).

14. S.-J. Lau, T. P. A. Kruck, and B. Sarkar, <u>J. Biol. Chem.</u> <u>249</u>, 5878 (1974).

15. V. Fawcett and D. Long, Biological Applications of Raman Spectroscopy, in <u>Molecular Spectroscopy</u>, Vol. 4, The Chemical Society, London, 1976.

16. O. Siiman, N. M. Young, and P. R. Carey, <u>J. Am. Chem. Soc.</u> <u>96</u>, 5585 (1974).

17. J. R. Kincaid, J. A. Larrabee, and T. G. Spiro, <u>J. Am. Chem. Soc.</u> <u>100</u>. 334 (1978).

APPLICATIONS OF RAMAN SPECTROSCOPY TO BIOMEMBRANE STRUCTURE

Bruce P. Gaber
Naval Research Laboratory, Wash., D.C. 20375
and
Dept. of Biochemistry, School of Medicine
University of Virginia, Charlottesville, VA 22908

Paul Yager and Warner L. Peticolas
Department of Chemistry
University of Oregon
Eugene, Oregon 97403

INTRODUCTION

In the seven years since the first observation of the Raman spectra of model membranes, the field has grown from early attempts at demonstrating the usefulness of the technique to the point at which Raman spectroscopy is now accepted as a powerful tool for the study of the structure of the components of membrane systems. This review offers selected examples which demonstrate the range of applications and future potential for the Raman spectroscopy of membranes.

MEMBRANES: THEIR CONSTITUTION AND STRUCTURE

The lipid constituents of biomembranes are distinguished by their amphiphilic character as manifested by a strong hydrophilic headgroup and a long hydrophobic tail. Typical phosphoglycerides have the general formula:

Theo M. Theophanides (ed.), Infrared and Raman Spectroscopy of Biological Molecules, 241–259.

```
            H
            |
    H-C-O-C(=O)-(CH2)n-CH3
            |
    H-C-O-C(=O)-(CH2)n-CH3
            |
    H-C-O-P(O2‾)-O-X
            |
            H
```

where X may be one of a variety of head-group moieties. For
example, if X = $-CH_2CH_2-N^+(CH_3)_3$, the lipid is a diacyl phosphatidyl choline, or lecithin. Other X groups include serine,
ethanolamine and glycerol. There are many variations of this
type of structure in which the acyl groups are changed. For
example, sphigomyelin has the structure,

```
             H
             |
    HO-C-CH=CH(CH2)12CH3
             |
     H-C-NH-C(=O)-(CH2)11-CH3
             |
     H-C-O-P(O2‾)-O-CH2CH2N+(CH3)3
             |
             H
```

Although the structural details are different, in each case the
general structure remains the same: "hydrophobic tail/hydro-
philic headgroup." Other common lipid components of membranes
include the steroids such as cholesterol. However we will not
discuss these latter compounds in this review. In water the
amphiphiles such as the lecithins with their hydrocarbon tails,
characteristically form bilayers (rather than micelles) with the
headgroups on the external surface and a layer of juxtaposed
hydrocarbon chains between them.
     Two varieties of phospholipid preparation have long served
as simple model biomembrane systems. Phospholipid dispersions
are prepared by mild agitation above the lipid phase transition
in excess water. Dispersions are large aggregates (typically
1-4μ diameter) and consist of multiple bilayers. Prolonged
ultrasonication followed by chromatographic or centrifugal
separation results in the small (radius 125 Å) unilamellar
vesicles.[2] Both unilamellar vesicles and multilamellar disper-

sions are useful model systems and, as we shall see below, have
been shown by Raman spectroscopy to have somewhat different
physical properties.

The lipid bilayer constitutes only about half of the membrane. An actual membrane includes about 50% constituent
proteins. Intrinsic proteins which are inserted directly into
the bilayer may be exposed on either the exterior or interior
surface of the bilayer or transverse the entire bilayer with
exposure on both surfaces. Extrinsic proteins, on the other
hand, are surface-bound and held largely by electrostatic
forces. When the proteins are inserted into the lipid bilayer
we arrive at a conceptual model of the membrane which may be
visualized schematically as in figure I. Detailed discussions
of lipid chemistry and biomembrane structure may be found in
references 3-5.

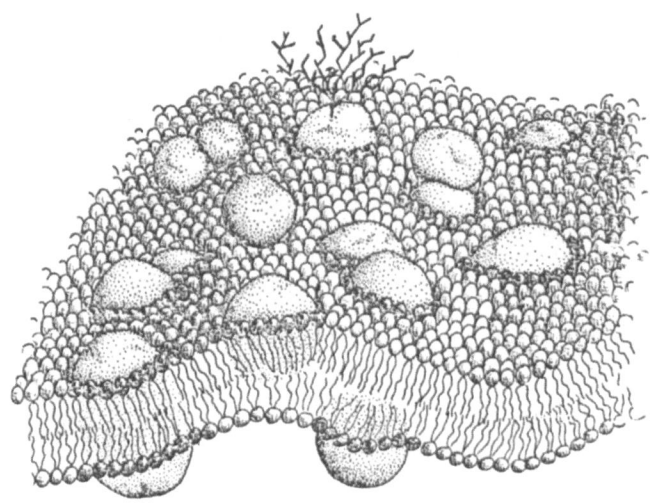

Figure I.   Diagrammatic representation of portion of a
            membrane indicating: Lipid bilayer; extrin-
            sic proteins (surface bound), intrinsic
            proteins (immersed in or transversing bi-
            layer), glyco protein (with branched car-
            bohydate), protein-associated lipid.

Raman Band Assignments

As depicted by the spectrum of solid dipalmitoyl phosphatidyl choline (figure 2), the Raman spectra of a phospholipid is
dominated by the vibrations of the long alkyl hydrocarbon portions of its fatty acyl chains superimposed upon which there
are bands from the headgroup. As a consequence assignment of
lipid spectra has benefitted enormously from the quite extensive
literature available on the vibrational spectroscopy of hydro-

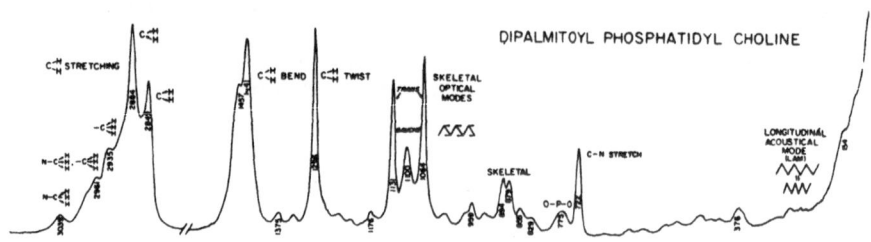

Figure 2.   Raman spectrum of solid dipalmitoyl phos-
             phatidylcholine (from Ref. 9).

carbons and polyethylenes. The conformationally sensitive vibrations of hydrocarbons may be divided into three general
regions:
     (1) In the low frequency range between 10 cm^{-1} and 300
cm^{-1} a band occurs which is the longitudinal acoustical mode
(LAM). This is an accordion-like motion of the entire hydrocarbon chain whose frequency is inversely related to the number
of all-trans bonds.[6,7] In Figure 2 the LAM may be seen as a
shoulder occurring at 154 cm^{-1}. A procedure for identifying
the LAM in hydrocarbons has been outlined.[8] Since this band is
known to split into two bands at higher frequency when the
hydrocarbon chain has the conformation of two all-trans portions
joined at a single gauche bond, this band should, in principle,
be very sensitive to conformations of lipids in which this type
of conformation occurs.[6] In actual fact because of its low frequency it is very difficult to observe this band, and many
phospholipid dispersions or vesicles simply do not show its
existence.
     (2) The skeletal-optical (-C-C-stretching) mode between
1000 cm^{-1} and 1500 cm^{-1} are particularly sensitive to the conformational state of hydrocarbons.[1,11] Of the three bands com-
prising this region the two bands at 1064 cm^{-1} and 1133 cm^{-1}
may be assigned[12-14] to the B_{1g} and A_g vibrational modes of the
all-trans chain segments while the third (1090 cm^{-1}) results
from structures containing gauche rotations. Detailed studies
of the origins assignments and conformational dependence of the
skeletal modes have been carried out in a number of laboratories.[15-18] Basic assignments are summarized in Table I.
However as we will discuss below this series of bands is useful
for obtaining an estimate of the relative number of trans and
gauche bonds along the hydrocarbon chain.
     (3) A number   of the complex bands in the C-H stretch-
ing region which extends from 2700 cm^{-1} to 3100 cm^{-1} have been
shown to change in shape and intensity distribution upon the
disruption of regular chain packing or lateral order.[21,22]

Table I   Assignments of the Raman bands of DPPC and Their Temperature Dependence in Dispersions.

Assignment	Trans	Gauche	Packing	Δcm^{-1} at 15°	Intensity rel. to 15°	Δcm^{-1} at 37°	Intensity rel. to 15°	Δcm^{-1} at 50°	Intensity rel. to 15°
$H_3C—N$ stretch				718	1.0	715	1.0	717	1.0
SOM	X			1,063	1.0	1,063	0.65	1,063	0.32
SOM *gauche*		X		1,080	1.0	1,070	~1.1	1,085	>5
SOM	X			1,098	1.0	1,090	0.95	?	?overlap with 1,080 cm^{-1}
SOM	X			1,128	1.0	1,127	0.49	1,122	0.04
CH_2 twist		X		1,296	1.0	1,296	0.54	1,301	0.28*
CH_2 sciss			X	1,436	1.0	1,436	0.76	1,439	0.70
CH_2 sym stretch	X		X	2,845	1.0	2,846	0.80	2,850	0.70
CH_2 triclinic marker			X	2,860	1.0	2,860	~0	2,860	~0
CH_2 asym stretch	X		X	2,881	1.0	2,880	0.73	2,888	0.47
CH_2		X		2,920	1.0	2,920	1.0	2,920	1.25 shoulder
CH_3 stretch				2,935	1.0	2,935	1.0	2,935	1.0 shoulder
CH_3 stretch				2,961	1.0	2,961	1.0	2,961	1.0
CH_3 stretch (choline)				3,039	1.0	3,039	1.0	3,039	1.0

SOM, skeletal optical modes.
*Peak broadens.

Table III .   Transition Temperatures for Dispersions and Vesicles of Diplamitoyl Phosphatidylcholine.

	Raman (°C)	Calorimetry * (°C)	Fluorescence * (°C)
Dispersion			
Pre-transition	34.2	35.4	25.2—33.9
Main transition	41.5	41.2	41.1
Vesicles	37	37	37

*Data from ref. 41.

Either melting or dissolution results in a decrease in intensity
of the 2890 cm^{-1} band relative to that at   2850 cm^{-1}. The
exact assignment of the various peaks of the bands of this complex of bands has been quite difficult. A great deal of work has
been carried out to explain exactly the origin of these band
intensities which involve Fermi resonance with the CH$_2$ bending
modes at half this frequency when the chains are packed in a
crystal lattice.[9,15,18-22]

     Experimental proof that it is changes in lateral packing
order and not changes in the <u>trans-gauche</u> conformation which
give rise to these intensity changes in the C-H stretching region is obtained by an experiment in which the hydrocarbon
chain is vibrationally decoupled from its neighbors while leaving its chemical and conformational properties unchanged. Thus
the preparation of a solution of hexadecane in a crystal matrix
of solid perdeutero-hexadecane results in a substantial decrease
in the peak height ratio of the asymmetric-symmetric CH$_2$ stretching bands.[9] Snyder et al.[20] have recently considered in detail
the physical origin of this observation; but we will use it here
as a semi-quantitative measure of lateral order as will be discussed below.

Deuterohydrocarbons as Non-disturbing Probes

     Raman spectra of single component phospholipid systems
such as DPPC can be interpreted to give highly detailed information about the structure of the component molecules. Multi-
component systems, however, give spectra in which many of the
structurally sensitive modes of different phospholipids or proteins may overlap. Isotopic substitution of one component
should, in principle, provide a convenient means by which to
separate those common bands arising from distinct species in a
mixture and to simultaneously observe their behavior. The use
of deuterocarbons in Raman spectroscopy of phospholipids was
suggested as early as 1973 by Chapman[25] and studies of specifically deuterated stearic acids by Sunder et al.[26] demonstrated
the utility of these molecules in Raman studies. A comparison
of the Raman spectrum of dipalmitoyl phosphatidylcholine with
its diperdeuterated analog (DPPC-d$_{62}$) is shown in Figure 3.
Arrows indicate how the major lipid bands change frequency upon
deuteration. Note that the CD$_2$ stretching modes of DPPC-d$_{62}$
appear in a spectral window free from interference from either
perhydro-lipids or proteins. Correspondingly a window is opened in the C-H stretching region permitting direct observation of
the lipid headgroup. Basic assignments for the perdeuterated
lipids are collected in Table II. Detailed discussions of these
assignments have been published recently.[27,28] The sensitivity
of the Raman spectrum of the perdeuterated lipid to changes in
bilayer conformation is demonstrated by the temperature difference spectra in Figure 4. We will see below how deuterated

Table II . Raman Assignments of DPPC-d_{62}

Δcm^{-1}	Change of band on melting	Assignment	Corresponding band in model compound
716		C—N stretch and CD_3 rock	715–716 in all choline-containing compounds
760	Increases	Phosphate symmetric diester stretch and CD_3 rock	772 in GPC solutions, weak at 760 in hexadecane-d_{34}
832	Decreases Broadens	CD* skeletal optical mode	828 in solid hexadecane-d_{34}
876		Head group and CD unassigned	877 in GPC, phosphocholine; weak in liquid hexadecane-d_{34}
918	Decreases Broadens Shifts to 940	CD_2 twist	918 in solid hexadecane-d_{34}, palmitic acid-d_{31}
954		CD unassigned	954 in hexadecane-d_{34}, palmitic acid-d_{31}
984	Decreases Broadens Shifts to 960	CD_2 scissoring	987 in solid palmitic acid-d_{31}, 991 in solid hexadecane-d_{34}
1,057		CD_3 symmetric bend	1,057 in hexadecane-d_{34}, palmitic acid-d_{31}
1,077	Increases	CD unassigned and head group	1,070 in GPC, 1076 in liquid hexadecane-d_{34}
1,100		O—P—O($-$) symmetric stretch	1,089 in GPC
1,125	Increases	CD unassigned	1,129 in liquid hexadecane-d_{34}
1,144	Decreases	CD skeletal optical mode and deformation	1,145 in palmitic acid-d_{31}, 1,150 in hexadecane-d_{34}
1,249	Decreases Broadens Shifts to 1,241	CD_2 wag	1,251 in solid hexadecane-d_{34}, palmitic acid-d_{31} 1,245 in liquid hexadecane-d_{34}
1,305		CD unassigned	1,305 in hexadecane-d_{34}, palmitic acid-d_{31}
1,451		Choline CH_2 scissoring	1,450 in phosphocholine solution
1,470		Glycerol and choline CH_2 scissoring	1,470 in α-glycerophosphate, 1,480 in phosphocholine solution
2,075	Decreases	CD_3 symmetric stretch	2,073 in hexadecane-d_{34}
2,101	Decreases	CD_2 " "	2,101 in solid hexadecane-d_{34}, 2,107 in liquid
2,135	Decreases	CD_3 " "	2,135 in solid hexadecane-d_{34}
2,173	Increases slightly	CD_2 " "	2,173 in solid hexadecane-d_{34}
2,194	Decreases	CD_2 asymmetric stretch	2,198 in hexadecane-d_{34}
2,210		CD_3 " "	2,218 in hexadecane-d_{34} (sharp)
2,765		Unassigned	
2,885	Decreases Broadens Shifts to 2,900	Glycerol and unchanging contributions from lone protons on the deuterocarbon chains	2,889 in glycerol at 25°C, 2,902 in GPC solution
2,935		Glycerol and perturbed head group methylene	2,921 in glycerol at 25°C, 2,938 in phosphocholine solution
2,980		Perturbed head group methylenes and choline CH_3 symmetric stretch	2,988 in solutions of phosphocholine, GPC
3,041		Choline CH_3 asymmetric stretch	3,044 in phosphocholine solution 3,042 in GPC solution

*CD denotes modes of the deuterocarbon chain.

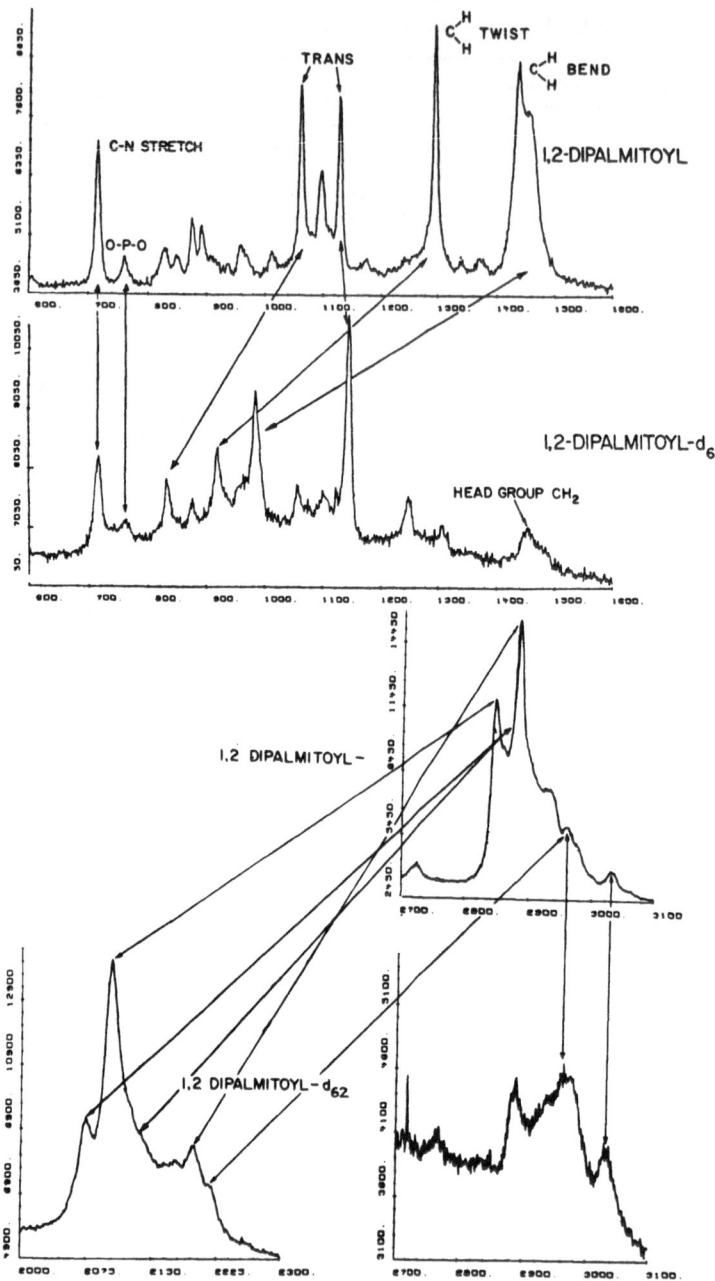

Figure 3. Raman spectra of solid samples of DPPC and DPPC-d₆₂, with arrows indicating correlations between the prominent bands.

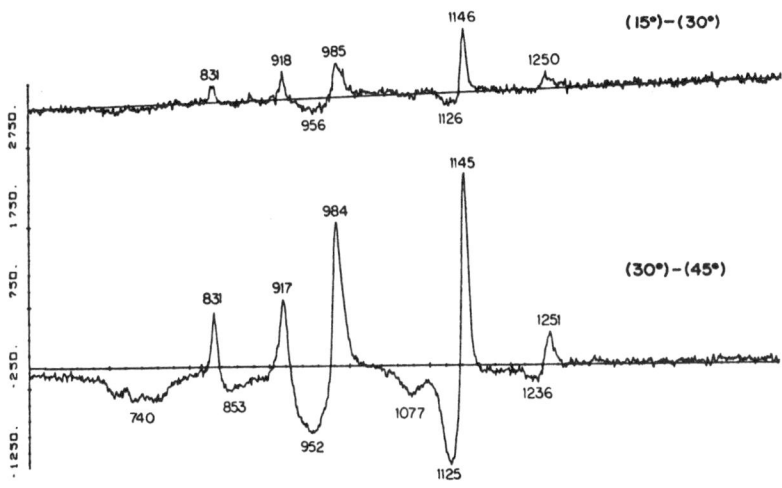

Figure 4.   Raman difference spectra for dipalmitoyl-d_{62}
             phosphatidylcholine. Difference spectra
             were created by computer subtraction of ab-
             solute spectra collected at the temperatures
             indicated above. Thus "(15°)-(30°)" denotes
             the data taken at 15° minus that taken at
             30°C. The difference spectra demonstrate
             the conformation-sensitive bands of DPPC-d_{62}
             (from Ref. 26).

lipids prove useful in studies of bilayer structure.
     Resonance Raman probes generally have proved to be rather
insensitive to the type of perturbation encountered in the lipid
bilayer. An exception is ruthenium red, a cationic dye which
can occupy calcium binding sites in phospholipids or proteins.
While the change in the electronic absorption spectrum of ruthenium red upon binding to a calcium site is quite small, the
alteration in the resonance Raman spectrum is substantial and
has permitted Friedman and co-workers[29] to identify the low affinity calcium binding site in mitocondria. More work needs to
be done on resonance Raman probes   of lipid phase transition.

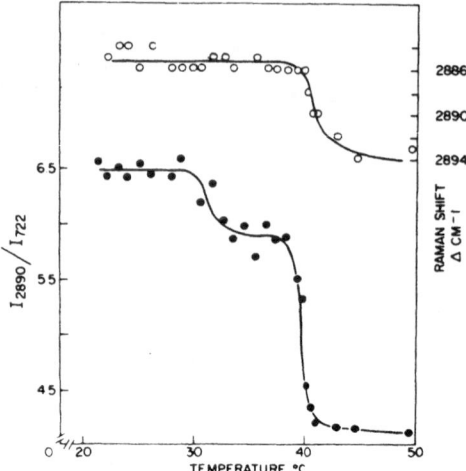

Figure 5.   Melting curve for a multilamellar disper-
            sion derived from: (lower), the change in
            the intensity ratio I_{2890}/I_{722} (closed
            circles); (upper) change in frequency of the
            asymmetric CH_2 stretch (ca. 2886 cm^{-1}) (open
            circles) (from Ref. 9).

APPLICATIONS

Lipid Phase Transitions

     Much of the early developmental work in the application of
Raman spectroscopy to biomembrane studies adopted the phenomen-
ological approach of monitoring the bilayer liquid crystal phase
transitions _via_ changes in peak amplitude and plotting the data
in the form of a melting curve. A set of such   "Raman melting
curves" (Figure 5), obtained at a resolution of 1°, show considerable detail. Two transitions may be noted (lower trace),
the well-documented main melting transition (Tm_2) at 41.5°C and
the premelting event (Tm_1) at 34.2. A favorable comparison of
the Raman data with that obtained by calorimetric and fluorescence procedures is evident in the data in Table 3. Not only
changes in peak intensity, but frequency shift[9,16] as well may
be monitored to generate melting curves (Figure 5).
     While it was the initial results of this sort which served
to validate Raman spectroscopy as a "sensitive probe," of membrane structure (1), vibrational spectroscopy has, of course,
the potential of providing substantially more than mere phenom-
enological information about the structure of the phospholipid
bilayer. As an example, the nature of the endothermic transitions in phosphatidylcholines may be studies in some detail

utilizing the Raman temperature difference procedures introduced
above.   Temperature difference spectra
created by computer subtraction of Raman data taken at various
temperatures show   evidence for three distinct bilayer structures characterizing temperatures below Tm_1, between Tm_1 and
Tm_2 and above Tm_2[18]. From these difference spectra it may be
concluded that below the pretransition temperature the hydrocarbon chains   assume a nearly all-<u>trans</u> conformation and are
well-packed. Between the pretransition and the melting temperature the number of <u>gauche</u> bonds increases slightly to about 1
or 2 per chain. The absence of a broad, strong <u>gauche</u> band at
1080 cm^{-1} is evidence for the absence of <u>gauche</u> rotations on
adjacent or nearby C-C bonds and suggests that the <u>gauche</u> bonds
are highly restricted in this phase and found only at the ends
of long all-<u>trans</u> segments. Above the melting transition chain-
chain interactions continue to decrease and the number of <u>gauche</u>
bonds increases sharply. The appearance of a strong broad band
at 1080 cm^{-1} indicates that the restriction on the placement of
<u>gauche</u> bonds is no longer present in this temperature range; the
<u>gauche</u> rotamers are free to migrate along the chain.

Structural alterations accompanying the phase transitions
are not restricted to the acyl chains. Taking advantage of the
"spectral window" in the CH_2 stretching region of DPPC-d_{62} spectrum (see Fig. 3), a temperature-induced conformational change
has been found in the headgroup.[26] A   lone band in the temperature difference spectrum is identified as arising from the lone
hydrogen atom on position SN2 of the glycerol moiety.[28] It is
observed that this H-atom is noticeably perturbed only above the
temperature region of the main phase transition.[28]

Radius of Curvature Effects

Another problem to which Raman spectroscopy has been successfully directed is the conjecture that differences should
exist between the molecular structure of large multilamellar
phospholipid dispersions and small single bilayered vesicles.
The proposed structural difference is postulated to arise from
disruption of the orderly hydrocarbon chain-packing induced by
the small radius of curvature of the vesicles. The problem has
been investigated in several laboratories[9,10,31] with similar
results and basic interpretive agreement.

The Raman spectra of phospholipid vesicles are compared
with those for solid dipalmitoyl phosphatidylcholine as well as
for multilamellar dispersions of the same lipid in Fig. **6.** In
the skeletal optical region the peak height ratio of 1130 cm^{-1}
to 1090 cm^{-1} (<u>trans</u> to <u>gauche</u>) is lower for the vesicles than
for either the solid or the dispersion. The pattern is confirmed in the C-H stretching region (Fig. 6) where the asymmetric stretch has lost intensity relative to the symmetric stretch.
Thus vesicles are found to be consistently more disordered than

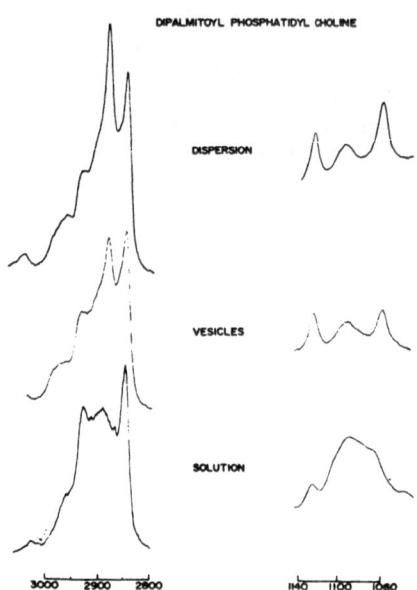

DIPALMITOYL PHOSPHATIDYL CHOLINE

DISPERSION

VESICLES

SOLUTION

Figure 6.   Raman spectra of dipalmitoyl phosphatidyl-
            choline as a multilamellar dispersion (20%
            W/W in H_2O), as single-bilayered vesicles
            and dissolved in $CHCl_3$ (or $CDCl_3$) (from
            Ref. 9).

dispersions both in the sense of their lateral or interchain
ordering as well as their intrachain ordering. The difference
is further demonstrated in the previously shown melting curve
(Figure 5), in which the melting of the vesicles is broader and
does not show the pretransition typical of the multilamellar
dispersion.

Estimate of Trans/Gauche Ratios

       Of intense interest to both theorists and experimentalists
is the development of a procedure which will permit direct experimental estimation of the proportion of trans and gauche bonds
in the acyl chain of a phospholipid. Several different procedures have been employed, and while still primitive, the results are in reasonable agreement with other experimental data
and with theoretical predictions. The general approach is to
assume that the relative intensity of the skeletal optical modes
near 1130 cm^{-1} and 1065 cm^{-1} provide a rough quantitative estimate of the number of bonds in an all-trans segment. Marsh[32]
originally proposed a linear relationship between the 1130 cm^{-1}
and 1090 cm^{-1} peak height ratio and the all-trans probability
P_T. The present authors[9] have developed a formalism for the

Raman intensity which makes the assumption that the C-C stretching vibration observed is the sum of intensities from individual all-trans segments three or more bonds in length. In assuming a unique intensity per "trans-unit" the formalism is, of course, approximate. Furthermore the specific sequence of trans and gauche bonds, which would give rise to different values to the total intensity, is not considered in this formalism, nor are the effects of end groups. However, by comparison with standard compounds (see below) it is possible to estimate a reasonable lower limit for the number of trans bonds in a palmitoyl chain of approximately 8 per chain.[9]

Yellin and Levin[17] have employed a somewhat different procedure for estimating the number of chains in the all-trans conformation. Applying the integrated form of the van't Hoff equation to their Raman data in the skeletal optical region at temperatures below the liquid crystal phase transition they estimate an enthalpy difference between the chains in the all-trans conformation and chain conformations containing gauche rotations. Comparing their data to the calorimetrically determined ΔH for the phase transition[33] leads to an estimate of 4-4.5 gauche bonds per chain in dipalmitoylphosphatidyl choline above T_M. This value compares favorably with our estimate of approximately 8 trans bonds per chain.

Estimates of Intra- and Inter-Chain Contributions to Bilayer Order

We have attempted to treat the Raman data so as to distinguish, insofar as possible, between order due to the intra-chain structure and that due to lateral crystalline interactions.[9] Order parameters have been defined such that $S = 1$ indicate the highest possible order and $S = 0$ no order (not necessarily the lowest possible). The trans-parameter is defined as

$$S_T = \frac{(I_{1133}/I_{REF})_{observed \cdot}}{(I_{1133}/I_{REF})_{DPPC, \; solid}}$$

and serves to reference the data to a standard of known all-trans chain length. Generation of an order parameter for the lateral interaction presents a more difficult problem since the change in intensity of the 2890 cm^{-1} is not simply a function of the loss of interchain interaction.[20] As a point of departure, however, an order parameter $S_{LATERAL}$ may be defined which compares the peak height ratios of the sample with those observed for crystalline samples of hexadecane. $S_{LATERAL}$ may then be defined as

$$S_{LATERAL} = \frac{I_{CH2} \; (observed)^{-0.7}}{1.5}$$

where $I_{CH2} = I(2890)/I(2850)$. The parameter is semiquantitative, but does provide insight into the amount of lateral interaction.

While acknowledging the empirical nature of these order parameters, use of normalization equations of this sort does provide a relative measure of the extent of intra- and inter-chain order. A chart may be prepared (Figure 7) which summariz-es the data in a form which provides for ready comparison bet-ween various phospholipid preparations. It may be concluded, for example, that at their respective transition temperatures the trans order for phospholipid vesicles and dispersions is very nearly the same, but at 30°C dispersions are substantially more ordered.

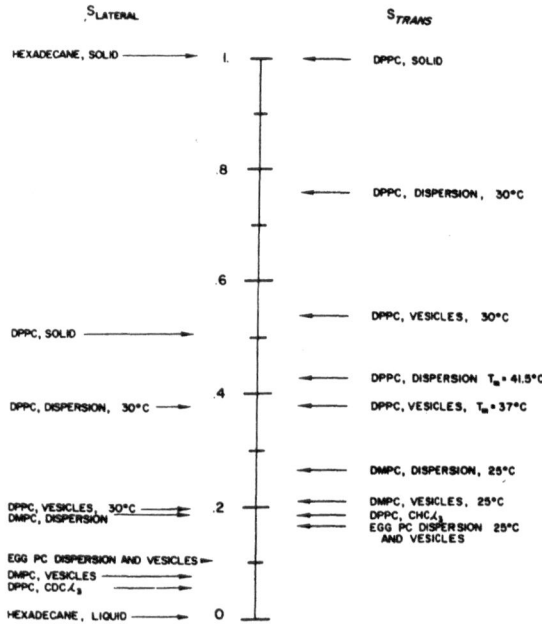

Figure 7. Various phosphatidylcholine preparations ranked in terms of their relative intrachain order (S_{trans}) and interchain order ($S_{LATERAL}$) (from Ref. 9).

Mixed Lipid Systems

One of the earliest Raman studies of mixed lipid systems was that of the cholesterol-phosphatidylcholine system in which it was shown that the phospholipid melting transition was broad-ened considerably in the presence of cholesterol;[1,34] suggesting that below T_M cholesterol has the effect of "softening" the bilayer, while above T_M it served to "stiffen" the bilayer. These studies led to the conclusion that cholesterol might

be considered a "plasticizer," permitting the bilayer to adjust
to changes in temperature.

More recently, interest has turned to the effect of chain
length on the phase separation properties of binary phospholipid
mixtures. When Raman melting curves are generated for mixtures
of dimyristoyl (C_{14}) and distearoyl (C_{18}) lecithin broad transitions are observed (Figure **8**).[35] Collecting these melting
curves at several mole fractions of C_{18} and C_{14}, the Raman data
may be used to construct a phase diagram , which then
may be compared with theoretical predictions.[36] As an example
of the utility of isotopic labeling, Mendelsohn[37] has performed
a similar experiment using a mixture of perdeuteromyristoyl
phosphatidylcholine and distearoyl phosphatidylcholine such that
the melting of each component is individually observable.

The study of mixed lipid systems need not be restricted to
mixtures of separate molecular species, but may include, for
example, phospholipids of heterogeneous chain composition.* The
spectrum of one such natural asymmetric lipid, sphingomyelein,
was first recorded and interpreted by Mendelsohn, et al.[34]
Raman melting curves of sphingomylein (J. P. Sheridan, P. E.
Schoent, R. Priest, unpublished) show multiple phase transitions
reminescent of the data in Figure 8.

The question of the conformation of the two hydrocarbon
chains of a single phospholipid can also be considered as a
"mixed lipid system" -- one which may be profitably explored
through use of isotopically-labeled phospholipids. The compounds 1-palmitoyl, 2-palmitoyl-d_{31} phosphatidylcholine and 1-
palmitoyl-d_{31}, 2-palmitoyl phosphatidylcholine were synthesized
and the Raman difference spectrum between the two compounds collected under identical conditions.[38] Here Raman spectroscopy
permits the simultaneous and independent observation of both
chains of the phospholipid. The inflections in the Raman difference spectrum are attributed to non-equivalent conformations
of the fatty acyl chains attached at glycerol positions Sn_1 and
Sn_2.[38] Below the pretransition temperature, the conformation of
the Sn_2 chain is, on the average, slightly less all-<u>trans</u> than
is the chain in position Sn_1. Differences in conformation per-
sist even above the melting temperature.[38]

Lipid-Protein Interactions

The nature of the interaction between lipid and protein in
membranes is an area of lively interest and one in which Raman
spectroscopy is making increasingly important contributions.
One such example is a recent study[39] of the interaction between
lipid and the coat protein from the filimentous virus Fd. The

---

*Compositional asymmetry is common   in natural lipids.

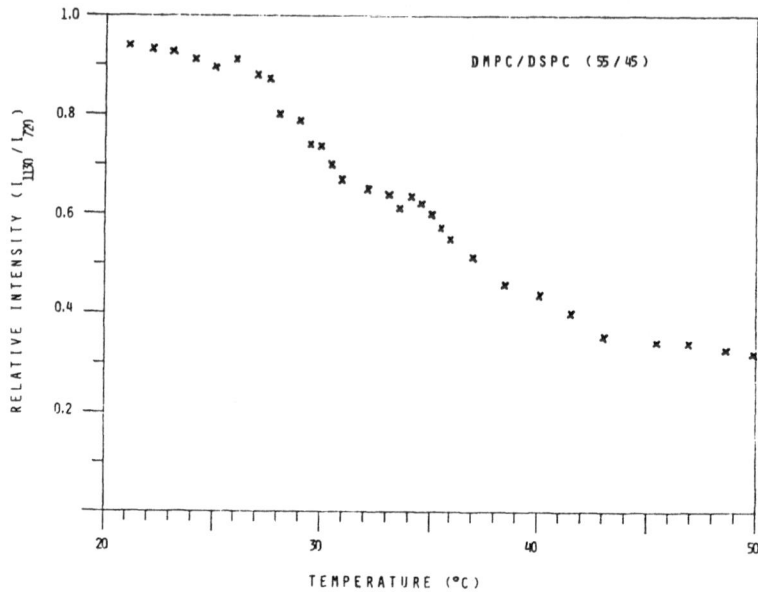

Figure 8.   Melting curve generated from Raman spectra
            of a 50:50 mole fraction mixture of dimy-
            ristoyl phosphatidylcholine (C_{14}) and dis-
            tearoyl phosphatidylcholine (C_{18}) (from
            Ref. 35).

coat protein can be isolated in pure form and is extraordinarily
hydrophobic. The protein, when sonicated with lipid forms
vesicles consisting of the lipid-protein complex. The Raman
spectra of such vesicles shows that a change in temperature
brings about a marked structural difference in the bands associated with the phosphilipid but no discernible change in protein
structure. When the Raman melting curve is determined for the
phospholipid component in the lipid protein complex, the lipid
is observed to melt some 20° below the temperature expected for
the pure lipid component. (Note that this experiment permits
the independent measurement of the lipid melting temperature --
a measurement which is not possible, for example, by calorime-
try.)

Membranes

The current focus in Raman applications to membrane studies has been to develop the technique to a point such that it could be applied directly to the elucidation of the structure of actual membranes. The Raman spectrum of red blood cells[40] demonstrates that high quality spectra may be collected from natural membranes spectra which are amenable to interpretation in terms of both lipid and protein conformation.

References

1. Lippert, J. L. and Peticolas, W. L., Proc. Acad. Sci. U.S. 68, 1572-76 (1971).

2. Huang, C-H., Biochem. 8, 344-51 (1969).

3. Ansell, G. B., Dawson, R. M. C., Hawthorne, J. N., eds. Form and Function of Phospholipids, B. B. A. Library, Vol. 3, Elsevier Scientific Publishing Company, Amsterdam, (1973).

4. Fox, C. F., and Keith, A., eds., Membrane Molecular Biology, Sinaver Associates, Inc., Stamford, Conn. (1972).

5. Kates, M., Techniques of Lipidology, North-Holland Publishing Co., Amsterdam (1972).

6. Schaufele, R. G., and Shimanouchi, T., J. Chem. Phys. 47, 3605-3610 (1967).

7. Schaufele, R. G., J. Chem. Phys. 49, 4168-4175 (1968).

8. Behroozi, F., Priest, R. G., and Schnur, J. M., J. Raman Specty. 4, 379-85 (1976).

9. Gaber, B. P. and Peticolas, W. L., Bioch. Biophys. Acta 465, 260-274 (1977).

10. Spiker, Jr., R. C., and Levin, I. W., Bioch. Biophys. Acta 455, 560-579 (1976).

11. Lippert, J. L., and Peticolas, W. L., Bioch. Biophys. Acta 282, 8-17 (1972).

12. Tasumi, M., Shimanouchi, T., and Miyazawa, T., J. Mol. Specty. 9, 261-287 (1962).

13. Snyder, R. G., and Schachtschneider, J. H., Spectrochim. Acta 19, 85-116 (1963).

14. Snyder, R. G., J. Chem. Phys. 47, 1316-1360 (1967).

15. Spiker, B. C., and Levin, I. W., Bioch. Biophys. Acta 388, 361-373 (1975).

16. Spiker, Jr., R. C., and Levin, I. W., Bioch. Biophys. Acta 433, 457-458 (1976).

17. Yellin, N., and Levin, I. W., Biochem. 16, 642-646 (1977).

18. Gaber, B. P., Yager, P., and Peticolas, W. L., Biophys. J. 21, 161-176 (1978).

19. Bunow, M. R., and Levin, I. W., Bioch. Biophys. Acta 487, 388-394 (1977).

20. Snyder, R. G., Hsu, R. M., and Krimm, S., Spectrochim. Acta (in press).

21. Larsson, K., and Rand, P., Bioch. Biophys. Acta 326, 245-255 (1973).

22. Brown, K. G., Peticolas, W. L. and Brown, E., Bioch. Biophys. Research Comm. 54, 358-364 (1973).

23. Schoen, P. E., Schnur, S. M. and Sheridan, J. P., Appl. Specty. 31 337-339 (1977).

24. Sheridan, J. P., Schnur, J. M. and Schoen, P. E., Proc. 5th Int. Conf. Raman Specty.,

25. Chapman, D., in Biological Membranes, Vol. 2., D. Chapman and D. F. H. Wallach, eds. Acad. Press, N.Y. (1973), pp. 91-144.

26. Sunder, S., Mendelsohn, R., Bernstein, H. J., Chem. Phys. Lipids 17 456-65 (1976).

27. Bunow, M. R. and Levin, I. W., Bioch. Biophys. Acta 489 191-206 (1977).

28. Gaber, B. P., Yager, P., and Peticolas, W. L., Biophys. J. 22, 191-207 (1978).

29. Friedman, J. M., Navon, G., Glynn, P., Lyons, K. B., Biophys. J. 21, 36a (1978).

30. Suurkuusk, J., Lentz, B. R., Barenholtz, Y., Biltonen, R. L., and Thompson, T. E., Biochem. 15, 1393-1401 (1976).

31. Mendelsohn, R., Sunder, S., and Bernstein, H. J., Bioch.
     Biophys. Acta 419, 563-569 (1976).

32. Marsh, D., J. Mem. Biol. 18, 145-162 (1974).

33. Hinz, H. T., and Sturtevant, J. H., J. Biol. Chem. 247,
     6071-75 (1972).

34. Mendelsohn, R., Bioch. Biophys. Acta 290, 15-21 (1972).

35. Sheridan, J. P., and Priest, R. G., Schoen, P. E. and
     Schnur, J. M. (in prep.).

36. Priest, B. G. and Sheridan, J. P., Liquid Crystals and
     Ordered Fluids, Vol. 3, pp. 209-224, Plenum Press, N.Y.
     1978.

37. Mendelsohn, R., et al. Bioch. Biophys. Acta ___ 000-000
     (1978).

38. Gaber, B. P., Yager, P., and Peticolas, W. L., Biophys. J.,
     (submitted for publication).

39. Dunker, A. K., Williams, R. W., Gaber, B. P., and Peticolas,
     W. L., Science, (submitted for publication).

40. Lippert, J. L., Gorczyca, L. E., and Melklejohn, G., Bioch.
     Biophys. Acta 382, 51-57 (1975).

41. Suurkuusk, J., Lentz, B. R., Barenholz, Y., Biltonen, B. L.,
     and Thompson, T. E. (1976) Biochemistry 15, 1393-1401.

Resonance Raman Spectra of Model Metallo Porphins

H. J. Bernstein

Division of Chemistry, National Research Council,
Ottawa, Canada

The discovery of the laser and its subsequent use as an irradiating source in Raman spectroscopy made it possible to investigate routinely highly-coloured gases[1], doped crystals[2], liquids[3,4,5] and solids[5]. In many cases, however, the incident light has a wavelength which lies within or near the contour of the electronic absorption band. Under such conditions the light is strongly absorbed with rapid heating of the sample which now refracts and diffuses the light into a cone. All the useful features of the laser irradiation are now destroyed and one is surrounded by the well known difficulties associated with scattering from a Hg arc lamp[4].

In order to restore the advantages of laser irradiation, it is necessary to cool the sample which can be readily accomplished by rotating the sample in the laser beam[5]. In this way, for example, it is possible to obtain spectra from CS_2 solutions of model compounds of the heme in heme proteins in relevant biological concentrations[6].

In order to analyse the Raman spectum of a molecule when the irradiation is in the electronic absorption, one recognizes that the Placzek condition[7] for ordinary Raman spectra is no longer fulfilled, namely that ν_0 the exciting frequency should be far from ν_e the effective absorption frequency of the molecule. For resonance, $\nu_0 - \nu_e$ is small. The intensity increases enormously due to the

261

Theo M. Theophanides (ed.), Infrared and Raman Spectroscopy of Biological Molecules, 261–265.

terms with resonance denominators:

$$I\alpha(\nu-\nu)^4\Sigma(\alpha_{ij})^2$$

or

$$I\alpha \sum \frac{<k|M_\sigma|r><r|M_\sigma|n>}{\nu_{rn} - \nu_o} + \frac{M_\sigma \rightarrow M_\sigma}{\nu_{rk}+\nu_o}$$

Thus the intensity of a Raman band depends strongly on the value of the exciting frequency and the variation of the Raman intensity with the frequency of the exciting line is often called the excitation profile.[8] The depolarization ratio depends on the invariants of the tensor of the polarizability derivatives where for the ordinary Raman effect $\alpha_{ij}= \alpha_{ji}$ for the resonance Raman effect $\alpha_{ij} = -\alpha_{ji}$. The derived polarizability tensor may be written

$$\alpha_{ij} = \begin{pmatrix} xx, & xy, & xz \\ yx, & yy, & yz \\ zx, & zy, & zz \end{pmatrix}$$

The depolarization ratio $\rho = \dfrac{3\gamma_s'^2+ 5\nu_a'^2}{45a'^2+4\nu_s'^2}$

where $a' = \dfrac{xx+yy+zz}{3}$

$$\nu_s'^2= 1/2 \ \{(xx-yy)^2+(xx-zz)^2+(yy-zz)^2\}$$

$$3/4 \ \{(xy+yx)^2+(xz+zx)^2\pm(yz+zy)^2\}$$

$$\nu_a'^2= 3/4 \ \{(xy-yx)^2+(xz-zx)^2+(yz-zy)^2\}$$

For ordinary Raman spectra $\alpha_{ij}= \alpha_{ji}$, and $\nu_a'^2\equiv0$ in resonance Raman spectra where $\alpha_{ij}=-\alpha_{ji}$, $\nu_a'^2=0$.

In the ordinary Raman effect there are two types of classification of modes 1) those with the depolarization ratio for polarized light, $\rho = 0 \longleftrightarrow 3/4$ values obtained for totally symmetric modes, and 2) $\rho = 3/4$ for the non totally symmetric modes. In the resonance Raman effect there is an additional class of totally symmetrical modes which shows dispersion in the depolarization ratio for which $\rho = 0 \longleftrightarrow \infty$. For non totally symmetric modes there are two additional classes for which $\rho = 3/4 \longleftrightarrow\infty$, and $\rho=\infty$.

A mode with $\rho = \infty$ is usually referred to as being inversely polarized, whereas those with $\rho = 3/4$ are referred to as being anomalously polarized. The ρ values for these classes have been assigned by McClain[9] to the different species of vibrations in the various molecular point group species. For molecular symmetry D_{4h} (Cu-porphin, for example) the vibronic coupling between two allowed electronic transitions is active for the vibrational species whose symmetry is contained in the direct product of their representations,

$$Eu \times Eu = A_{1g} + B_{1g} + B_{2g} + A_{2g}$$

The intensity of the A_{1g} modes arises from changes in bond and angle dimensions in the excited state; the intensity in the A_{2g}, B_{1g}, B_{2g} modes arises from vibronic coupling between two electronic transitions of Eu symmetry whereas the Eg modes which are out of plane are weak.

The A_{1g} modes have ρ values between 1/8 and 3/4: the B_{1g} and B_{2g} modes have ρ values of 3/4 and for the A_{2g} modes $\rho = \infty$. With our instrumentation and slits of 1 cm^{-1} the value measured[10] for one of the A_{2g} modes of Cu porphin is 50. This is in part due to the fact that the resonance denomination is never zero because of the damping term Γ viz

$$I\alpha \ \prod_0^M \frac{a_n}{(\nu_0 - \nu_e \pm n\nu_R)^2 + \Gamma^2}$$

When the metallo porphin is distorted as it is in some heme proteins we could have D_{4h} becoming C_{4h} which could become C_{2h} and eventually C_S. One could follow this symmetry by observing the values of the depolarization ratios of the anomalously polarized modes, anticipating that ρ would decrease as the molecular point group symmetry became less, i.e. from D_{4h} to only a plan of symmetry C_S. After analysing the spectra for several types of methyl and ethyl substituted Cu-porphins of different symmetries[11] we arrived at a partial assignment from which it was possible to make a normal co-ordinate calculation. Because of the many force constants required for a reasonably accurate treatment some simplifying assumptions were made: (1) the stretching force constants were obtained from the plot of the valence field force constants and the observed bond distance (2) the valence force bending

constants were obtained from the product of the bond
distances of the bonds containing the angle
(3) methyl groups were replaced by particles of
appropriate mass and the newly introduced stretch bond
constants obtained from the previously described
correlations.

The NCT was used to compute 140 frequencies for
the molecules Cu-porphin, Cu-porphin-d_4, Cu-octa-
methy porphin, all of D_{4h} symmetry, also Cu-tetrameth-
porphin of C_{4h} symmetry with an estimated error of
1-2% in the agreement between observed and calculated
frequencies[12].

These calculations are compared with those in the
literature[13] from which one can estimate a statistical
value for each mode. Our calculations are as useful
as any other compared,[13-16] and has the distinguishing
feature that it depends on only 12 parameters in the
potential function whereas for all other calculations
at least 18 are required. Finally our results for Cu
and Ni octamethyl porphin are correlated with the
persistent Raman modes in heme proteins[14], to indicate
the origin and metal sensitivity of the modes.

References

1) W. Holzer, W.F.Murphy and H.J. Bernstein, J. Chem.
   Phys. 52, 399 (1970).

2) W. Holzer, W.F. Murphy and H.J. Bernstein, J. Mol.
   Spectr. 32, 13 (1969).

3) a. T.G. Spiro and T.L. Strekas, Proc. Nat. Acad.
      Sci. USA 69, 2622 (1972).

   b. L. Riman, M.E. Heyde, H.C. Heller and D. Gill,
      Chem. Phys. Lett. 10, 207 (1971).

   c. W.L. Peticolas, L. Nafie, P. Stein and B. Fani-
      coni, J. Chem. Phys. 1576 (1970).

   d. A.L. Verma and H.J. Bernstein, J. Ram. Spectr. 2,
      163 (1974).

4) See, for example, J. Behringer.

5) W. Kiefer and H.J. Bernstein, J. Appl. Spectr. 25,
   500, 609 (1971).

6) 3a, 3d.

7) G. Placzek, Handbuch der Radiologie V12, 205 (1934).

8) 3b, 3a.

9) W.M. McClain, J. Chem. Phys. 55, 2789 (1971).

10) Cu-porphin, A.L. Verma and H.J. Bernstein, J. Chem. Phys. 61, 2560 (1974).

11) R. Mendelsohn, S. Sunder, A.L. Verma and H.J. Bernstein, J. Chem. Phys. 62, 37 (1975); S. Sunder and H.J. Bernstein, J. Chem. Phys. 52, 2851 (1974).

12) S. Sunder and H.J. Bernstein, J. Raman Spectr. 5, 351 (1976).

13) a. Solovyov et al. XII Int. Mol. Symposium (1977), Wroclaw, Poland.

   b. H. Susi, J.S. Ard, Spectrochem. Acta 33A, 561 (1977).

   c. A. Warshel, private comm.

14) Spiro and Strekas, etc. Hemeproteins.

15) Kitagawa et al., J. Phys. Chem. 60, 1181 (1976), Chem. Lett. p. 249, 1976.

16) H. Ogoshi, V. Saito and K. Nakomoto, J. Chem. Phys. 57, 4194 (1972).

RESONANCE RAMAN SPECTROSCOPY OF HEME PROTEINS

Thomas G. Spiro

Department of Chemistry, Princeton University,
Princeton, New Jersey 08540

ABSTRACT. Resonance Raman spectroscopy can monitor the vibrational
modes of the heme prosthetic group in situ. Porphyrin ring
vibrations are dominant, enhanced by the strong π-π* transitions.
Some of them are sensitive to the competition for iron d_π electrons between the π* orbitals on porphyrin and on π-acceptor
axial ligands. In this context O_2 is shown to be as good a π-
acceptor as NO. Other modes are sensitive to iron spin-state,
an effect attributable to expansion of the porphyrin core in the
high-spin complexes. Although weak, axial ligand vibrational
modes can sometimes be detected, and confirmed via isotope shifts.
These include ν_{Fe-O} of O_2Hb, whose frequency is consistent with
appreciable Fe-O_2 multiple bonding.

1. GENERAL FEATURES

Since the visible and near-UV spectra of heme proteins contain
the intense absorption bands of the heme prosthetic group, their
Raman spectra are dominated by resonance enhanced heme vibrational
modes. These are readily detectable in the concentration range
10^{-3} - 10^{-5} M. Heme vibrational frequencies can therefore be
monitored in dilute solution, without interference from polypeptide
vibrational modes. Indeed, the latter cannot be observed at all
with ordinary laser sources, since heme protein solutions that
are sufficiently concentrated for non-resonance Raman studies
absorb too much of the light. Heme protein resonance Raman (RR)
spectra resemble those of protein-free metalloporphyrins. Model-
ling of heme protein RR spectra with iron porphyrins, in various
oxidation and ligation states, is very useful, since the chemistry
and structures of the porphyrin complexes have been well studied.

Theo M. Theophanides (ed.), Infrared and Raman Spectroscopy of Biological Molecules, 267–275.

RESONANCE RAMAN SPECTROSCOPY OF HEME PROTEINS

Thomas G. Spiro

Department of Chemistry, Princeton University,
Princeton, New Jersey 08540

ABSTRACT. Resonance Raman spectroscopy can monitor the vibrational
modes of the heme prosthetic group in situ. Porphyrin ring
vibrations are dominant, enhanced by the strong π-π* transitions.
Some of them are sensitive to the competition for iron d_π electrons between the π* orbitals on porphyrin and on π-acceptor
axial ligands. In this context O_2 is shown to be as good a π-
acceptor as NO. Other modes are sensitive to iron spin-state,
an effect attributable to expansion of the porphyrin core in the
high-spin complexes. Although weak, axial ligand vibrational
modes can sometimes be detected, and confirmed via isotope shifts.
These include ν_{Fe-O} of O_2Hb, whose frequency is consistent with
appreciable Fe-O_2 multiple bonding.

1. GENERAL FEATURES

Since the visible and near-UV spectra of heme proteins contain
the intense absorption bands of the heme prosthetic group, their
Raman spectra are dominated by resonance enhanced heme vibrational
modes. These are readily detectable in the concentration range
10^{-3} - 10^{-5} M. Heme vibrational frequencies can therefore be
monitored in dilute solution, without interference from polypeptide
vibrational modes. Indeed, the latter cannot be observed at all
with ordinary laser sources, since heme protein solutions that
are sufficiently concentrated for non-resonance Raman studies
absorb too much of the light. Heme protein resonance Raman (RR)
spectra resemble those of protein-free metalloporphyrins. Model-
ling of heme protein RR spectra with iron porphyrins, in various
oxidation and ligation states, is very useful, since the chemistry
and structures of the porphyrin complexes have been well studied.

Theo M. Theophanides (ed.), Infrared and Raman Spectroscopy of Biological Molecules, 267–275.

3. ELECTRONIC EFFECTS ON PORPHYRIN FREQUENCIES

Systematic examination of heme derivatives has revealed that certain porphyrin frequencies are appreciably influenced by chemical changes at the iron atom.[5,14] Three of these frequencies, labelled I, III and V,[14] are observed to correlate well with iron oxidation state and the π acidity of the axial ligands. This correlation is demonstrated in Table I for a series of 6-coordinate hemes.

TABLE I

RR PORPHYRIN FREQUENCIES (cm^{-1}.)
CORRELATING WITH OXIDATION STATE AND AXIAL LIGAND π ACIDITY

	I	III	V
$(Im)_2Fe^{II}MP^{a,14}$	1358	1534	1617
$(P\ Et_3)_2Fe^{II}MP^{14}$	1358	1544	1620
$(Py)_2Fe^{II}MP^{14}$	1365	1555	1622
$(P\ OEt_3)_2Fe^{II}MP^{14}$	1368	1559	1629
$(CO)(Py)Fe^{II}MP^{17}$	1374	1570	1633
$(NO)(Py)Fe^{II}MP^{17}$	1375	1570	1640
$(Im)_2Fe^{III}MP^{14}$	1375	1572	1640
$CoHb^{41}$	1373	1564	1635
O_2Hb^5	1377	1564	1640
$NOHb^{17,26}$	1375	1564	1633

[a]Symbols: MP - mesoporphyrin IX, dimethyl ester; Im - imidazole; Et - ethyl; Py - pyridine; Hb - hemoglobin.

All three modes show appreciably lower frequencies for Fe^{II} than for Fe^{III} when both axial ligands are imidazole. This can be attributed to increased back-donation of Fe^{II} d_π electrons into the π^* acceptor orbitals of the porphyrin ring, with resultant decrease in the π bond order. As imidazole is replaced, in the Fe^{II} complex, with ligands that themselves have low-lying π^* acceptor orbitals, the back-donation to the porphyrin ring is relieved and the frequencies increase. The increases correlate

quite well with the expected π acidity of the axial ligands.
The best π acid, NO, produces the highest porphyrin frequencies;[17]
they are fully as large as those found for low-spin Fe^{III}, suggesting that π acceptance by NO is equivalent, with respect to the
porphyrin frequencies, to the removal of one electron from the
iron d_π orbitals.

In this context it is of considerable interest that anomalously
low values of the band I frequency have been observed for cytochrome P450, in both oxidized[18] and reduced[19] forms. One of the
axial ligands is believed to be a cysteine thiolate group, which
should be a good π donor. Thus the low frequencies are fully
consistent with the transmission of π effects between porphyrin
and axial ligands via the iron atom.

It is likely, however, that σ as well as π effects influence the
porphyrin frequencies. Thus band I is found at 1382 cm^{-1} for
$(Im)_2Co^{III}MP$,[20] 7 cm^{-1} higher than for $(Im)_2Fe^{III}MP$. Co^{III}, with
a d_π^6 configuration is unlikely to be a poorer π donor than Fe^{III},
with a d_π^5 configuration, but it does have one unit higher nuclear
charge. The resultant increase in polarization of the pyrrole
N atoms might explain the increased frequency of band I, which
is known from its ^{15}N shift to be primarily a C-N breathing mode.[21]
Horseradish peroxidase Compound II, which is believed to contain
Fe^{IV}, also shows an elevated band I frequency,[22] 1382 cm^{-1}, but
·bands III and V are at frequencies characteristic of low-spin
Fe^{III}. It may be that bands III and V are sensitive primarily
to π effects, which may be already saturated for Fe^{III}, while
band I is sensitive as well to polarization in the metal-pyrrole
σ bonds.

4. BONDING IN OXYHEMOGLOBIN

The early observation that the band I frequency shown by O_2Hb was
characteristic of Fe^{III} led to the suggestion by Yamamoto, et
al.,[23] that the RR evidence supported Weiss' characterization[24]
of O_2Hb as a superoxide adduct of (low-spin) Fe^{III}, as against a
neutral dioxygen adduct of Fe^{II}.[25] Indeed bands I, III and V all
have essentially the same frequencies in O_2Hb as they do in
metHb cyanide (Table II). This is true as well for NOHb,[17,26]
and the frequencies for COHb are also closer to those of Fe^{III}
than of Fe^{II} derivatives. (Band III falls 6 cm^{-1} lower in Hb
derivatives than in the MP analogs, as seen in Table I, probably
reflecting an influence of the vinyl groups in protoporphyrin IX;
the trends are the same for Hb and MP, however.) In the light
of the preceding discussion, these facts are fully in accord with
the π-back donation model, with O_2 acting as an effective π acid,
stronger than CO, and about as strong as NO.

Multiple bonding between Fe and O_2 is also suggested by the relatively high frequency of ν_{Fe-O}, 567 cm^{-1}, observed[9] in O_2Hb. The O-O stretch, on the other hand, has been determined by infrared spectroscopy[27] to be at 1109 cm^{-1}, a frequency characteristic of superoxide. Similar frequencies $\nu_{O-O} = 1159$ cm^{-1} [28] and $\nu_{Fe-O} = 577$ cm^{-1} [29] have been observed for a dioxygen adduct of Collman's "picket-fence" porphyrin.[30] These data confirm the analogy between O_2Hb and the picket-fence adduct, whose crystal structure[30] shows the O_2 to be bound end-on, with an Fe-O-O angle of 135°. The short Fe-O distance, 1.75 Å, is again suggestive of Fe-O multiple bonding.

Taken together, these observations support the bonding model shown pictorially in Figure 1, involving both σ and π bonding between

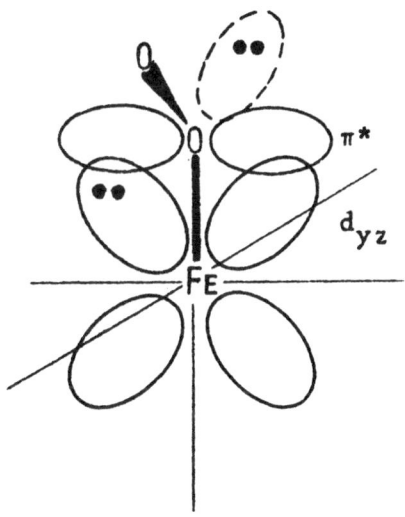

Figure 1. Model of π-bonding in O_2Hb

Fe and O_2. The partial occupancy of the O_2 π* orbital lowers the O-O bond order, and ν_{O-O}, to that expected for superoxide. Equal division of the electron pair in the π bonding orbital between Fe and O_2 would indeed produce a formal $Fe^{3+}O_2^-$ electron distribution, but of course there is substantial forward donation from O_2 to Fe through the σ bond. Electronic calculations[31] indicate that O_2 does not build up appreciable charge on binding.

5. OUT-OF-PLANE HEMES: DEOXYHb AND METHb

Three porphyrin frequencies, labelled II, IV and V, change apprecia-
ciably between planar six-coordinate (low-spin) hemes and out-of-
plane five-coordinate (high-spin) hemes, as demonstrated in
Table II. Five-coordinate $(2\text{-MeIm})Fe^{II}MP$ shows frequency

TABLE II

RR PORPHYRIN FREQUENCIES (cm^{-1}) SHIFTING WITH SPIN-STATE

	II	IV	V
$(Im)_2Fe^{II}MP^{14}$ (ls)[a]	1490	1583	1617
$(2\text{-MeIm})Fe^{II}MP^{14}$ (hs)	1472	1558	1606
$Fe^{II}Hb^5$ (hs)	1473	1552	1607
$Fe^{II}HRP^5$ (hs)	1472	1553	1605
$Fe^{III}Hb^{14}$ (hs)	1482	1555	1608
$Fe^{III}HRP^{36}$ (hs)	1500	1575	1630
$\left[(DMSO)_2Fe^{III}PP\right]^{+33}$ (hs)	1475	1559	1609
$\left[(2\text{-MeIm})Fe^{III}MP\right]^{+17}$ (hs)	1495	1572	1632
$\left[Im_2Fe^{III}MP\right]^{+17}$ (ls)	1505	1588	1640

[a]Symbols, as in Table I: ls - low-spin; hs - high-spin;
2-MeIm-2-methylimidazole; HRP - horseradish peroxidase;
PP - protoporphyrin IX dimethyl ester.

decreases of 18, 25 and 11 cm^{-1} for bands II, IV and V respectively, relative to six coordinate $(Im)_2Fe^{II}MP$.[14] Although
these shifts were originally attributed to doming of the porphyrin
ring in the out-of-plane structures,[32] the best current interpretation,[33] following the suggestion of Spaulding, et al.,[34]
is that the shifts result from expansion of the porphyrin ring,
resulting in a stretching of the methine bridge bonds. The
Fe-pyrrole bonds are lengthened for high-spin Fe^{II} because of
the partial occupancy of antibonding orbitals. This repulsion
is largely taken up by movement of the iron atom out of the porphyrin plane, but there is also a significant expansion of the
porphyrin cavity, with respect to six-coordinate Fe^{II} porphyrins.[35]
A similar but smaller expansion is observed for high-spin five-

coordinate Fe^{III} porphyrins; the spin-marker frequency shifts are likewise in the same direction but smaller in magnitude (Table II). It has recently been demonstrated[36] that high-spin Fe^{III} can be forced into the porphyrin plane by two weak-field ligands (e.g., H_2O or dimethylsulforide) with a further expansion of the porphyrin core. As shown by the entry in Table II for $(DMSO)_2Fe^{III}PP +$ (PP = protoporphyrin IX dimethyl ester), the spin-marker frequencies once again shift down.[33]

For the heme proteins studied so far, those with high-spin Fe^{II} (deoxyHb and Mb[5], reduced horseradish peroxidase (HRP),[37] reduced cytochrome c'[38]) show spin-marker frequencies that are essentially the same as those of the five-coordinate model, $(2\text{-MeIm})Fe^{II}MP$.[14] High-spin Fe^{III} heme proteins fall into two classes: (1) metHb and metMb, which show spin-marker frequencies close to six-coordinate $(DMSO)_2Fe^{III}PP +$;[33] (2) HRP[37] and cytochrome c',[38] which show spin-marker frequencies close to five-coordinate Fe^{III} porphyrins.[14] X-ray crystallography gives iron out-of-plane displacements of 0.40 Å for met Mb[39] and 0.23 and 0.07 Å for the α and β chains of metHb.[40] This variability is not reflected in the spin-marker frequencies. However, the iron atom is six-coordinate in both metMb and metHb, a water molecule providing the second axial ligand. Perhaps the non-bonded interactions between the axial ligands and the pyrrole nitrogen atoms force the porphyrin expansion which appears to be characteristic of six-coordinate Fe^{III} hemes. The higher spin-marker frequencies observed for HRP and cytochrome c', close to those of five-coordinate Fe^{III} hemes, suggest that a sixth ligand is not bound in these proteins.

REFERENCES

1. M. Gouterman, J. Chem. Phys. <u>30</u>, 1139 (1959); J. Mol. Spectrosc. <u>6</u>, 138 (1961).
2. T.G. Spiro, Biochim. Biophys. Acta <u>416</u>, 169 (1975).
3. T.C. Strekas, A. Packer and T.G. Spiro, J. Raman Spec. <u>1</u>, 197 (1973).
4. T.G. Spiro and T.C. Strekas, Proc. Natl. Acad. Sci. <u>69</u>, 2622 (1972).
5. T.G. Spiro and T.C. Strekas, J. Amer. Chem. Soc. <u>96</u>, 338 (1974).
6. S. Sunder, R. Mendelsohn and H.J. Bernstein, J. Chem. Phys. <u>63</u>, 573 (1975).
7. A. Warshel, Chem. Phys. Lett. <u>43</u>, 273 (1976).
8. J.M. Burke, J. Kincaid and T.G. Spiro, J. Amer. Chem. Soc., in press.
9. H. Brunner, Naturwiss <u>61</u>, 129 (1974).
10. G. Chottard and D. Mansey, Biochim. Biophys. Res. Commun. <u>77</u>, 1333 (1977).

11. J. Kincaid and K. Nakamoto, Spectrosc. Lett. 9, 19 (1976).

12. S.A. Asher, L.E. Vickery, T.M. Schuster and K. Sauer, Biochemistry, 16, 5849 (1977).

13. W.A. Eaton and R.M. Hochstrasser, J. Chem. Phys. 49, 985 (1968).

14. J.M. Burke and T.G. Spiro, J. Amer. Chem. Soc. 98, 5482 (1976).

15. P. Wright, J.M. Burke and T.G. Spiro (to be published).

16. S. Asher and K. Sauer, J. Chem. Phys. 64, 4115 (1975).

17. J.M. Burke, P. Daly, J. Stong and T.G. Spiro (to be published).

18. P.M. Champion and I.C. Gunsalus, J. Amer. Chem. Soc. 99, 2000 (1977).

19. Y. Ozaki. T. Kitagawa, Y. Kyogoku, H. Shimada, T. Iizuka and Y. Ishimura, J. Biochem.

20. W.H. Woodruff, D.H. Adams, T.G. Spiro and T. Yonetani, J. Amer. Chem. Soc. 97, 1695 (1975).

21. T. Kitagawa, M. Abe, Y. Kyogoku, H. Ogoshi, H. Sugimoto and Z. Yoshida, Chem. Phys. Lett. 48, 55 (1977).

22. G. Rakhit, T.G. Spiro and M. Uyeda, Biochim. Biophys. Res. Commun. 71, 803 (1976).

23. T. Yamamoto, G. Palmer, D. Gill, I.T. Salmeen and L. Rimai, J. Biol. Chem. 248, 5211 (1973).

24. J. Weiss, Nature 203, 83 (1964).

25. L. Pauling, Nature 203, 182 (1964).

26. A. Szabo and L. Barron, J. Amer. Chem. Soc. 97, 660 (1975).

27. C.H. Barlow, J.C. Maxwell, W.J. Wallace and W.S. Caughey, Biochim. Biophys. Res. Commun. 58, 166 (1973).

28. J.P. Collman, J.I. Brauman, T.R. Halbert and K.S. Suslick, Proc. Natl. Acad. Sci. 73, 3333 (1976).

29. J.M. Burke, J. Kincaid, S. Peters, R. Gagne, J.P. Collman and T.G. Spiro, J. Amer. Chem. Soc., in press.

30. J.P. Collman, R.R. Gagne, C.A. Reed, W.T. Robinson and G.A. Rodley, Proc. Natl. Acad. Sci. 71, 1326 (1974).

31. B.H. Huynh, D.A. Case, M. Karplus, J. Amer. Chem. Soc. 99, 6103 (1977).

32. P. Stein, J.M. Burke and T.G. Spiro, J. Amer. Chem. Soc. 97, 2304 (1975).

33. T.G. Spiro and J. Stong, submitted for publication.

34. L.D. Spaulding, C.C. Chang, N.T. Yu and R.H. Felton, J. Amer. Chem. Soc. 97, 2517 (1975).

35. J.L. Hoard in "Porphyrins and Metalloporphyrins," K.M. Smith, Ed., Elsevier, New York, N.Y., 1975, pp. 317-376.

36. a) M.E. Kastner, W.R. Scheidt, T. Mashiko and C.A. Reed, J. Amer. Chem. Soc., 100, 666 (1978).
    b) T. Mashiko, M.E. Kastner, K. Spartalian, W.R. Scheidt, and C.A. Reed, J. Amer. Chem. Soc., in press.

37. G. Rakhit and T.G. Spiro, Biochem. 13, 5317 (1974).

38. T.C. Strekas and T.G. Spiro, Biochim. Biophys. Acta 351,
     237 (1974).
39. T. Takano, J. Mol. Biol. 110, 533 (1977).
40. R.C. Ladner, E.J. Herdner and M.F. Perutz, J. Mol. Biol.
     114, 385 (1977).
41. L. Rimai, I.T. Salmeen and D.H. Petering, Biochem. 14, 578
     (1975).

HYDROGEN BONDED SYSTEMS

VIBRATIONAL SPECTROSCOPY OF HYDROGEN BONDED SYSTEMS

A. Novak

Laboratoire de Spectrochimie Infrarouge et Raman - CNRS
2, rue Henri Dunant - 94320 - Thiais - France

ABSTRACT. The effect of hydrogen bonding on infrared and Raman
spectra of donor (AH) and acceptor (B) molecule and in particular
on AH stretching and bending vibrations, is discussed. Different
correlations between the AH stretching frequency and other spectroscopic and non-spectroscopic parameters is given. The intermolecular hydrogen bond vibrations yielding direct information about the
force and dynamics of hydrogen bonds are analysed. A few applications of vibrational spectroscopy to structure and equilibrium
constant determination of hydrogen bonded entities in gases and
liquids are given. The problem of assignment of internal and lattice modes of hydrogen bonded crystal based on low temprature
single crystal study with polarised light and the isotopic mixed
crystal technique is discussed. Factors determining the AH stretching band breadth and structure of weak and of strong hydrogen
bonds are analysed.

1. INTRODUCTION

The hydrogen bond A-H...B is a molecular interaction of the donor-
acceptor type where a Brönsted acid AH is bonded to a base B, i.e.
one covalently bound hydrogen atom forms a second bond to another
atom. A hydrogen bond is distinctly directional and specific and
stronger than the Van der Waals type interactions. However, it is
much weaker than the usual chemical bonds and can be easily broken
by a temperature increase or by dilution even with an inert solvent. The hydrogen bond (H.B.) plays thus an important role in
biological systems including water ; the conformation of biopolymers containing proton donors and acceptors, for instance, is
frequently determined by making and breaking of H.B. In fact, in

279

Theo M. Theophanides (ed.), Infrared and Raman Spectroscopy of Biological Molecules, 279–303.
All Rights Reserved. Copyright © 1979 *by D. Reidel Publishing Company, Dordrecht, Holland.*

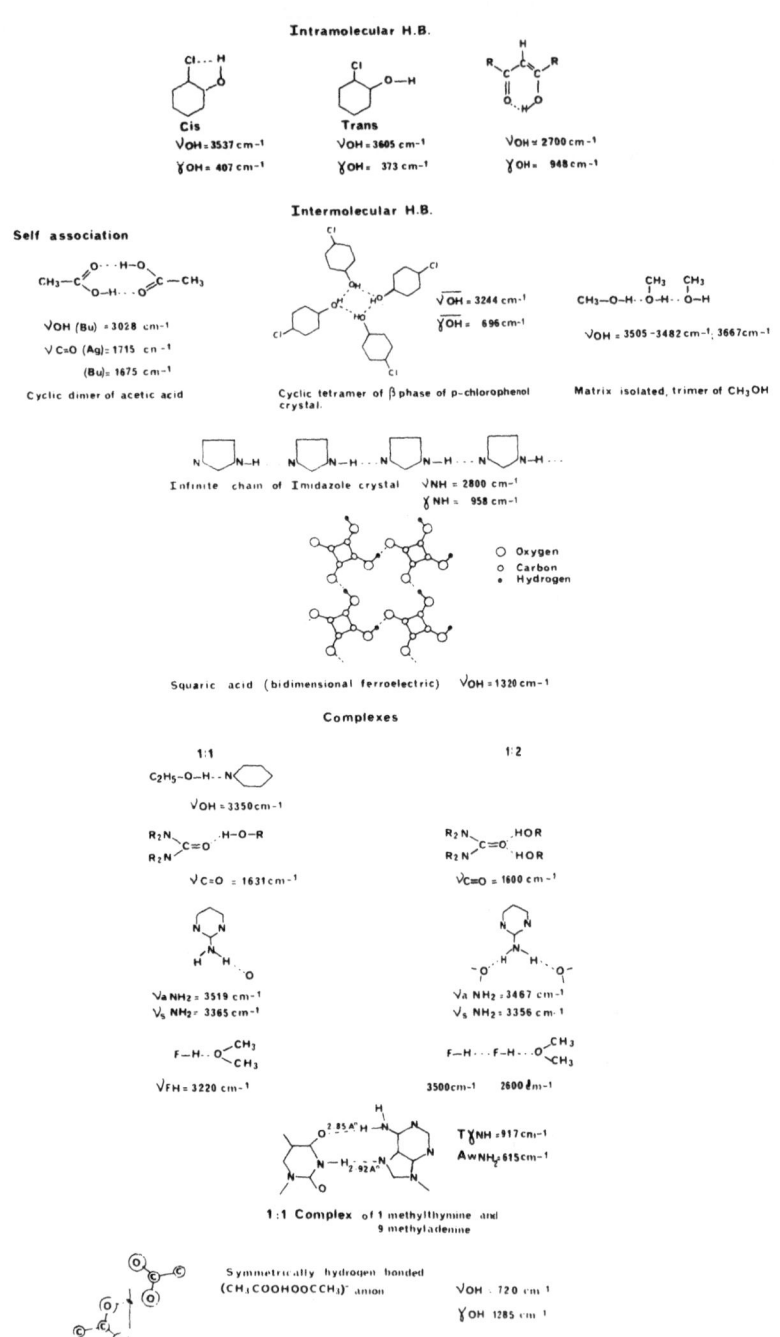

Fig. 1. Structures of some hydrogen bond systems

the chemistry of living systems the H.B. is as important as the
carbon-carbon bond (1-5).

1.1 Classification of hydrogen bonds

H.B. can be classified according to 1. the nature of the atoms A
and B, 2. the structure of the H.B. entities, 3. the strength and
4. the type of the interaction.

A H.B. can occur between atoms with electronegativity greater
than that of hydrogen. Oxygen, nitrogen, and halogen (X) atoms are
the most frequent partners and may give rise to strong H.B., in
particular O-H...O, O-H...N, N-H...N and X-H...X. Carbon, phosphorus, sulfur, and selenium atoms, on the other hand, are usually
involved in weak, if any, H.B. (6). However, some relatively strong
C-H...O (7) and N-H...S (8) H.B. have also been observed.

A H.B. is called intramolecular when it is formed between
groups within a single molecule (fig. 1) and intermolecular when
it involves association of two or more molecules. The intermolecular H.B. between the molecules of the same substance leads to
self-association. The resulting species may form dimer or polymer
rings the carboxylic cyclic dimer (RCOOH)$_2$ being a well known
example, or chains, open dimers or oligomers in gases or solutions
and infinite chains (spirals) in crystalline solids. If there are
more than one proton donor group in the molecule three dimensional
networks can be created such as found in the structure of ices (3)
or glycerol (9). Molecules of different substances are linked by
H.B. into AH...B complexes which are more or less complicated depending on the number of proton donors and acceptors. The 1:1 and
1:2 complexes may thus be distinguished for 2-aminopyrimidine (10)
(Fig. 1), and 1:1, 1:2 and 1:3 complexes for NH$_3$ (11).

In order to classify H.B. according to the strength of the
interaction we must first recall that the basic group B in the
A-H...B system represents a supplementary attraction potential for
the proton and can create a second potential well as illustrated
diagrammatically in Fig. 2. The potential energy function of the
free AH group is thus modified by the potential B ; it becomes
broader and the vibrational levels become closer as reflected by
the shift of the AH stretching band toward lower frequencies shown
by infrared and Raman spectroscopy. At the same time, the proton
shifts toward B and the equilibrium r A-H distance increases while
the intermolecular R A...B distance discreases to a value less
than the sum of the van der Waals radii of the A and B atoms as
shown by neutron and X-ray diffraction methods (2, 12). We have
thus in addition to the enthalpy of the H.B. formation ΔH, two
main criteria for estimating the H.B. strength : 1. AH stretching
frequency ν or its relative shift $(\nu_o-\nu)/\nu_o$ where ν_o is the
frequency of the "free" AH group and, 2. r A-H, R A...B and R H...B

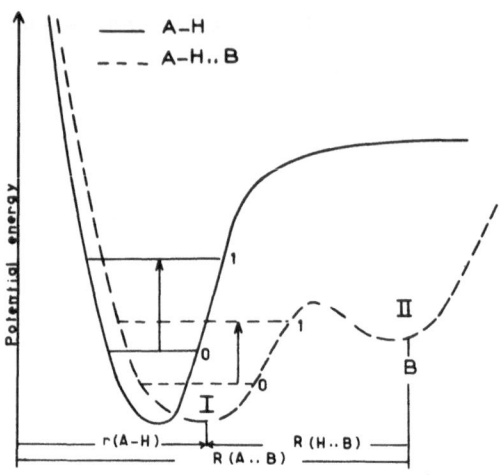

Fig. 2. Qualitative potential curves for the free (——) and
hydrogen bonded AH group.

distances. Table I gives an approximate classification of the
O-H...O H.B., the system most studied. Here we can distinguish
weak (long), strong (short), and intermediate H.B. (13).

The type of H.B. may also be classified according to the
shape of the potential curve of the proton of an A-H...B system
or the position of the proton (2, 13, 14). In fact, an acid-base
reaction can be represented by a series of equilibria :

$$AH + B \rightleftarrows A-H...B \rightleftarrows (A-H-B) \rightleftarrows A^-...B^+ \rightleftarrows A^- + HB^+.$$

Table I. Approximative classification of O-H...O hydrogen bonds

H.B.	ν OH cm^{-1}	$\Delta\nu/\nu_0$ (%)	RO...O (A°)	ΔH (kcal/mole)	Example
weak	> 3200	< 12	> 2.70	< 5	H_2O, R-OH
intermediate	3100-2800	12-22	2.70-2.60	6-8	RCOOH
strong	2700- 600	25-83	2.60-2.40	> 8	$MH(RCOO)_2$

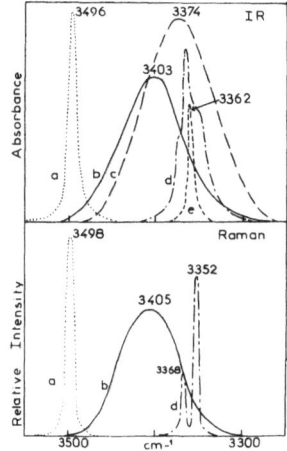

Fig. 3. NH stretching region of pyrrole ; (a) diluted solution in
CCl$_4$; (b) pure liquid ; (c) vitreous state at 90 K ;
(d) crystal at 90 K ; (e) isotopically diluted crystal
(NH/ND ≃ 5 %) at 90 K.

The formation of a H.B. is followed by a transition state of
the proton transfer which can give a complex between the BH$^+$ acid
and its conjugated base A$^-$ and finally "free" ions. The A-H...B
system containing a weak or moderately strong H.B. has a potential
curve similar to that of the fig. 2. For the A$^-$...H$^+$B system the
potential curve is similar but with the well II deeper than well I.
Finally, the potential curve may be symmetric. In all three cases
the potential barrier separating the two minima is high and the
proton is always closer to one atom than to the other. However,
in the (A-H-B) system and more particularly in those represented
by (A-H-A)$^-$ and (B-H-B)$^+$ containing strong H.B., other types of
potentials may exist. This is the case of "hesitating proton"
characterised by a small or zero potential barrier. We can distinguish 1. asymmetric double minimum, 2. symmetric double minimum,
and 3. symmetric single minimum with r A-H = 1/2 R A...A.

2. EFFECT OF HYDROGEN BONDING ON INFRARED AND RAMAN SPECTRA

The formation of an intermolecular H.B. complex will modify to a
varying degree the intramolecular vibrations of donor and of
acceptor molecule and will give rise to new -intermolecular degrees
of freedom- hydrogen bond vibrations. The most H.B. sensitive
vibrations of donor molecule are obviously the AH stretching and
bending modes. However, in polyatomic systems other vibrations

will also be affected. This happens both through kinematic effects
and by changes of force constants due to the electronic effects of
H.B. (5). A complete account of the effects of H.B. on the whole
molecule can be obtained in principle by a full normal coordinate
analysis but only a few polyatomic systems have been studied in
detail so far (5, 15).

2.1. AH stretching vibration

With increasing strength of H.B. the absorption bands due to ν AH
vibration are progressively 1. lowered in frequency, 2. enhanced
in integrated intensity, and 3. broadened in contour (16). This
simple rule describes trends which have been observed in an extremely large number of cases and it can be said that the first two
effects are principally due to changes in electron distribution in
the AH region of the molecule brought about by H.B. formation (17,
18). The origin of broadening and generally of band shape is more
complicated and will be discussed later. The AH stretching band
yields the most information and has been more widely investigated
in H.B. studies than any other experimental effect (3).

The spectroscopic manifestation of a weak H.B. is illustrated
in fig. 3. The so called "free" NH stretching band of a diluted
solution of pyrrole in CCl_4 is observed near 3497 cm^{-1} in infrared
as well as in Raman spectrum with the half-width $\nu_{1/2} = 12$ cm^{-1}.
The "bonded" band of pure liquid containing N-H...π H.B., on the
other hand, shifts to 3404 cm^{-1} and broadens to $\nu_{1/2} = 84$ cm^{-1}.
The integrated infrared intensity increases from 1.1 to 2.2 10^{-4}
liter mole^{-1} cm^{-2} under the same conditions. The frequency shift
is thus about 3 % while the intensity change is of the order of
200 % and that of band-width nearly 700 % (19, 20).

The infrared spectra of medium strong and strong O-H...O H.B.
are shown in fig. 4. The "free" OH stretching frequency of the acetic
acid molecule in the gaseous state at 430 K is observed at 3583 cm^{-1}
It decreases to 2875 cm^{-1} in the acetic acid crystal near 90 K, to
1400 cm^{-1} in potassium hydrogen diacetate KH(CH$_3$COO)$_2$, and finally
to 720 cm^{-1} in sodium hydrogen diacetate NaH(CH$_3$COO)$_2$ containing
a symmetric H.B. The corresponding RO...O distances are 2.63, 2.49
and 2.44 Å, respectively. The changes in the intensity, breadth,
and structure of the ν OH bands are no less spectacular (12).

There are many studies of H.B. by infrared but only few by
Raman techniques. Yet we anticipate that Raman scattering will play
an important role in the investigation of H.B. Fig. 5 shows Raman
spectra of some crystal containing medium strong and strong H.B.
It can be concluded that the Raman ν AH bands are in may ways
similar to those observed in infrared as far as frequency and shape
is concerned when this is allowed by selection rules. The Raman
bands are usually less broad and the band structure is thus better

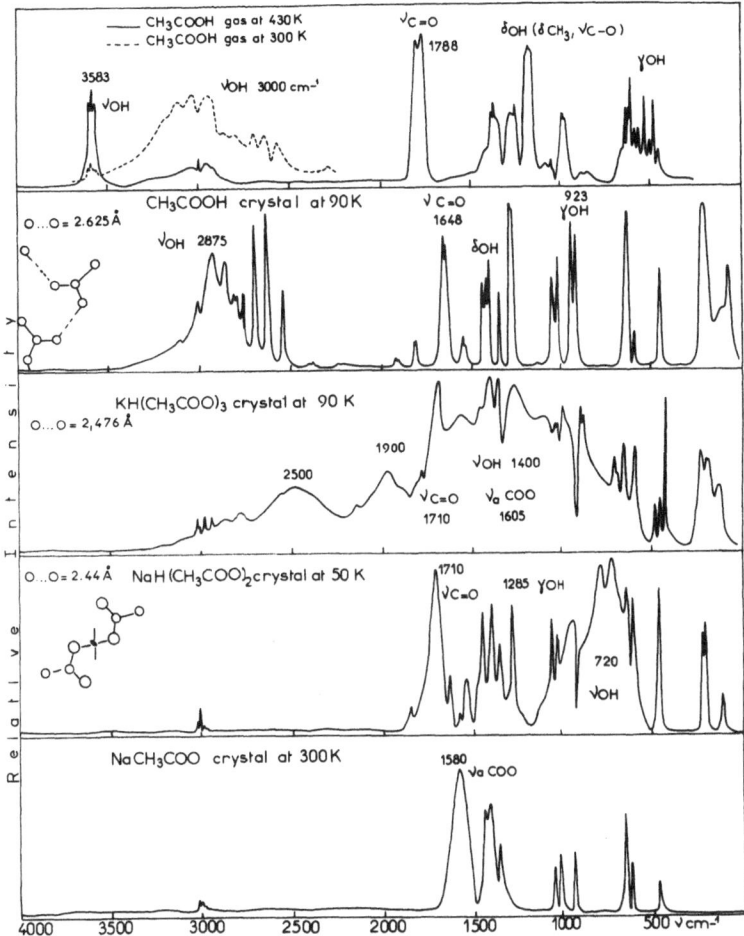

Fig. 4. Infrared spectra of acetic acid and of some of its acid
and neutral salts.

resolved. However, the most important difference concerns the intensity. The √ AH vibration gives usually rise to the strongest absorption band but is quite weak in Raman. This low peak intensity probably explains the scarcity of Raman data on H.B. systems. Nevertheless, Raman studies of single crystals at low temperature appear
promising and can give more information than infrared spectra
alone (9, 15, 21).

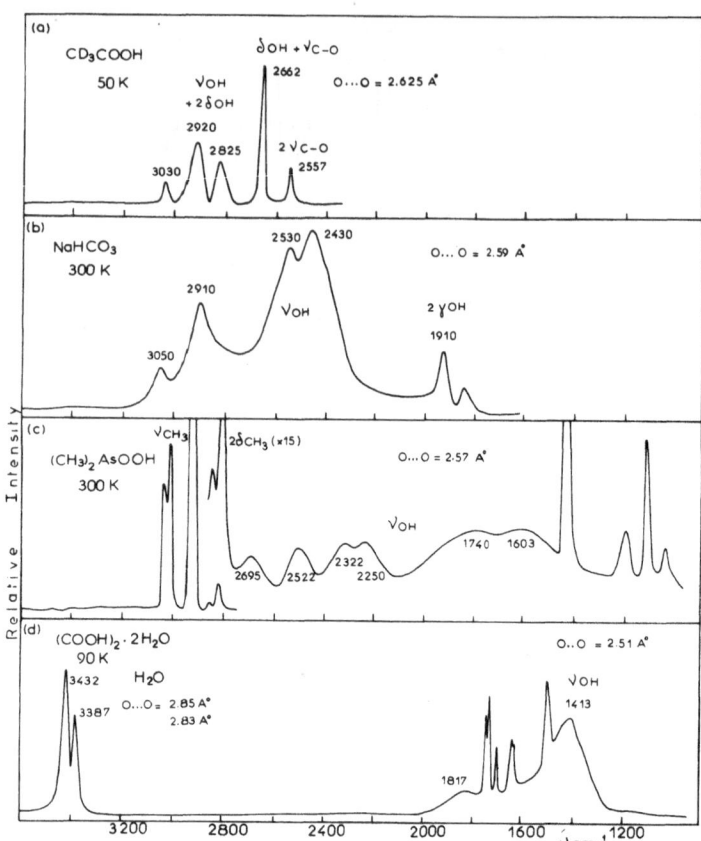

Fig. 5. Raman spectra of (a) CD₃COOH crystal at 50 K ; (b) sodium
hydrogen carbonate at 300 K ; (c) dimethylarsinic acid
(CH₃)₂AsOOH at 300 K ; (d) oxalic acid dihydrate at 90 K.

<u>Correlations involving AH stretching frequency.</u>   The ν AH
frequency or its relative shift $\Delta\nu/\nu_0$ has been widely used for
correlation purposes with other parameters such as enthalpy of H.B.
formation, acidity and basicity of donor and acceptor, ionisation
potential of the base, R A...B and r A-H distances, AH bending
frequency and H.B. stretching force constant (1, 3, 5, 12, 13).
However, accurate determination of the ν AH frequency is not always
an easy task, particularly in the case of strong H.B., and the results of infrared and Raman spectra combined with some special
techniques must be used (12). One of the most important correlations
is that between ν AH frequency and R A...B distance, which is
already implied in the fig. 2, and is shown for the O-H...O system

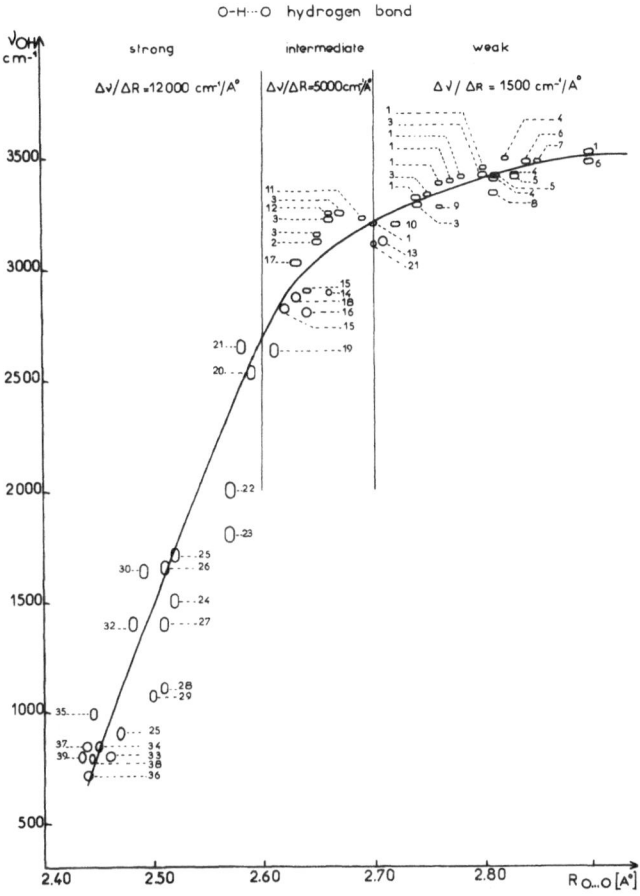

Fig. 6. Relation between OH stretching frequency and R (O...O)
         distance

in fig. 6. This relationship is useful for the estimation of intermolecular H.B. distances in the compounds which have not been or
cannot be studied by diffraction methods and has been used successfully. Another characteristic of this curve is that its steepness
$\Delta\nu/\Delta R$ increases sharply in going from weak to strong H.B. This is
significant for the understanding of the band width. $\Delta\nu/\Delta R$ increases
in the same direction as the intrinsic band width of the OH stretching band -from 10 to 10^2 and 10^3 cm^{-1} for weak, medium-strong,
and strong H.B., respectively- and appears to be the most important
parameter determining the ν AH band breadth (22). Other curves
concerning N-H...N and N-H...O H.B. have been published (1, 2, 12, 23).

Another important correlation exists between the isotopic frequency ratio ν AH/ν AD and the ν AH frequency and is related to the anharmonicity and isotope effect on H.B. (1, 2, 5, 12). This ratio is about 1.35 for the "free" AH group and decreases with decreasing ν AH frequency. According to Savelev and Sokolov (24) the isotopic frequency ratio is determined by the $\nu°$ AH/ν AH ratio of the stretching frequency of the free ($\nu°$ AH) and bonded AH group following the relation ν AH/ν AD = $2/\{[(2^{1/2} - 1)\,\nu°$ AH/ν AD] + 1$\}$. The main assumption in deriving the above expression involves strong anharmonic coupling between ν AH and ν A...B stretching vibrations. Low ν AH/ν AD ratios are usually observed for strong H.B. with double minimum exhibiting positive isotope effect, i.e. the substitution of the acid hydrogen with deuterium lengthens the A...B distance $R_{AD...B} > R_{AH...B}$ (12). ν AH/ν AD $\leq$ 1 have been observed for O-H...O H.B. shorter than about 2.55 Å which implies that ν OD frequency is similar or higher than ν OH frequency (12, 25). The symmetric single minimum H.B., on the other hand, shows negative or zero isotope effect, $R_{AD...B} \leq R_{AH...B}$ (2) and the corresponding spectroscopic manifestation is an increase of the ν AH/ν AD ratio which can be higher than that of the free AH group. In the case of $NaH(CH_3COO)_2$ (fig. 4) the ν OH/ν OD ratio is 1.44 (26) while it is 0.5 for $LiHC_2O_4$ (25). The ν AH/ν AD ratios can thus be used in order to distinguish the type of hydrogen bond.

AH stretching band intensity. The investigated IR intensity of ν AH band is at least as important a characteristic of H.B. as the frequency shift. It increases with the strength of the H.B. and the ratios of the integrated intensities of the bonded to the monomer ν AH bands vary from 2 to 40 (3). In the case of some weak H.B. the intensity is much more sensitive than the frequency. The ν CD frequency of chloroform, for instance, decreases only 4 cm^{-1} (0.1 %) when going from a diluted CCl_4 solution to that of dioxane while the integrated IR intensity increases from 190 to 2600 ł mole^{-1} cm^2, i.e. almost 14 times (27).

The intensity enhancement of ν AH absorption band is very important for the theory of H.B.. It does not depend only on the change of polarity of the A-H bond but on the charge distribution in the whole H. bonded complex with normal coordinate (5). Pimentel first suggested that polarisation of the base (acceptor) molecule contributes to the intensity changes in H.B. (1). Boobyer and Orville-Thomas (28) have extended this idea in their pulsed charge cloud model. In the case of carboxylic acid dimers, for instance, the polar O-H bond (μ_o) will induce a dipole (μ_i) in the highly polarisable lone pair electrons of the carbonyl group. As the O-H bond vibrates during the bond stretching vibration, so also will the induced moment, thus contributing an appreciable component to overall change of dipole moment during the vibration (3, 28).

The effect of H.B. on the intensity of Raman bands has been very little studied. Perchard and Perchard have shown that the OH

stretching band of alcohols (29) and the NH stretching band of
amines (30) also increase in intensity upon association. They
interpret the intensity increase using the above pulsed charge
cloud model. In fact, it can be assumed that during the OH (NH)
stretching vibration the fluctuation of polarisability involves
not only the O-H (N-H) bond but also the lone pair of the oxygen
(nitrogen) atom. Nevertheless, the ν AH Raman intensity change
appears to be less spectacular than the IR one.

2.2. AH bending vibration

Model calculations for the triatomic linear A-H...B system show
that the AH bending frequency should increase with the decreasing
AH stretching frequency on H.B. (13) in agreement with the experiment. There is only one doubly degenerate bending mode in the above
system and can be compared to proton donors of high symmetry such
as $CHCl_3$ or X-H which have been used for correlations between
frequencies and intensities of the AH stretching and bending modes.
However, in most cases AH molecules are less symmetric and we have
to distinguish two AH bending modes of rather different frequency :
in-plane, δ AH, and out-of-plane, γ AH, bending (deformation)
vibration. H.B. increases both frequencies but their diagnostic
value is different (12).

The δ AH mode is, unlike ν AH, more or less coupled with one
or several other in-plane vibrations. The type and degree of coupling depends on the molecule and the H.B. strength. The δ OH mode
of CH_3COOH monomer strongly mixes with the CH_3 symmetrical bending
and ν C-O vibrations (31) while in the crystal the coupling is
limited mainly to the C-O stretch (32) ; in the sodium hydrogen
diacetate crystal containing much stronger H.B. the δ OH frequency
increases considerably, decouples from the ν C-O mode and mixes
with C=O stretch (33). Another example is that of liquid ethanol
-where δ OH is coupled with CH_2 scissoring and CH_2 twisting modes
(34). Finally, the δ NH vibration of imidazole (35) and purine (36)
is mixed with three or four heterocyclic ring vibrations. Therefore,
vibrational coupling makes the δ AH frequency less specific, less
sensitive, and thus less suitable for correlation studies.

The γ AH vibrations, on the other hand, appears frequently
pure and very sensitive towards molecular interactions. The corresponding infrared band (γ AH Raman bands are usually quite weak) can
thus be used as a criterion for H.B. strength. It seems useful to
distinguish two types of vibration : 1. the γ AH mode of H.B.
systems which transforms itself into a low-frequency torsional
motions for the free monomer. This is the case of many alcohols
and phenols, in particular, and is characterised by a big relative
γ AH shift. The torsional frequency of gaseous methanol (τ OH)
observed at 197 cm^{-1} (37) increases to 730 cm^{-1} when it becomes a
γ OH vibration of H.B. solid CH_3OH (38) and the $(\gamma_o-\gamma)/\gamma_o$ shift

amounts to about 270 %. 2. the second γ AH type does not change
its character when going from free to bonded species -such as the
γ OH vibration of carboxylic acids and γ NH vibration of heterocyclic nitrogen molecules (12).

A few correlations have been established between ν AH and
γ AH frequencies for the first type -mostly for phenols (39, 40)-
as well as for the second type involving O-H...O and N-H...X
systems (12). For carboxylic acids and their acid salts, for
instance, there is a simple relationship between the ν OH and γ OH
frequencies : the slope of the line $\Delta\nu$ OH/$\Delta\gamma$ OH = 5.8, i.e. the
variation of γ OH frequency is almost six times less than that of
ν OH frequency. The $\Delta\gamma/\gamma_o$ shift, on the other hand, is always
greater than $\Delta\nu/\nu_o$ shift and can reach a 100 % for the shortest
H.B. of the second type. The analytical importance of γ OH bands
appears to be considerable, especially for strong H.B. where the
ν OH bands are very broad and ill-defined while γ OH are narrow
and show a strong high frequency shift on temperature lowering.

2.3. Other vibrations

Influence of H.B. on other vibrations can be analysed in principle
when a complete normal coordinate calculation is performed (5, 41)
such as in the phenol-pyrimidine complex (42) or self associated
1,2,4-triazole (15). However, there are few complete calculations
and the distinction between the effect of mechanical coupling and
electronic charge distribution is usually made by using suitable
isotopic species. In the case of CH_3COOH for instance, the δ_s CH_3
frequency is shifted only because of different coupling for free
and bonded molecules while the skeletal frequencies are changed by
both mechanical and electronic effects (32). In the series
CH_3COOH monomer, dimer, polymer the C-O and C-C stretching frequencies increase while the ν C=O frequency decreases, the frequencies
of the polymer being closer to those of acetate ion than to the
monomer frequencies. This is correlated with the corresponding C-O
and C-C distances and implies that the ionic contribution CH_3COO^- H^+
to the structure of associated acetic acid becomes important (32).

Examples of heterocyclic nitrogen bases such as pyridine or
pyrimidine shows that the effect of H.B. on some planar ring
vibrations in much the same as that of charge transfer complexes.
The ν_{8a} vibration observed at 1583 cm^{-1} for pure pyridine increases
to 1594, 1603 and 1630 cm^{-1} for the O-D...N, Ni-N, and Cl^-...NH^+
systems, respectively the $\Delta\nu$ increase being related to the charge
density modification on N and C atoms (43). The ν_{19b} frequency
increases in the same order from 1439 cm^{-1} for pure substance to
1444, 1447, and 1533 cm^{-1}. However, the largest shift observed for
the proton transfer H.B. is due as much to the electronic effect
as to mechanical coupling of the ν_{19b} mode with the new (δ NH)
degree of freedom introduced by protonation (43).

Table II. H.B. frequencies of formic acid cyclic dimer

Mode	Sym.	Calc. (cm^{-1})	Obs. (cm^{-1}) IR	Raman
γ (OH...O)	B_g	260		245
ν (OH...O)	B_u	249	248	
ν (OH...O)	A_g	224		
γ (OH...O)	A_u	167	173	
δ (OH...O)	A_g	91		
γ (twist)	A_u	68	68	

2.4. Hydrogen bond vibrations

When an A-H...B complex or an (AH)$_2$ cyclic dimer is formed with
polyatomic donor and acceptor molecules, six rotational and translational degrees of freedom are transformed into vibrational
degrees of freedom usually designated as hydrogen bond vibrations.
Their frequencies are generally low, lower than 300 cm^{-1}, since
the hydrogen bond is a weak force and masses and moments of inertia
of participating molecules are large. These vibrations, in particular the stretching H.B. vibrations, ν (AH...B), are of great
interest since they can give direct information about the force
and dynamics of H.B. However, there are few thorough studies of
H.B. vibrations because of experimental difficulties -the H.B.
frequencies being expected in the far-infrared region in absorption
and near the exciting line in Raman scattering- and because of the
assignment problem : these modes may mix with some intramolecular
modes and, in the case of crystals, with other external vibrations.
The main criterion for assigning the H.B. bands in gas phase or
solution is that their intensity is expected to decrease with the
decreasing intensity of the "associated" ν AH band by dilution or
temperature increase. Table II gives the six H.B. frequencies of
the cyclic dimer (HCOOH)$_2$ as an example. The vibrations are assigned to their symmetry species of C_{2h} point group symmetry and
described in terms of stretching (ν), in-plane (δ) and out-of-plane
(γ) bending H.B. vibrations. Four of the six frequencies have been
observed experimentally (44).

The criteria for assigning H.B. vibrations in solid state are
frequently based on single crystal study with polarised light and
on temperature and isotopic shifts (cf. 4.1 and 4.3). The stretching

H.B. frequencies in general are observed mostly in the 200 to 100
cm^{-1} range (3, 4) and there is no simple relationship between
ν (AH...B) and ν AH frequencies. However, such relationship may
exist between $F_{H...A}$ and F_{A-H} force constants (12). The H.B. force
constants can vary from 0.14 to 0.25 md/Å for weak H.B. such as
C-H...N (15) or O-H...O of ice (45) and increase to typical
0.35 md/Å value for medium-strong H.B. of acetic acid crystal (44)
and purine (46). Strong H.B. yield much higher values estimated to
0.5 md/Å for $(HCO_3^-)_2$ dimer of $KHCO_3$ and 1.8 md/Å for the symmetrical H.B. of $NaH(CH_3COO)_2$ (12).

3. HYDROGEN BONDED GASES AND LIQUIDS

Infrared and Raman spectroscopy is particularly useful in studies
of structure or conformation of H.B. entities in gases and liquids
where diffraction studies are few or non existent. Another important application concerns the determination of thermodynamic
quantities from equilibrium constants and of acid-base properties.

3.1. Intramolecular H.B.

Intramolecular H.B. form internal rings which have usually five,
six or seven atoms in them. The spatial requirements produce an
isomeric effect, the ortho or cis isomers correspond to intramolecular H.B. while the para, meta or trans isomers do not (fig. 1).
Intramolecular H.B. may be weak like in o-chlorophenol where the
"associated" band is very close to the "free" one, $\Delta\nu = 68$ cm^{-1},
or strong like in the enol form of β-diketones, ν OH = 2700 cm^{-1}
and in nickeldimethylglyoxime, ν OH $\simeq$ 1000 cm^{-1} (5). The most
direct and readily obtainable test to distinguish intramolecular
H.B. from intermolecular ones is the observation of changes in
the infrared and Raman spectra at low concentration and pressure.
Intermolecular H.B. and its spectral characteristics, associated
ν OH band in particular, disappear at low concentration in an
inert solvent, whereas intramolecular H.B. do not (1).

There are not many thermodynamic data on intramolecular H.B.
The ΔH enthalpy values for ortho-halophenols (ΔH between cis and
trans isomers) have been obtained from OH torsional frequencies
(fig. 1) for both isomers by Carlson et al. (47). The relative
order of H.B. strength is F = Cl (3.41) > Br (3.13) > I (2.75
kcal/mole) in gas phase. In most other thermodynamic studies comparison of ν OH frequency shifts and proton magnetic resonance
have been used. Such comparisons can be misleading since $\Delta\nu$ OH
and PMR chemical shifts are affected by geometric effects as well
as by the acidity of the AH and the basicity of B (4).

3.2. Intermolecular H.B. Thermodynamics (3)

An important aspect of understanding the interaction between a proton donor A-H and proton acceptor B is based on the experimental determination of the thermodynamic parameters, the changes in free ΔG, enthalpy ΔH, and entropy ΔS. The free energy is derived from the equilibrium constant governing the interaction, while enthalpy and entropy are obtained from the variation of the equilibrium constant with temperature. For the simplest case of a 1:1 complex, the equilibrium is represented as follows : AH + B $\rightleftharpoons$ A-H...B. The association equilibrium constant is K = (AH...B)/(AH)(B) and can thus be calculated from the equilibrium concentrations of (A-H...B), (AH), and (B). Usually we know the total concentration of the donor $c°_{AH}$ = (AH) + (AH...B) and the total concentration of the acceptor $c°_B$ = (B) + (AH...B). Therefore, if we can measure any one of (AH...B), (B) or (AH) we can compute the others from the known values of $c°_{AH}$ and $c°_B$ and hence compute K. Infrared spectroscopy allows one frequently to determine directly (AH) or (AH...B) or both using the Beer-Lambert law. This law states that for any solution absorbing radiation at a given frequency ν the absorbance $A\nu$ is given by A_ν = log (I/I_0) = a_ν cl where a is the absorptivity (extinction coefficient) of the given group, c is the concentration in moles per liter and l is the thickness of the absorbing solution in centimeters. I_0 and I are the intensities of incident and emerging radiation, respectively. The free energy change ΔG is given by :

$$\Delta G = \Delta H - T\Delta S = - RT\ln K \text{ and } \ln K = - \Delta H/RT + \Delta S/R$$

lnK is thus a linear function of T^{-1} if ΔH and ΔS are constants independent of temperature. The ΔH and ΔS parameters can thus be calculated from the plot of log K versus l/T. The great majority of intermolecular H.B. in gases and solutions are weak or medium strong and the corresponding thermodynamic parameters are as follows : ΔH = - 1 to - 10 kcal mole^{-1}, ΔS = - 3 to - 25 e.u., and ΔG = 0 to - 4 kcal mole^{-1}. The absolute values of ΔG are relatively small and indicate that these H.B. are fragile. Free molecules in equilibrium with the associated species of different type can thus be observed simultaneously in a solution.

3.3. Complexes

Intermolecular complexes may be of different types. In a ternary solution of ethanol and pyridine in CCl_4 only the 1:1 complex containing O-H...B H.B. is observed. There is a "free" ν OH band at 3630 cm^{-1} and the "associated" one at 3350 cm^{-1} (3). A solution of tetramethylurea in butanol on the other hand, contains two types of complexes shown in fig. 1. They give rise to different C=O stretching frequencies near 1660, 1631, and 1600 cm^{-1} corresponding to "free" molecules, 1:1 and 1:2 complexes, respectively (48).

Another type is illustrated by 2-aminopyrimidine (fig. 1) whose NH_2 group can be engaged to one or two acceptor molecules. The antisymmetric and symmetric NH_2 stretching frequencies observed at 3543 cm^{-1} and 3436 cm^{-1} for free molecule shift to 3519 and 3365 cm^{-1} for 1:1 complex and to 3467 and 3356 cm^{-1} for 1:2 complex with dioxane (10). Finally, binary solution containing small molar fraction of HF in $(CH_3)_2O$ shows ν FH infrared band at 3220 cm^{-1} assigned to 1:1 F-H...O complex. New ν FH bands near 2600 and 3500 cm^{-1} appear with the increasing molar fraction of FH and are ascribed to a 1:2 complex (fig. 1). F-H...O H.B. (2600 cm^{-1}) is thus much stronger than F-H...F H.B. (3500 cm^{-1}) of this complex (49).

3.4. Self-association

Molecules containing both donor and acceptor groups can self-associate through H.B. in one or more different ways into a number of different species which may coexist with one another in a series of complicated equilibria. The simplest case is a molecule such as imidazole (fig. 1) which can form only linear structures such as open dimer or trimer. The main difficulty inherent in the calculation of the corresponding association constants from IR measurements utilising the monomer ν NH absorption band at 3485 cm^{-1} is the possibility that the terminal NH group may absorb at the monomer frequency (3).

The most complicated cases of selfassociation are the ones where the cyclic and linear H.B. species coexist. The best known examples are water, alcohols, and phenols. Analysis of such systems requires the determination of many equilibrium constants. The usual approach in studies of selfassociation using IR and PMR techniques has been to develop a model which assumes that only a few H.B. species are present and then to carry out computation from the experimental data to ascertain whether consistent K values are obtained. In a typical example, ethanol was assumed to exist in CCl_4 solution as monomer, open dimer, and trimer, and cyclic tetramer with $K_{dimer} = 0.95$ (1 mole^{-1}), $K_{trimer} = 95$ (1 mole^{-2}) and $K_{tetramer} = 650$ (1 mole^{-3}) (3).

A very useful technique for studying types of association is so called matrix isolation which involves a rapid condensation of a mixture of the absorbing species and a diluent gas at liquid helium or liquid hydrogen temperature (50). The main advantage of this technique is that the bands are much narrower than those of gases and solutions at room temperature and it is possible to resolve features separated by as little as 1 cm^{-1}. In the case of methanol, for instance, the well-known broad features of the ambient-temperature ν OH absorption has been resolved into about twenty different frequencies between 3679 and 3200 cm^{-1} characteristic of monomer, open chain, dimer, trimer (fig. 1) and tetramer, cyclic tetramer and higher polymers (50).

Some carboxylic acids such as HCOOH and CH_3COOH may form cyclic dimers as well as linear structures depending on physical state, temperature, and pressure. The spectroscopic evidence for cyclic structure is based on the non-coincidence of IR and Raman frequencies required by the mutual exclusion rule of the centrosymmetric dimer ring (fig. 1). The largest difference between symmetric (A_g) and antisymmetric (B_u) frequency is observed for the C=O stretching vibration. In the case of acetic acid in CCl_4 solution the former gives rise to a Raman band at 1675 cm^{-1} and the latter to an infrared band at 1715 cm^{-1} (52). Pure liquid CH_3COOH contains some open dimers and multimers which coexist with the dominant cyclic dimers. However, when the temperature increases to 200°C, liquid CH_3COOH consists mainly of open dimers characterised by two ν C=O bands near 1755 and 1720 cm^{-1} observed both in IR and Raman. The former is assigned to the terminal "free" C=O group while the latter corresponds to the bonded one (52).

3.5. Correlations between spectroscopic and thermodynamic parameters

A linear relationship between $\Delta\nu$ shift of the AH stretching frequency and the enthalpy of H.B. formation was first suggested by Badger and Bauer (53). It appears reasonable to expect a greater shift for stronger H.B., that is larger ΔH (table I), provided special factors such as solvation and steric repulsions are not involved. Sokolov has shown theoretically that for relatively weak H.B. a relationship $\Delta\nu/\nu_o \simeq - \Delta H.K$ (K = 1.16 . 10^{-2} for O-H...O system) is expected (13). $\Delta\nu$ - ΔH correlations have been extensively discussed by Joesten and Shaad (4) : there is lack of general $\Delta\nu$ - ΔH correlation which, however, does not preclude the use of frequency shifts to predict H.B. enthalpies. If frequency shifts are measured in inert solvents with low concentration of base appropriate equations for the given system can be used : - ΔH (Kcal/mole) = $m\Delta\nu AH$ (cm^{-1}) + b where m and b are different constants for each acid. These equations will provide a reasonable estimate of ΔH (4).

Another relationship has been proposed by Iogansen (54) between ΔH and the difference between the square roots of integrated intensities of the free and bonded ν OH absorption bands

$$- \Delta H = 5.3 \ \Delta\Gamma^{1/2} \text{ where } \Delta\Gamma^{1/2} = \Gamma^{1/2}_{complex} - \Gamma^{1/2}_{free}$$

is defined by (cl^{-1}) $\int$ ln (l_o/I) dlnν and is related to the integrated band intensity B, $\Gamma \simeq B/\nu$. This linear relationship is apparently obeyed by a number of H.B. systems such as complexes of phenols and alcohols with different bases as well as by selfassociated systems such as carboxylic acids and alcohols (3). Gramstad et al. (55) have reported a linear correlation between $\Delta\nu$ OH and ΔG for adducts of phenols with bases for a given base type :
- ΔG(Kcal/mole) = m $\Delta\nu$ OH + b while Pullin and Werner (56) proposed

the equation $\Delta G = K \log \Delta \nu + c$ for N-H...X H.B. of amines and amides.

Relative shift $\Delta \nu / \nu_o$ can be used to predict pK_A values. Cutmore and Hallam have related the slopes of $\Delta \nu / \nu_o$ lines obtained by plotting $\Delta \nu / \nu_o$ for various solvents relative to $\Delta \nu / \nu_o$ of pyrrole for the same solvents (57). When these slopes are plotted against pK_A, the dissociation constant of the conjugated acid, a straight line is obtained for each structural family of bases. They claim that the slope of $\Delta \nu / \nu_o$ line is a measure of the intrinsic proton accepting power of the base (57, 4).

4. HYDROGEN BONDED SOLIDS

The main advantage of studying solids is that these are usually well defined systems structures of which have been or can be deter- mined by X-ray and neutron diffraction methods. Therefore, spectros- copic and crystallographic data can be usefully correlated (figs. 2 and 6). Moreover, some special techniques such as study of single crystals with polarised radiation in infrared and Raman and that of isotopically diluted crystals at low temperature bring a wealth of information about H.B. in crystals.

4.1. General considerations

The number of fundamental vibrations of a non-linear molecule is determined by $3n - 6$, where n is the number of atoms, and the selection rules are derived from the point group symmetry of the molecule. The number of fundamentals of a molecular crystal, in the wave-vector-equel-to-zero-approximation, is given by $3p - 3$ with $p = q.r$ where q is the number of atoms in the molecule and r the number of molecules in the smallest unit cell. The intramole- cular forces being much stronger than intermolecular forces it is convenient to distinguish $(3q - 6) r$ internal vibrations and $6r - 3$ external or lattice modes. The latter can be divided into 3r rota- tional vibrations (R') and $3r - 3$ translational vibrations (T'), this distinction being rigorous only when imposed by the symmetry. The selection rules of the crystal are derived from the factor group analysis which consists of treating a primitive cell of a crystal as if it was a single polyatomic molecule (58). In order to predict the optically active vibrations of a molecular crystal we must know its space group and the number of molecules in the primitive cell (Z).

Example. Factor group analysis.   Imidazole (fig. 1) crystallises into a monoclinic lattice of space group $P2_1/C \equiv C_{2h}^5$ and with four molecules in the unit cell. The molecules are linked into infinite chains by strong N-H...N H.B. parallel to the c axis of the crystal. There are thus 3 (9 x 4) - 3 = 105 optically active vibrations which can be divided into 21 x 4 = 84 internal and 6 x 4 - 3 = 21 external modes. Under the C_{2h} factor group symmetry the former can be represented by 21 A_g + 21 B_g + 21 A_u + 21 B_u and the latter by 6 A_g + 6 B_g + 5 A_u + 4 B_u the subscripts g and u indicating Raman and IR activity, respectively. In other words, each of the 21 intramolecular vibrations of imidazole molecule will split in the crystal (Davydov or correlation field splitting) into four components -two IR and two Raman active. The lattice vibrations can be further divided into 12 R' (3 A_g + 3 B_g + 3 A_u + 3 B_u) and 9 T' (3 A_g + 3 B_g + 2 A_u + 1 B_u) motions.

Band assignment.   The assignment of the bands to their symmetry species can be made by measuring a single crystal with polarised light. In absorption the intensity of a band will be maximum when the electric vector of the incident light $\vec{E}$ is parallel to the transition moment $\vec{M}$ of the crystal for the given vibration. The A_u species gives rise to $\vec{M}_{A_u}$ parallel to the b axis of the crystal, while that of the B_u species will be situated in the a, c plane. In Raman scattering, the form of the polarizability tensors corresponding to A_g and B_g vibrations indicate that A_g species should be observed only in polarised spectra of types (XX), (YY), (ZZ), and (XZ) while those of B_g species should appear only in (YY) -and (YZ)-polarised spectra (58).

An assignment of the bands to their approximate type of motion can also be given on the grounds of the spectra of totally and specifically deuterated derivatives and by comparison of the spectra of imidazole in gaseous, liquid, and solid state (35). In particular, lattice modes can be identified since their frequencies are much lower, < 200 cm^{-1}, than those of internal modes, > 600 cm^{-1}. The interpretation of lattice modes in terms of R' and T' motions is helped by examining the shifts on total deuteration. A considerably larger isotopic frequency ratio is expected for rotational ($\nu/\nu' \simeq 1.08$) than for translational ($\nu/\nu' = 1.03$) vibrations, the former being proportional to the square root of moments of inertia and the latter to the square root of reduced masses of the heavy and light imidazole molecule (59).

4.2. Intramolecular vibrations

Internal vibrations of a molecule in a H.B. crystal are modified with respect to those of a free molecule in two ways : (i) their frequencies change because of electronic effect of H.B. and different vibrational coupling and (ii) their number increases since there are usually several molecules (Z) in the unit cell. An intramolecular vibration is split to at most three infrared active and four Raman active components for any space group and Z if the molecules are crystallographically equivalent. More components may be observed if there are non-equivalent molecules. Finally, Fermi resonances may increase the band multiplicity and complicate the

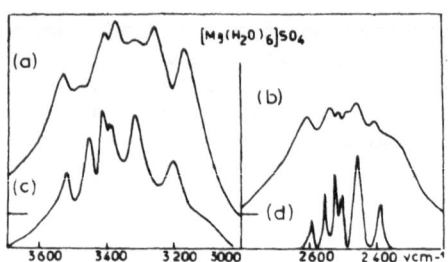

Fig. 7. Infrared spectra of OH (OD) stretching region of hexahydrate
        of magnesium sulphate : (a) H_2O ; (b) D_2O ; (c) HOD diluted
        in D_2O ; (d) HOD diluted in H_2O.

vibrational analysis. In order to separate these factors polarised
light study and isotopical dilution are used.

Isotopically mixed crystals.   This very useful technique was introduced by Hornig and Hiebert (60) and consists of studying the vibrational spectra, in particular ν AH absorption bands, of crystals
containing a small amount (< 5 %) of AH groups in AD (> 95 %)
environment or vice versa. Under these conditions, each A-H oscillator is surrounded by A-D oscillators, the vibrational frequencies
of which are very different and there is no longer any intermolecular vibrational coupling any more. The same applies to the intramolecular coupling, i.e. H_2O, NH_2, NH_3 molecules or groups are
substituted by HOD, NHD, and NHD_2 species. In other words, in the
AH stretching region of an isotopically (D) diluted crystal only
one ν AH band is expected if the AH molecules or groups are all
crystallographically equivalent. In the case of crystalline pyrrole
(fig. 3) the ν NH vibration gives an infrared triplet and a Raman
doublet while the infrared spectrum of an isotopically diluted
crystal shows a single sharp component. There are thus at least
four crystallographically equivalent molecules in the unit cell.
Another example concerns the β phase of 1,4 chlorophenol which
contains two types of non-equivalent molecules characterised by
ν OH frequencies at 3290 and 3200 cm^{-1} and γ OH frequencies at
685 and 702 cm^{-1} of an isotopically diluted crystal (61). Finally,
$Mg(H_2O)_6SO_4$ shows a strong and broad ν OH absorption band which is
resolved by isotopical dilution into seven well defined ν OH (ν OD)
bands of HOD species indicating at least seven different O-H...O
bonds between H_2O molecules and $SO_4^=$ anions (20) (fig. 7). The
multiplicity of bands in the ν OH region of acetic acid crystal
(figs. 4 and 5), on the other hand, is due neither to correlation
field splitting nor to different hydrogen bonds, but to Fermi resonances between the ν OH fundamental and combinations involving δ OH,
ν C-O and δ CH_3 (20).

No simple relationship appears to exist between the magnitude of the crystal field splitting and the strength of the H.B. Weakly or non-hydrogen bonded molecular crystals such as pyrrole or benzene show splittings up to 24 (ν NH, γ NH) and 27 cm^{-1}, respectively, while strongly H.B. heterocycles of similar size as imidazole or 1,2,4 triazole give splittings less than 11 cm^{-1} (γ NH) (15). A very large splitting has been observed for both phases of anhydrous oxalic acid for almost all internal modes. However, the splittings of α-$H_2C_2O_4$ are bigger (ν OH-174, ν C=O-76, ν C-O-29, δ OH-72, ν C-C-33, γ OH-28 and δ COO-30 cm^{-1}) than those of β phase in spite of the stronger H.B. of the latter (62). An exceptionally large splitting exists for γ OH vibration of p.chlorophenols and amounts to 130 cm^{-1} for the β phase (61) containing cyclic tetramers (fig. 1).

4.3. Lattice vibrations

Lattice frequencies are important for understanding intermolecular forces, molecular motions, and phase transitions in crystals. They can be assigned to their symmetry species and R' or T' vibrations using polarised radiation and isotopic shifts.

In the case of H.B. crystals all lattice frequencies do not necessarily involve H.B. and it is thus interesting to assign these also in terms of H.B. vibrations (cf. 2.4.). It can be shown that in pyridinium chloride crystal, for instance, only three translational vibrations T' of each complex $C_5H_5NH^+...Cl^-$ correspond to H.B. vibrations while the R' modes do not. The former give rise to strong IR bands at 192, 116, and 92 cm^{-1} and are assigned to ν (NH...Cl), δ (NH...Cl), and γ (NH...Cl) vibrations, respectively. Their frequencies decrease rapidly when Cl$^-$ is substituted by Br$^-$ and I$^-$ in the same order as the corresponding NH stretching frequencies increase indicating weakening of the H.B. The rotational vibrations, on the other hand, are strong in Raman and their frequencies are not H.B. sensitive (63). In the case of imidazole crystal (fig. 1) an isolated chain model is used in order to distinguish H.B. vibrations from other lattice modes and it can be argued on symmetry grounds that the infrared bands correspond to intra-chain and thus H.B. modes while the Raman spectra contain both intra and inter-chain lattice vibrations.

However, a recent calculation of 1,2,4-triazole crystal containing also infinite chains of NH...N (R N...N = 2.82 Å) H.B. molecules could not confirm this approximation. This crystal for which 32 out of 45 lattice modes have been assigned experimentally has four parallel chains in the unit cell and one would thus expect 32 "intra" and 13 "inter chain" vibrations. The results of the calculation show that there are only six vibrations which do not involve H.B. motions. Finally, a complete normal coordinate calculation of this crystal gives an overall H.B. stretching force

constant of 0.35 md/Å ; this value may be further decomposed into
$F_{H...N} = 0.20$ md/Å and $F_{N...N} = 0.08$ md/Å. The remainder represents
the long-range interactions in the chain. The overall force-constant
accounts for about 90 % of the potential energy of the H.B. stretching. The remaining 10 % are due to inter-chain interactions (15).

5. AH STRETCHING BAND BREADTH AND STRUCTURE

AH stretching "bonded" bands are generally broad and have more or
less complicated structure as shown in figs. 3, 4, 5, and 7. These
unusual features have been considered as closely related to the
theory of the H.B., in particular to the dynamical behaviour and
the type of the potential energy curve of the proton and have been
widely discussed in the literature (1, 3, 5, 20, 22, 64, 65). There
are different types of theories (5, 22, 64) and without entering
into details, we give the main factors determining the ν AH band
shape as follows :
(a) Anharmonic coupling of the ν AH mode with low frequency H.B.
    vibrations ("frequency modulation").
(b) Anharmonic coupling of the ν AH fundamental with combinations
    or overtones of intramolecular vibrations, in particular from
    those having AH bending character, giving rise to Fermi reso-
    nances.
(c) Double-well potential governing the motion of the AH stretching
    vibration with subsequent tunnelling and splitting of the cor-
    responding vibrational bands.
(d) Correlation field splitting.
(e) Non-equivalent hydrogen bonds
(f) Structural disorder .

    All these factors may be responsible for the breadth and/or
the structure but their relative importance depends on the physical
state and the strength of H.B.

    The "frequency modulation" is certainly a decisive factor in
determining the band breadth. Fig. 6 shows that the $\Delta\nu/\Delta R$ parameter
can be correlated with the intrinsic band width : in cases of strong
H.B. a variation of the O...O distance of about 0.01 Å corresponds
to a $\Delta\nu$ change of about 120 cm^{-1}. In other words, any H.B. vibration
which involves a change of the O...O distance will contribute to
the ν OH band breadth. The responsability of the "frequency modu-
lation" for the band structure, on the other hand, seems much less
obvious and the only example where a subband could be ascribed to
a combination with a H.B. vibration is that of $(CH_3)_2O...HCl$ complex (66). The double minimum potential has so far not been shown
as a factor determining the ν AH band structure. The other factors
are analysed separately for weak, medium-strong, and strong H.B.

    Weak H.B.   The ν AH band width of weak H.B. is usually due
to structural disorder and/or intermolecular coupling. It may
attain a few hundred wavenumbers but its intrinsic band width is

small, $\nu_{1/2}$ = 3 to 30 cm^{-1}, as observed in the matrix isolated species or isotopically diluted crystals (fig. 3). The band structure can be in most cases ascribed to correlation field splitting (fig. 3) and different hydrogen bonds (fig. 7).

Medium-strong H.B.   The multiplet structure of medium-strong H.B. such as selfassociated carboxylic acids (figs. 4 and 5) or N-H...N H.B. of nitrogen bases can usually be explained in terms of Fermi resonances between ν AH and combinations involving δ AH vibrations and vibrations coupled with this mode. The intrinsic band width varies from 30 to 150 cm^{-1}, is very temperature and deuteration sensitive indicating a high anharmonicity of the ν AH oscillator.

Strong H.B.   There are different types of band shapes illustrated by NaHCO$_3$ ($\nu_{1/2} \simeq$ 440 cm^{-1}), (CH$_3$)$_2$AsOOH (fig. 5), KH(CH$_3$COO)$_2$ and NaH(CH$_3$COO)$_2$ (fig. 4). The main features of the first two can be interpreted as combination and overtones of δ OH and γ OH vibrations in Fermi resonance with ν OH. The ν OH bands of KH(CH$_3$COO)$_2$ ($\nu_{1/2} \simeq$ 1000 cm^{-1}) and NaH(CH$_3$COO)$_2$ ($\nu_{1/2} \simeq$ 200 cm^{-1}) containing asymmetric and symmetric H.B., respectively, behave rather differently towards temperature and deuteration. The band width of the former decreases strongly upon deuteration but is almost temperature insensitive while the opposite occurs to the latter. These differences are closely related to the shape of the corresponding potential curves.

REFERENCES

1. G. Pimentel and A.L. Mc Clellan, Hydrogen Bond, W.H. Freeman and Co, San Francisco and London, 1960 ; Ann. Rev. Phys. Chem. 22, 347, 1971.
2. W.C. Hamilton and J.A. Ibers, Hydrogen Bonding in Solids, W.A. Benjamin Inc., New-York, 1968.
3. S.N. Vinogradov and R.H. Linnell, Hydrogen Bonding, Van Nostrand Reinhold Company, p. 59, 1971.
4. M.D. Joesten and L.J. Schaad, Hydrogen Bonding, Marcel Dekker Inc., New-Yord, 1974.
5. D. Hadzi and S. Bratos, in P. Schuster, G. Zundel and C. Sandorfy, The Hydrogen Bond II, North Holland Publishing Company, p. 565, 1976.
6. R.D. Green, Hydrogen Bonding by C-H Groups, Macmillan, 1974.
7. L.A. Leites, N.A. Ogorodnikova and L.I. Zakharkin, J. Organometal. Chem., 15, 287, 1968.
8. Unpublished results of this laboratory.
9. J.-L. Beaudoin, Thesis, Reims, 1977.
10. A. Lafaix and M.-L. Josien, J. Chim. Phys., 62, 684, 1965.
11. J. Corset and J. Lascombe, J. Chim. Phys., 64, 665, 1967.
12. A. Novak, Structure and Bonding, 18, 177, 1974.
13. N.D. Sokolov, Vodorodnaya Svyaz, Moscow, p. 7, 1964.

14. R. Blinc, D. Hadzi and A. Novak, Zt. Electrochem., 64, 567, 1960

15. D. Bougeard, N. Le Calvé, B. Saint Roch and A. Novak, J. Chem.
     Phys., 64, 5152, 1976.

16. N. Sheppard, Hydrogen Bonding, Editor D. Hadzi, Pergamon,
     London, p. 85, 1959.

17. C.A. Coulson and G.N. Robertson, Proc. R. Soc. London, A, 337,
     167, 1974.

18. P.A. Kollman and L.C. Allen, J. Chem. Phys., 51, 3286, 1969.

19. A. Lautié and A. Novak, J. Chem. Phys., 56, 2479, 1972.

20. A. Novak, J. Chim. Phys., 72, 981, 1975.

21. G. Lucazeau and A. Novak, J. Raman Spectroscopy, 1, 573, 1973.

22. S. Bratos, J. Chem. Phys., 63, 3499, 1975.

23. A. Lautié, F. Froment and A. Novak, Spectroscopy Lett., (5), 9,
     289, 1976.

24. V.A. Savel'ev and N.D. Sokolov, Chem. Phys. Lett., 34, 281, 1975

25. J. de Villepin and A. Novak, J. Mol. Struct., 30, 255, 1976.

26. D. Hadzi, B. Orel and A. Novak, Spectrochim. Acta, 29 A, 1745,
     1973.

27. J. Lascombe, J. Devaure and M.-L. Josien, J. Chim. Phys., 1271,
     1964.

28. G.J. Boobyer and W.J. Orville-Thomas, Spectrochim. Acta, 22,
     147, 1966.

29. C. Perchard and J.-P. Perchard, J. Raman Spectroscopy, 3, 277,
     1975.

30. C. Perchard and J.-P. Perchard, J. Raman Spectroscopy, 6, 74,
     1977.

31. M. Haurie and A. Novak, J. Chim. Phys., 62, 137, 1965.

32. M. Haurie and A. Novak, Spectrochim. Acta, 21, 1217, 1965.

33. A. Novak, J. Chim. Phys., 69, 1615, 1972.

34. J.-P. Perchard and M.-L. Josien, J. Chim. Phys., 65, 1856, 1968.

35. C. Perchard, A.-M. Bellocq and A. Novak, J. Chim. Phys., 62,
     1344, 1965.

36. A. Lautié and A. Novak, J. Chim. Phys., 65, 1359, 1968.

37. R.F. Lake and H.W. Thompson, Proc. Roy. Soc. London, 291, 469,
     1966.

38. M. Falk and E. Whalley, J. Chem. Phys., 34, 1554, 1952.

39. R.A. Nyquist, Spectrochim. Acta, 19, 1655, 1963.

40. P.V. Huong, M. Couzi and J. Lascombe, J. Chim. Phys., 64, 1056,
     1967.

41. J.L. Wood, in J. Yarwood Spectroscopy and Structure of Mole-
     cular Complexes, Plenum Press, 1973.

42. D.L. Cummings and J.L. Wood, J. Mol. Struct., 20, 1, 1974.

43. R. Foglizzo and A. Novak, J. Chim. Phys., 66, 1539, 1969.

44. R.J. Jakobsen, J.W. Brasch and Y. Mikawa, Applied Spectroscopy,
     22, 641, 1968.

45. E. Whalley, Developments in Applied Spectroscopy, vol. 6,
     p. 277, Plenum Press, 1968.

46. D. Bougeard, A. Lautié and A. Novak, J. Raman Spectroscopy,
     6, 80, 1977.

47. G.L. Carlson, W.G. Fateley, A.S. Manocha and F.F. Bentley,
     J. Phys. Chem., 76, 1553, 1972.

48. M.-L. Josien, Pure Applied Chemistry, 4, 33, 1962.
49. J.-C. Lassègues and P.V. Huong, Chem. Phys. Letters, 17, 444, 1972.
50. H.E. Hallam, in P. Schuster, G. Zundel, and C. Sandordy, The Hydrogen Bond III, North Holland, p. 1065, 1976.
51. M. Haurie and A. Novak, J. Chim. Phys., 62, 120, 1965.
52. M. Haurie and A. Novak, C.R. Acad. Sci., 264 A, B, 694, 1967.
53. R.M. Badger and S.H. Bauer, J. Chem. Phys., 5, 839, 1937.
54. A.V. Iogansen, Dokl. Akad. Nauk. SSSR, 164, 610, 1965.
55. T. Gramstad, Spectrochim. Acta, 19, 829, 1963.
56. J.A. Pullin and R.L. Werner, Spectrochim. Acta, 21, 1257, 1965.
57. E.A. Cutmore and H.E. Hallam, Trans. Faraday Soc., 58, 40, 1962.
58. G. Turrell, Infrared and Raman Spectra of Crystals, Academic Press, 1972.
59. C. Perchard and A. Novak, J. Chem. Phys., 48, 3079, 1968.
60. D.F. Hornig and G.L. Hiebert, J. Chem. Phys., 27, 752, 1957.
61. N. Le Calvé, M.-H. Limage, S. Parent and B. Pasquier, J. Chim. Phys., 74, 917, 1977.
62. J. de Villepin and A. Novak, Spectrochim. Acta, in press.
63. R. Foglizzo and A. Novak, J. Chem. Phys., 50, 5366, 1969.
64. G.L. Hofacker, Y. Maréchal and M. Ratner, in P. Schuster, G. Zundel and C. Sandorfy, The Hydrogen Bond I, North Holland, 1976.
65. C. Sandorfy, in P. Schuster, G. Zundel and C. Sandorfy, The Hydrogen Bond I, North Holland, 1976.
66. J.E. Bertie and D.J. Millen, J. Chem. Soc., 497, 514, 1965 ; J.E. Bertie and M.V. Falk, Can. J. Chem., 51, 1713, 1973.

OVERTONES AND COMBINATION TONES:
APPLICATION TO THE STUDY OF MOLECULAR ASSOCIATIONS

C. Sandorfy

Département de Chimie
Université de Montréal
C.P. 6210, Succ. A
Montréal (Québec) Canada   H3C 3V1

ABSTRACT. Overtones and combination tones can be useful for the study of H-bonding and other molecular associations in biological systems. The more favorable free: associated ratio in H-bond systems and bands of simultaneous excitation could be of value in such investigations. Some simple facts about overtones and combination bands are summarized.

INTRODUCTION

The transition moment for a vibrational transition is, for any given normal vibration,

$$R^{v'v''} = \int \psi_{v'}^{*} \, M \, \psi_{v''} \, d\tau \qquad \ldots \; [1]$$

where $\psi_{v'}$ and $\psi_{v''}$ are the vibrational wave functions for the upper and lower vibrational states and M is the variable dipole moment. Provided the oscillator is exactly harmonic and the variation of the dipole moment with the normal coordinate is exactly linear, the $\Delta v = \pm 1$ selection rule applies and only fundamentals can appear in the spectrum. ($v'' = 0 \to v' = 1$). Hot bands of the $v'' = 1 \to v'' = 2$, --- type would coincide with the fundamental since the energy levels of the harmonic oscillator are equidistant. These conditions, however, are never exactly met. Overtones and combination tones do appear in the spectra. In what follows we shall recall some elementary facts about these and then see if they can yield information about hydrogen bonds and other types of molecular association.

305

Theo M. Theophanides (ed.), Infrared and Raman Spectroscopy of Biological Molecules, 305–318.

Diatomic oscillators

For vibrations of diatomic molecules or good diatomic group vibrations (like carbonyl or hydroxyl stretching motions) the potential function, for small displacements, can be written as

$$V = kQ^2 + k_3Q^3 + k_4Q^4+... \qquad [2]$$

where Q is the normal coordinate and k_1, k_3, k_4, --- are the harmonic, cubic, quartic, --- potential constants respectively with

$$k = \frac{1}{2}\left(\frac{d^2V}{dQ^2}\right)_e$$

$$k_3 = \frac{1}{3!}\left(\frac{d^3V}{dQ^3}\right)_e \quad, \quad k_4 = \frac{1}{4!}\left(\frac{d^4V}{dQ^4}\right)_e \quad,... \qquad [3]$$

all taken at the equilibrium distance.

To the second order we can consider the cubic and quartic parts as the perturbation neglecting higher terms. (This is equivalent to the usual Morse-curve approximation). Then perturbation calculation yields for the vibrational terms:

$$G_v = \omega(v + \frac{1}{2}) + X(v + \frac{1}{2})^2... \qquad [4]$$

where ω is the harmonic frequency and X the anharmonicity constant, both in cm^{-1}. Furthermore

$$X = \frac{15}{4\omega_e} k_3^2 - \frac{3}{2} k_4 \quad... \qquad [5]$$

and

$$\nu^{ov} = \omega_e v + Xv(v+1) \quad... \qquad [6]$$

For the first, second and third overtones this gives

$$\nu^{01} = \omega_e + 2X$$

$$\nu^{02} = 2\omega_e + 6X$$

$$\nu^{03} = 3\omega_e + 12X$$

so that

$$X = \frac{\nu^{02}}{2} - \nu^{01} = \frac{\nu^{03}}{3} - \frac{\nu^{02}}{2} =... \qquad [7]$$

(If the frequency of the ν^{02} is less than twice ν^{01} which is normally the case, X is given the negative sign). [7] offers a test for the validity of the second order approximation: from the fun-

damental and the first overtone or from the first and second overtones etc. the same value should be obtained for the anharmonicity
constant. The perturbed wave-function will be a linear combination
of the "original" harmonic wave functions leading to non-zero (but
usually small) transition moments for the overtones. This anharmonicity which is due to the non-quadratic part of the potential is
called mechanical anharmonicity.

It is perhaps good to remember that while the harmonic frequency, ω_e, depends on the square root of the reduced mass, the
anharmonicity constant, X, depends on the reduced mass itself.

There is, however, another reason for the appearance of overtones in the infrared spectrum. The dipole moment in [1] does not
vary quite linearly with the normal coordinate:

$$M(Q) = M_e + M_1 Q + M_1 Q^2$$

where

$$M_1 = \left(\frac{dM}{dQ}\right)_e \ , \quad M_2 = \left(\frac{d^2 M}{dQ^2}\right)_e \ , \quad \cdots \qquad [8]$$

The transition moment is

$$R^{v'v''} = M_e \int \psi_{v'}^{*} \psi_{v''} d\tau + \left(\frac{\partial M}{\partial Q}\right)_e \int \psi_{v'}^{*} Q\psi_{v''} d\tau$$

$$+ \frac{1}{2}\left(\frac{\partial^2 M}{\partial Q^2}\right)_e \int \psi_{v'}^{*} Q^2 \psi_{v''}^{*} d\tau + \cdots \qquad \cdots [9]$$

The first term in [9] is zero because of the orthogonality condition. The integral in the second term is non-zero for $\Delta v = \pm 1$ and
the term itself is non-zero if, in addition, the dipole moment can
vary with the normal coordinate. The integral in the third term
is non-zero for $\Delta v = \pm 2$ and the term itself is non-zero if the second derivative of the dipole moment is non-zero and so forth. The
The third and higher terms is [9] are called the electrical anharmonicity. It contributes to the intensity of the fundamental and
it can give intensity to the overtones even if the oscillator is
exactly harmonic. (The ψ_v in [9] are the harmonic wave functions).

Thus we see that both mechanical and electrical anharmonicities can give intensity to overtones. In actual cases both of them
contribute and the expression for the transition moment becomes
rather more complecated. Retaining only the cubic and quartic
terms in the potential [2] and the quadratic term in the expansion
of the dipole moment [8] we obtain:

$$R^{v'v''} = M_1 \, R_1^{v'v''} + M_2 \, R_2^{v'v''} \ldots \qquad [10]$$

where

$$R_1^{v'v''} = \int \psi_{v'}^{*} \, Q \, \psi_{v''} \, d\tau$$

and

$$R_2^{v'v''} = \int \psi_{v'}^{*} \, Q^2 \psi_{v''} \, d\tau$$

and the ψ_v are now wave functions perturbed to the second order. Formulas for the integrated intensities were given by a number of authors. According to Herman and Shuler (1) those for the fundamental and for the first overtone are:

$$A^{1,0} = \frac{8\pi^3 N}{3hc}$$

$$\times \, \omega^{1,0} \left\{ M_1 \left(\frac{1}{\sqrt{2}} - \frac{3}{2\sqrt{2}} \, g + \frac{11}{4\sqrt{2}} b^2 \right) \right.$$

$$\left. + \, M_2 \left(\frac{5b}{\sqrt{2}} + \frac{88}{\sqrt{2}} \, bg - \frac{715}{12\sqrt{2}} b^3 \right) \right\}^2$$

$$A^{2,0} = \frac{8\pi^3 N}{3hc}$$

$$\times \, \omega^{2,0} \left\{ \sqrt{2} M_1 \left(\frac{b}{2} - \frac{111}{8} \, bg - \frac{71}{48} b^3 \right) \right.$$

$$\left. + \, \frac{M_2}{\sqrt{2}} \left(1 - \frac{15}{4} \, g + \frac{3}{8} b^2 \right) \right\}^2 \qquad \ldots \, [11]$$

where ω is the wave number in cm^{-1}: $b \equiv k_3/\omega_e$, $g \equiv k_4/\omega_e$, and N is the number of molecules per cm^3 in the lower vibrational level. Equation [11] throws some light on the joint operation of mechanical and electrical anharmonicity. Concerning $A^{1,0}$ the intensity of the fundamental, normally the largest terms are the first terms in each bracket. Provided M_1 and M_2 have the same sign and since k_3 (b) is usually negative these two terms will reinforce each other. The anharmonic contribution will depend on the product of the mechanical and electrical anharmonicities. It has been shown with model calculations (2) that, except for strong perturbations, for allowed infrared bands this contribution does not exceed ±20%. Conditions are very different for $A^{2,0}$, the intensity of the first overtone. There, naturally, all intensity comes from anharmonic contributions. Again the largest terms are normally the first terms in each bracket. If M_1 and M_2 have the same sign, then because of the negative sign of b these two terms will have opposite

signs and can partly or entirely cancel and distroy the intensity
of the overtone. With perturbations of the order of a medium strong
hydrogen bond, for example, a combination of a strongly polar bond
with a fairly high mechanical anharmonicity could compensate for
the effect of electrical anharmonicity and the overtone could become very weak. On the other hand, if M_1 and M_2 have opposite signs
the anharmonic contribution to $A^{1,0}$ and the intensity of $A^{2,0}$
would be expected to run parallel to each other for reasonable
combinations of b and M_2. The consequences of these conditions will
be dealt with below.

Polyatomic oscillators

The potential function (to the second order) has the form
(see (3), (4)):

$$V = V_o + \sum_{i=1}^{3N-6} \left(\frac{\partial V}{\partial Q_i}\right)_e Q_i + \frac{1}{2} \sum_{i,j=1}^{3N-6} \left(\frac{\partial^2}{\partial Q_i \partial Q_j}\right)_e Q_i Q_j$$

$$+ \frac{1}{3!} \sum_{i,j,k=1}^{3N-6} \left(\frac{\partial^3 V}{\partial Q_i \partial Q_j \partial Q_k}\right)_e Q_i Q_j Q_j +$$

$$\frac{1}{4!} \sum_{i,j,k,l=1}^{N-6} \left(\frac{\partial^4 V}{\partial Q_i \partial Q_j \partial Q_k \partial Q_l}\right)_e Q_i Q_j Q_k Q_l + \ldots \qquad [12]$$

The first term can be absorbed in the zero level of the energy,
the second term (the force) is zero at equilibrium, the third therm
is the quadratic term. Subsequent terms represent the mechanical
anharmonicity. The cubic and quartic potential constants are usually abbreviated as k_{ijk} and k_{ijkl}, respectively.

We are going to illustrate the problem on a triatomic model
with no degeneracy. A vibrational term will have the form

$$G(v_1,v_2,v_3)=\omega_1(v_1+\tfrac{1}{2})+\omega_2(v_2+\tfrac{1}{2})+\omega_3(v_3+\tfrac{1}{2})+X_{11}(v_1+\tfrac{1}{2})^2+X_{22}(v_2+\tfrac{1}{2})^2$$

$$+X_{33}(v_3+\tfrac{1}{2})^2+X_{12}(v_1+\tfrac{1}{2})(v_2+\tfrac{1}{2})+X_{13}(v_1+\tfrac{1}{2})(v_3+\tfrac{1}{2})$$

$$+ X_{23}(v_2+\tfrac{1}{2})(v_3+\tfrac{1}{2}) \qquad \ldots \quad [13]$$

It contains six anharmonicity constants of which three belong to
one of each of the three normal vibrations and the other three are
coupling constants. The wavenumber of a given vibrational transition is:

$$\nu = G'(v_1',v_2',v_3')-G''(v_1'',v_2'',v_3'') \qquad \ldots \quad [14]$$

and for bands other than hot bands

$$\nu = G'(v_1, v_2, v_3) - G''(0,0,0) \qquad \text{... [15]}$$

It is instructive to go into some detail. From [15] the fundamental of vibration Q_1 is

$$\nu_1^{01} = \omega_1 + 2X_{11} + \frac{1}{2}X_{12} + \frac{1}{2}X_{13} \qquad \text{... [16]}$$

For a diatomic oscillator only the first two terms would appear but now we see that the coupling constants between Q_1 and the other two vibrations also have a bearing on the frequency, even if it is a fundamental. The first overtone of the same vibration is:

$$\nu_1^{02} = 2\omega_1 + 6X_{11} + X_{12} + X_{13} = 2\nu_1^{01} + 2X_{11} \text{ ...} \qquad [17]$$

so that if we measure the overtone and the fundamental we can compute X_{11} just as for a diatomic oscillator.

Polyatomic molecules also have combination bands. A few examples may be useful. Let $(\nu_1 + \nu_3)$ be the (observed) wavenumber of the binary combination of vibrations Q_1 and Q_3. Then

$$(\nu_1 + \nu_3) = G'(1,0,1) - G''(0,0,0)$$

$$(\nu_1 + \nu_3) = \omega_1 + \omega_3 + 2X_{11} + 2X_{33} +$$

$$+ 2X_{13} + \frac{1}{2}X_{12} + \frac{1}{2}X_{23} = \nu_1^{01} + \nu_3^{01} + X_{13} \qquad \text{... [18]}$$

Thus if we measure the combination band and the two fundamentals we can compute the coupling constant X_{13}.

This is a summation tone. The difference tone of ν_1 and ν_3 is obtained if ν_3 is at level $\nu_3 = 1$ when the photon strikes. (Hot band).

$$(\nu_1 - \nu_3) = G'(1,0,0) - G''(0,0,1)$$

$$(\nu_1 - \nu_3) = \omega_1 - \omega_3 + 2X_{11} - 2X_{33} + \frac{1}{2}X_{12} - \frac{1}{2}X_{23}$$

$$= \nu_1^{01} - \nu_3^{01} \qquad \text{... [19]}$$

Interestingly, the coupling constant cancels out and the wavenumber of the difference band is simply the difference of the wavenumbers of the two fundamentals. This can be useful. If for some reason we cannot measure one of the fundamentals, say ν_3^{01}, but we have the other fundamental and a sum and a difference tone we can still compute the coupling constant:

$$X_{13} = [(\nu_1 + \nu_3) - \nu_1^{01}] - [\nu_1^{01} - (\nu_1 - \nu_3)] \text{ ...} \qquad [20]$$

This might occur if one of the fundamentals is too weak to be observed, or falls into a crowded part of the spectrum and cannot be assigned with certainty or if its wavenumber is altered by Fermi resonance, etc. In cases of weak or medium strong hydrogen bonds, if ν_1 is the X-H stretching frequency and ν_3 the X...Y bridge stretching frequency, ν_3 is much lower than ν_1. Then the spectral location of $(\nu_1+\nu_3)$ and $(\nu_1-\nu_3)$ is roughly determined by $\nu_1{}^{01}$ which might be more favorably located. We can transpose the bands to higher frequencies by using ternary combinations, some of which can be relatively intense:

$$(2\nu_1+\nu_3) = G'(2,0,1) - G''(0,0,0)$$

$$= 2\omega_1 + \omega_3 + 6X_{11} + 2X_{33} +$$

$$+ X_{12} + \frac{7}{2}X_{13} + \frac{1}{2}X_{23} = \nu_1{}^{02} + \nu_3{}^{01} + 2X_{13}$$

$$= 2\nu_1{}^{01} + 2X_{11} + \nu_3{}^{01} + 2X_{13}$$

$$= (\nu_1+\nu_3) + \nu_1{}^{01} + 2X_{11} + X_{13} \qquad \text{... [21]}$$

Eq. [21] offers some possibilities for cross-checking. For example, we can compute the coupling constant X_{13} from a) the measured frequencies of $(2\nu_1+\nu_3)$, the overtone $\nu_1{}^{02}$ and the fundamental $\nu_3{}^{01}$; b) the measured frequencies of $(2\nu_2+\nu_3)$, the fundamentals $\nu_1{}^{01}$ and $\nu_3{}^{01}$ and X_{11} measured from $\nu_1{}^{01}$ and $\nu_1{}^{02}$; c) the measured frequencies of $(2\nu_1+\nu_3)$, the binary combination $(\nu_1+\nu_3)$, the fundamental $\nu_1{}^{01}$ and X_{11}.

The quaternary combination also offers various possibilities:

$$(2\nu_1+2\nu_3) = G'(2,0,2) - G''(0,0,0)$$

$$= 2\omega_1 + 2\omega_3 + 6X_{22} + 6X_{33}$$

$$+ X_{12} + 6X_{13} + X_{23}$$

$$= \nu_1{}^{02} + \nu_3{}^{02} + 4X_{13}$$

$$= 2\nu_1{}^{01} + 2\nu_3{}^{01} + 2X_{11} + 2X_{33} + 4X_{13}$$

$$= 2(\nu_1+\nu_3) + 2X_{11} + 2X_{33} + 2X_{13} \text{ ... [22]}$$

A complete list of anharmonicity and potential constants for the water molecule is found in the classic paper of Darling and Dennison (5).

For the "free" OH stretching band of an alcohol X_{11} is about -80 cm^{-1}. It is about -70 cm^{-1} for the free NH band of secondary amines and -50 cm^{-1} for the SH band of thiols. In H-bonded amine dimers it changes little. In H-bonded alcohol polymers it might increase to -120 cm^{-1} or more. In the ether+hydrogen fluoride system it probably reaches -200 cm^{-1} for the F-H stretching vibration. Carbonyl stretching bands are not very anharmonic ($X_{11} \sim 5-10$ cm^{-1}). The vital X_{13} coupling constant is probably about ~ 70 cm^{-1} for the ether+HF system, a high value.

A rather interesting type of combination band is obtained when a, say, $v_1''=0 \rightarrow v_1''=1$ transition of a vibration of high frequency is combined with $v_3''=1 \rightarrow v_3''=1$, etc. that is 1-1, 2-2, 3-3,... jumps of a vibration of low frequency. The series of bands obtained in this way are analogous to the "sequences" of electronic spectroscopy and are, of course, hot bands. For example,

$$(\nu_1+\nu_3-\nu_3) = G'(1,0,1) - G''(0,0,1)$$

$$= \omega_1 + 2X_{11} + \frac{1}{2}X_{12} + \frac{3}{2}X_{13}$$

$$= \nu_1{}^{01} + X_{13} \qquad \ldots [23]$$

The combination is just X_{13} apart from the $\nu_1{}^{01}$ fundamental. Furthermore since the coupling constant might be either positive or negative we might find this hot band at either the high or the low frequency side of $\nu_1{}^{01}$. Other examples need little comment:

$$(\nu_1+2\nu_3-2\nu_3) = G'(1,0,2) - G''(0,0,2)$$

$$= \omega_1 + 2X_{11} + \frac{1}{2}X_{12} + \frac{5}{2}X_{13}$$

$$= \nu_1{}^{01} + 2X_{13} \qquad \ldots [24]$$

$$(\nu_1+2\nu_3-\nu_3) = G'(1,0,2) - G''(0,0,1)$$

$$= \omega_1 + \omega_3 + 2X_{11} + 4X_{33}$$

$$+ \frac{1}{2}X_{12} + 3X_{13} + \frac{1}{2}X_{23}$$

$$= \nu_1{}^{01} + \nu_3{}^{01} + 2X_{33} + 2X_{13} \qquad \ldots [25]$$

$$(\nu_1 + \nu_3 - 2\nu_3) = G'(1,0,1) - G''(0,0,2)$$

$$= \omega_1 - \omega_3 + 2X_{11} - 4X_{33}$$

$$+ \frac{1}{2}X_{12} + X_{13} - \frac{1}{2}X_{23}$$

$$= \nu_1{}^{01} - \nu_3{}^{01} - 2X_{33} + X_{13} \qquad \cdots \quad [26]$$

From [25] and [26] X_{13} could be calculated.

Electrical anharmonicity is just as important for polyatomic as for diatomic molecules and it is, of course, much more difficult to handle. If we develop the dipole moment into series we obtain:

$$M = M_e + \sum_{i=1}^{3N-6} \left(\frac{\partial M}{\partial Q_i}\right)_e Q_i + \frac{1}{2} \sum_{i,j=1}^{3N-6} \left(\frac{\partial^2 M}{\partial Q_i \partial Q_j}\right)_e Q_i Q_j$$

$$+ \frac{1}{3!} \sum_{i,j,k=1}^{3N-6} \left(\frac{\partial^3 M}{\partial Q_i \partial Q_j \partial Q_k}\right)_e Q_i Q_j Q_k + \qquad \cdots \quad [27]$$

The terms higher than the linear ones constitute electrical anharmonicity. Even if we neglect all but the quadratic terms we obtain, in addition to the $(\partial^2 M/\partial Q_i)_e Q_i{}^2$ mixed terms of type $(\partial^2 M/\partial Q_i \partial Q_j)_e Q_i Q_j$. These are the ones which give intensity to the combination bands. Their contribution is probably just as important as that of mechanical anharmonicity. Little is known about these derivatives. Yet, their knowledge would be quite important for theories of H-bonding or molecular associations among others. It might invalidate many calculations on the intensities of combination bands which have been published.

Combination bands of the [18] – [26] types are essential for the understanding of the vibrational spectra of hydrogen bonded species. The gas phase works of Millen (6,7), Couzi et al (8), and Thomas (9,10) have shown that binary combinations of the X-H and X...Y stretching vibrations (like [18], [19]) and higher combinations of the $(\nu_1 + 2\nu_3)$, $(\nu_1 - 2\nu_3)$,... types explain the main satellite bands in the spectra of acetone+HF, or ether+HF and other complexes whereas the finer structure of the bands can be explained through the sequences in ν_3 and other low frequency vibrations. (Cf. Stepanov's Franck-Condon scheme (11, 12). In solution randomization of the intermolecular forces broadens out the bands producing the familiar diffuse spectra of H-bonded systems. It does not exclude at all that overtones and combinations of other fundamentals

boosted by Fermi resonance with the X-H fundamental contribute to
the breadth and complexity of the IR region of absorption of H-
bonded systems. This is just another consequence of anharmonicity.
This simple interpretation based on anharmonicity and disorder is
probably sufficient to explain the main facts about H-bonds except,
possibly, the very strong ones. For the latter the particular shape
of the potential (double well, or single but shallow well) might
introduce additional features such as tunnelling (13) and very
high polarisability (14) which in certain cases lead to even
broader regions of infrared absorption. However, for the most commonly occurring H-bonds anharmonicity is the key quantity. In particular all theoretical treatments attempting to calculate the interaction between the "fast" X-H and "slow" X...Y motions are approximations to the anharmonic problem with the coupling constant
X_{13} as the main target. This applies to nearly all biologically
important H-bonds. (For reviews of H-bond theories see references
13,14 and 15; on the anharmonicity of vibrations affected by H-
bond formation see ref. 16. The most recent comprehensive treatments are from Sokolov and Savelev (17) and from Robertson (18).

The solution of many chemical and biological problems can
depend on the ratio of free and associated (H-bonded) molecules
and this is usually obtained from the relative intensities of the
free and associated infrared OH, NH, SH bands. Now, it is a remarkable fact that this ratio differs for the fundamentals, for the
overtones and for the various combination tones. Luck and Ditter
(19) were the first to observe that the relative intensity of the
free OH stretching band is much greater for the overtones than for
the fundamentals. This has been confirmed later on a great number
of systems including OH, NH, and SH bands. It has been also observed
(20) that the relative intensities of the bands due to different associated species also differ for the fundamentals and for the overtones so that the less associated species have the stronger overtones.
        It follows that measurements on overtones or combination tones
can make it possible to ascertain the presence and to estimate the
amount of free or weakly associated apecies even if the corresponding bands are absent at the level of the fundamental. This helps
in the study of concentrated solutions and liquids, for example.

An application of combination bands made it possible for us
to obtain a correlation between the free/associated ratio of OH
and NH bands and the potency of halogenated anesthetics (21). Such
anesthetics shift the ratio in favor of the free bands and this
has been studied on model systems containing O-H---O, N-H---N
and N-H---O=C type H-bonds. At room temperature and at concentrations which have to be used, the free band of the fundamental is
usually too weak or entirely absent in such systems. The region
of the first overtone is often complicated by overlap between different bands. In some cases we encounter the opposite difficulty:

it is the associated band which is too weak (22). Thus we have to
look for combination bands. Alcohols and phenols possess bands
in the near IR centered at about 5200-5300 cm^{-1} for which the free
and associated bands overlap relatively little and they have an
equitable intensity ratio. This band is a combination of the OH
stretching and in-plane bending modes. Secondary amides have a band
in the 5000-4900 cm^{-1} range, a combination of the NH stretching and
in-plane bending modes. These bands made it possible for us to carry
out quantitative studies at room temperature. (See the following
paper in this volume).

Simultaneous excitation in associated molecules

Many cases were reported in the literature where two vibrations, one in each of two associated molecules are excited simultaneously by one photon. (For a theory of these we refer to Hooge
and Ketelaar (23)). Let us consider the two associated molecules as
one supermolecule (24) (25). Then if Q_a and Q_b are the normal coordinates of two vibrations "originally" located in molecule _a_ and _b_
respectively, the variable dipole moment can be written as

$$M = M_e + \left(\frac{\partial M}{\partial Q_a}\right)_e Q_a + \left(\frac{\partial M}{\partial Q_b}\right)_e Q_b$$

$$+ \frac{1}{2}\left(\frac{\partial^2 M}{\partial Q_a^2}\right)_e Q_a + \frac{1}{2}\left(\frac{\partial^2 M}{\partial Q_b^2}\right)_e Q_b + \left(\frac{\partial^2 M}{\partial Q_a \partial Q_b}\right)_e Q_a Q_b + \dots \quad [28]$$

where now the non-linear terms represent the electrical anharmonicity of the supermolecule. Three cases can be distinguished.
a) No mechanical coupling between the two molecules so that Q_a is
localized in molecule _a_ and Q_b in molecule _b_. If they are two identical vibrations of two identical molecules $\nu_a^{01}=\nu_a^{01}=\nu_b^{01}$ and only

one fundamental will be observed. The first overtones of Q_a and Q_b

will also coincide with wavenumber $\nu^{02}=2\nu^{01}+2X$. Its intensity will
naturally depend on both mechanical and electrical anharmonicity
within molecules _a_ and _b_. In addition, however, there will be a
combination band, $(\nu_a+\nu_b) = 2\nu^{01}$. In the absence of mechanical coupling (either harmonic or anharmonic) the intensity of this combination band will be due entirely to the mixed term in [28]:
$(\partial^2 M/\partial Q_a \partial Q_b)_e Q_a Q_b$ which is a part of the electrical anharmonicity

of the supermolecule. The wavenumber of the combination band will
not be the same as that of the overtone ν^{02}. While the latter is
affected by mechanical anharmonicity $(X_{aa} = X_{bb})$, the combination

is not. Its wavenumber will be the double of that of the fundamental. (or the sum of ν_a^{01} and ν_b^{01} if the two vibrations are dif-

ferent). b) If there is mechanical coupling between Q_a and Q_b they

will couple to give two normal vibrations (in-phase and out-of-phase) of the supermolecule. There will be two fundamentals, $\nu_{a'}^{01} \neq \nu_{b'}^{01}$. Each of them will have an overtone, $\nu^{02} = 2\nu_a^{01} +$

$2X_{a'a'}$ and $\nu_b^{02} = 2\nu_b^{01} + 2X_{b'b'}$, their intensity depending on both

mechanical and electrical anharmonicity. (the second derivatives).

There will be a combination band too. $(\nu_{a'} + \nu_{b'})$. If the

mechanical coupling is purely harmonic its intensity still comes entirely from the mixed term in the expression of the electrical anharmonicity and its wavenumber will be the sum of $\nu_{a'}$ and $\nu_{b'}$.

If we have both harmonic and anharmonic coupling the wavenumber of the combination band $(\nu_a + \nu_b)$ will depend on the (inter-

molecular) anharmonic coupling constant: $\nu_a^{01} + \nu_b^{01} + X_{ab}$. Its

intensity will be due to both X_{ab} and $(\partial^2 M/\partial Q_a \partial Q_b)_e Q_a Q_b$.

Band splittings due to intermolecular coupling and bands due to simultaneous excitation might be of value for the study of molecular associations, including biologically important systems.

Bands due to simultaneous excitation have been known for a long time in high pressure gases. (See (23). More recently many such bands have been identified in solution spectra by Burneau and Corset (26). The spectra of H-bonded complexes formed by chloroform and various bases provide good examples. In the spectrum of the $CHCl_3 + CD_3CN$ complex they found a band at 5287 cm^{-1} which is neither a $CHCl_3$ nor a CD_3CN band. Its wavenumber is exactly the sum of the $\nu(CH)$ of $CHCl_3$ (3023 cm^{-1}) and $\nu(C\equiv N)$ of CD_3CN (2264 cm^{-1}) showing that no mechanic coupling is involved. Chloroform and acetone – d_6 have a band of simultaneous excitation at 4720 cm^{-1}. ($\nu(CH)$: 3020 cm^{-1}; $\nu(C=O)$: 1702 cm^{-1}). The water bands of 3544 and 3629cm^{-1} also give bands of simultaneous excitation with CD_3CN [26]. They are at 5804 and 5883 cm^{-1} respectively and show a slight amount of mechanical coupling.

Rossarie et al (27) found a $\nu(OH)+\nu(CN)$ band at 5808 cm^{-1} for alcohol+acetonitrile. The sum of the two unperturbed fundamentals is equal to 5814 cm^{-1}. Earlier Ron and Hornig (28) reported the case of crystalline HCl. The case of dissolved alcohols is an example for more mechanical coupling. (Asselin and Sandorfy (29)). Perchard and Perchard (30) found the two fundamentals in the Raman spectrum of isopropanol at 3290 and 3170 cm^{-1} while the combination is at 6428 cm^{-1}. The importance of electrical anharmonicity is obvious in most cases.

Conclusions

Applications of overtones and combination tones to biologically important systems have not been very numerous in the past. It is believed, however, that such bands, especially those which fall into the near IR have potentiality for the solution of biological problems.

There are, of course, some practical advantages: water is a lesser obstacle in the near IR than in IR proper. Silica cells and long path lengths can be used. (Some solvents like CCl_4 can be used with over a meter).

Our principal conclusion is, however, that the main promise lies with the possibility to vary the free/associated ratio in hydrogen bonded systems and to use bands of simultaneous excitation to detect molecular associations.

REFERENCES

1. R.C. Herman and K.E. Shuler, J. Chem. Phys. 22, 481 (1954).
2. T. Di Paolo, C. Bourdéron, and C. Sandorfy, Can. J. Chem. 50, 3161 (1972).
3. E.B. Wilson, Jr., J.C. Decius, and P.C. Cross, Molecular Vibrations. McGraw-Hill, New York, 1955.
4. P. Barchewitz, Spectroscopie infrarouge. Vol. 1. Gauthier-Villars, Paris, 1961.
5. B.T. Darling and D.M. Dennison, Phys. Rev. 57, 128 (1940).
6. J. Arnold and D.J. Millen, J. Chem. Soc. 503, 510 (1965).
7. J.E. Bertie and D.J. Millen, J. Chem. Soc. 497, 514 (1965).
8. M. Couzi, J. Le Calvé, P.V. Huong, and J. Lascombe, J. Mol. Struct. 5, 363 (1976).
9. R.K. Thomas, Proc. Roy. Soc. London, A322, 137 (1971).
10. R.K. Thomas, Proc. Roy. Soc. London, A325, 133 (1971).
11. B.I. Stepanov, Zh. Fiz. Khim. 19, 507 (1945).
12. B.I. Stepanov, Zh. Fiz. Khim. 20, 408 (1946).
13. G.L. Hofacker, Y. Maréchal, and M.A. Ratner, Chapter 6, vol. 1, in The Hydrogen Bond; P. Schuster, G. Zundel, and C. Sandorfy, Editors. North-Holland, Amsterdam, 1976.
14. G. Zundel, ibidem. Chapter 15, Vol. 2.
15. D. Hadži and S. Bratos, ibidem. Chapter 12, Vol. 2
16. C. Sandorfy, ibidem. Chapter 13, Vol. 2.
17. N.D. Sokolov and V.A. Savelev, Chem. Phys. 22, 383 (1977).
18. G.N. Robertson, Phil. Trans. Roy. Soc. London, 286, 25 (1977).
19. W.A.P. Luck and W. Ditter, J. Mol. Struct. 1, 261 (1967-1968).
20. C. Bourdéron and C. Sandorfy, J. Chem. Phys. 59, 2527 (1973).

21. G. Trudeau, K.C. Cole, R. Massuda, and C. Sandorfy, Can. J. Chem. $\underline{56}$, 000 (1978).

22. M.-C. Bernard-Houplain and C. Sandorfy, J. Chem. Phys. $\underline{56}$, 3412 (1972).

23. F.N. Hooge and J.A.A. Ketelaar, Physica $\underline{23}$, 423 (1957).

24. J.-J. Péron, C. Bourdéron, and C. Sandorfy, Chem. Phys. Letters $\underline{33}$, 212 (1975).

25. J.-J. Péron and C. Sandorfy, Chem. Phys. Letters $\underline{36}$, 150 (1975).

26. A. Burneau and J. Corset, J. Chem. Phys. $\underline{56}$, 662 (1972).

27. J. Rossarie, J.C. Lavalley, and R. Romanet, Comptes Rendus Acad. Sci. (Paris) $\underline{B273}$, 185 (1971).

28. A. Ron and D.F. Hornig, J. Chem. Phys. $\underline{39}$, 1129 (1963).

29. M. Asselin and C. Sandorfy, J. Chem. Phys. $\underline{52}$, 6130 (1970).

30. C. Perchard and J.-P. Perchard, Chem. Phys. Letters $\underline{27}$, 445 (1974).

INFRARED STUDIES ON ANESTHETIC COMPOUNDS

C. Sandorfy and G. Trudeau

Département de Chimie
Université de Montréal
C.P. 6210, Succ. A
Montréal (Québec) Canada   H3C 3V1

ABSTRACT. Anesthesia is an ideal field of application for our knowledge on molecular associations. Vibrational spectroscopic studies on model systems give insight into the way anesthetics can perturb molecular associations, in particular hydrogen bonds which determine the conformation and permeability of the macro-molecules forming the nerve cell membrane.

INTRODUCTION

The proper functioning of the nerve cell is linked to the permeability to ions of the cell membrane. The latter consists of layers of lipids and proteins so that its permeability depends on the conformation of these macromolecules. This, in turn, depends on steric factors and a number of weak intermolecular associations including Van der Waals interactions, hydrogen bonding and, possibly, charge transfer. Then, perturbation of these inter-molecular interactions might lead to partial or complete break-down of the nervous system (For reviews see (1-4)). This is what seems to happen in a state of anesthesia. Therefore the phenomenon of anesthesia is a par excellence field of application of our knowledge on molecular associations. It follows that infrared, Raman, NMR and other forms of spectroscopy should be able to contribute to the elucidation of the mechanism of anesthesia. The subsequent discussion refers to inhalation, or general anesthetics.

These anesthetics have widely differing chemical structures; it has never been possible to classify them on chemical ground. Xenon, nitrous oxide, ethylene, cyclopropane, ether, chloroform, C_2F_6, SF_6, CF_3-CHCl_2, $CF_3-CHClBr$ (halothane), for example, all

Theo M. Theophanides (ed.), Infrared and Raman Spectroscopy of Biological Molecules, 319–324.

have anesthetic potency. Their anesthetic action does not involve chemical reactions. Then it is natural to try to divide
them into categories according to the types of association into
which they might enter and consider their action in terms of changes they might bring about in the pattern of associations affect-
ing the macromolecules of the nerve cell membrane.

There are anesthetic molecules which can have Van der Waals
interactions only; others can form, in addition, hydrogen bonds
as proton acceptors, like ether, or as proton donors like chloroform or halothane; still others could, perhaps, enter interactions
closer to the charge-transfer type, like fluorocarbons containing
higher halogens.

The mechanism of inhalation anesthesia constitutes an important bio-medical problem. The theory that carries most credit a-
mong anesthesiologists is the "hydrophobic site theory". It is
based on the Meyer-Overton rule which is a relationship between
oil-water partition coefficients (or lipid solubility) and anesthetic potency and dates back to the turn of our century. On it
is based the"unitary hypothesis" meaning that the mechanism of
action of all inhalation anesthetics is essentially the same and
it implies only interactions at hydrophobic sites of the membrane.
(See 1-4) In actual fact lipid solubility while it gives a measure of the accumulation of the anesthetic in the membrane does
not reveal its mode of action (5). (Nor are, of course, lipids
entirely hydrophobic).

Infrared studies

Several years ago it has been observed in our laboratory that
fluorocarbon anesthetics containing higher halogens "break" H-
bonds of the N-H---N, O-H---O or N-H---O=C types (6,7). At a given
concentration and temperature the IR spectrum of a solution containing one of these systems exhibits a free and an associated
(or more) NH or OH bands. Upon lowering the temperature normally
the free band gradually disappears and the associated bands become preponderant. In the presence of halofluorocarbons, however,
the opposite happens: the associated bands are weakened and eventually might disappear entirely while the free band "comes back".
That halogenated solvents shift the free: associated equilibrium
in favor of the free species has been known for some time (8-12).
Our contribution consisted in showing that this effect can be
studied to advantage through low temperature solution spectroscopy
and that it can be related to the anesthetic power of halogenated
molecules. We measured the IR spectra of many H-bonded systems in
the presence of such anesthetics demonstrating the great general-
ity of the phenomenon (13,14). Many examples were given. In view
of these observations it would seem to be surprising if dissocia-

tion or perturbation of certain H-bonds was not involved with the mechanism of anesthesia, for the halo-fluorocarbon type anesthetics at least.

Most of our work mentioned so far has been done at low temperatures. The reason for this was the necessity to magnify the effect. In most (but not all) cases, the weakness of the free fundamental NH or OH band prevented us from demonstrating the H-bond "breaking" effect convincingly at room temperature. Thus, more recently, we have carried out (15) a room temperature study using stretching-bending combination bands for which the free: associated intensity ratio favors the free band (See the preceding contribution in this volume). Our model systems were water-dioxane, self-associated N-ethylacetamide and tert-butanol. These contain H-bonds which commonly occur in the living organism. Table I (from (15)) illustrates the results of these measurements on the example of N-ethylacetamide. The solutions contained equimolar amounts of the anesthetics. Integrated intensities have been computed on the bases of a curve analysis using Cauchy-Gauss functions to fit the observed spectrum. (16-18) The increase in the intensity of the free NH band and the decrease of the intensity of the associated band with increasing anesthetic potency appear clearly. (p is defined as the effective anesthetic pressure in atmospheres required to suppress the righting reflex of mice in half of the experimental animals. It is widely used as a measure of anesthetic potency).

Now, since anesthetic compounds are seen to have a potency to "break" H-bonds there must exist an association which competes with H-bonding. This might be through Van der Waals forces or a charge transfer mechanism or indeed, the formation of another H-bond. It is very significant in this respect that many of the most potent anesthetics possess an acidic hydrogen atom, like chloroform, halothane (CF_3-CHClBr), methoxyflurane ($CH_3OCF_2CHCl_2$), enflurane CHF_2OCF_2CHFCl). Thus another series of measurements has been undertaken (Massuda and Sandorfy (19)) in which one of these anesthetics has been mixed into a solution containing an amide or a self-associated amine or alcohol. As temperature was lowered the intensity of the <u>associated</u> NH or OH bands decreased and at the same time the intensity of the associated CH (or CD) band increased in every case. Further evidence for C-H---N or C-H---O hydrogen bond formation came from the NMR work of Brown and Chaloner (20) and Koehler *et al* (21).

The role of the acidic hydrogen has been emphasized by Hansch *et al* (22) who examined the partition coefficients of a number of gaseous anesthetics in the octanol-water system. They came to the conclusion that the relative anesthetic potency depends on hydrophobicity of the anesthetic and on a polar factor. Di Paolo, Kier and Hall (23) arrived at similar conclusions through connectivity

index calculations as did Davies *et al* (24,25) who approximated
the activity coefficient of anesthetics as a function of the dominant intermolecular interactions in a given phase. The latter
found that the dominant ones for halogenated anesthetics are Van
der Waals and H-bond donor (C-H---O) properties.

    Table 1. Computed relative band area for the NH
stretching + in-plane bending combination. The band areas are on
different relative scales for each system and should only be compared within a given column.

TABLE 1

Anesthetic	log 1/p	Band Area self-associated N-ethylacetamide free NH	Band Area self-associated N-ethylacetamide ass. NH
C_7H_{14}	~0	37.78	46.68
$CFCl_3$	0.82	37.89	45.42
CH_2Cl_2	1.52	39.19	42.15
CF_3CHCl_2	1.57	40.90	---
$CHCl_3$	2.08	41.73	33.60
$CF_3CHClBr$	2.11	42.95	30.75
$CH_3OCF_2CHCl_2$	2.66	44.70	29.20

DISCUSSION

    Not all halocarbon anesthetics possess an acidic hydrogen.
We have ascertained the H-bond "breaking" property of halocarbons
containing no hydrogen at all. (CF_3Br, CF_2Br-CF_2Br, and many others). It is remarkable, however, that the most potent anesthetics
do have this acidic hydrogen. Their action is likely to be through
formation of H-bonds as proton donors. Since, however, available
proton acceptor sites in the membrane are quite certainly already
engaged in H-bond formation the anesthetic must dissociate or, at
least, perturb such existing H-bonds. This is well in line with
our original suggestion. (13,14) It is certainly not the only type
of interaction that contributes to anesthesia. Halocarbons containing no acidic hydrogen could act through Van der Waals forces,
both non-polar and polar and partial charge transfer. These forces
can provide additional interactions when the molecule does have
an acidic hydrogen.

There are, of course, anesthetics which do not contain either Cl, Br, I or acidic hydrogen, like ethylene, cyclopropane, xenone. These must act mainly through dispersion forces although dipole-induced dipole interactions might be important in certain cases. (26).It is quite possible that these interactions lead indirectly to perturbations of H-bonds in the membrane macromolecules but as yet we have no proof for this.

Much remains to be done. It appears clearly, however, that the mechanism of inhalation anesthesia cannot be explained on the basis of hydrophobic interactions only. It is a very important biomedical phenomenon into whose nature vibrational spectroscopy can provide insight.

REFERENCES

1. K.W. Miller and E.B. Smith, in A Guide to Molecular Pharma-cology - Toxicology, part 2, chapter 11. R.M. Featherstone, editor. M. Dekker, New York, 1973, pp. 427-475.
2. M.J. Halsey, in Anesthetic Uptake and Action. E.I. Eger, edi-tor. Williams and Wilkins, Baltimore, 1974, pp. 45-76.
3. J.C. Miller and K.W. Miller in Physiological and Pharmaco-logical Biochemistry. H.K.F. Blaschko, Editor. Butterworths, London (1975), pp. 33-75.
4. R.D. Kaufman, Anesthesiology 46, 49 (1977).
5. D.D. Denson, Chemtech 7, 446 (1977).
6. M.-C. Bernard-Houplain, C. Bourdéron, J.-J. Péron and C. Sandorfy, Chem. Phys. Letters 11, 149 (1971).
7. M.-C. Bernard-Houplain and C. Sandorfy, J. Chem. Phys. 56, 3412 (1972).
8. K.B. Whetsel, Spectrochim. Acta 17, 614 (1961).
9. A. Allerhand and P. von R. Schleyer, J. Am. Chem. Soc. 85, 371, 1233 (1963).
10. D.P. Stevenson and G.M. Coppinger, J. Am. Chem. Soc. 84, 149 (1962).
11. M. Gomel Ann. Chim. 3, 415 (1968).
12. L.J. Bellamy and R.J. Pace, Spectrochim. Acta 22, 535 (1966).
13. T. Di Paolo and C. Sandorfy, J. Med. Chem. 17, 809 (1974).
14. T. Di Paolo and C. Sandorfy, Can. J. Chem. 52, 3612 (1974).
15. G. Trudeau, K.C. Cole, R. Massuda and C. Sandorfy, Can. J. Chem. 56, 0000 (1978).
16. R.N. Jones, R. Venkataraghavan and J.W. Hopkins, Spectrochim. Acta 23A, 925 (1967).
17. R.N. Jones et al. NRC Bulletin No 15, National Research Coun-cil of Canada, Ottawa (1968).
18. J. Pitha and R.N. Jones, NRC Bulletin No 12, National Re-search Council of Canada, Ottawa (1968).
19. R. Massuda and C. Sandorfy, Can. J. Chem. 55, 3211 (1977).
20. J.M. Brown and P.A. Chaloner, Can. J. Chem. 55, 3380 (1977).

21. L.S. Koehler, W. Curley and K.A. Koehler, Mol. Pharmacol. 13, 113 (1977).

22. C. Hansch, A. Vittoria, C. Silipo and P.Y.C. Jow, J. Med. Chem. 18, 546 (1975).

23. T. Di Paolo, L.B. Kier and L.H. Hall, Mol. Pharmacol. 13, 31 (1977).

24. R.H. Davies, R.D. Bagnall and W.G.M. Jones, Int. J. Quant. Chem. Quant. Biol. Symp. 1, 201 (1974).

25. R.H. Davies, R.D. Bagnall, W. Bell and W.G.M. Jones, Int. J. Quant. Chem. Quant. Biol. Symp. 3, 171 (1976).

26. G. Leroy, G. Louterman-Leloup, D. Peeters and M. Sana, Ann. Soc. Sci. Bruxelles, 89, 115 (1975).

EXPERIMENTAL TECHNIQUES

EXPERIMENTAL TECHNIQUES OF RESONANCE RAMAN
SPECTROSCOPY APPLIED TO BIOLOGICAL MOLECULES

M. Berjot[*], M. Manfait[*] and T. Theophanides[**]

[*]Université de Reims (France)
[**]Université de Montreal (Canada)
National Hellenic Research Foundation (Greece)

INTRODUCTION

The authors have endeavoured to present the material in a style understandable to the average reader. However, by means of the refs.(1 to 7) the reader will easily gain access to the original special reports on Resonance Raman pratice and to reviews of biological applications published in recent years.

The central challenge of modern biochemistry is to elucidate the biological function in terms of molecular structure. A considerable catalog of information on biomolecular architecture is already available from X-ray diffraction studies on crystalline or partially ordered materials. Numerous spectroscopic methods have been introduced to monitor structural features and detect changes which accompany the biological function. Among them, vibrational spectroscopy offers high promise, since vibrational frequencies, available from Raman or infrared spectra, are sensitive to geometric and bonding arrangements of localized groups of atoms in a molecule.

Several points can be mentioned in favor of vibrational spectroscopy vis-à-vis other physico-chemical techniques. First, in

Theo M. Theophanides (ed.), Infrared and Raman Spectroscopy of Biological Molecules, 327–353

contrast to methods of X-ray diffraction, Raman spectroscopy offers speed, simplicity and versatility. Second, Raman spectroscopy imposes no requirement of sample crystallizability. Spectra
of satisfactory quality may be obtained from both amorphous solids and aqueous solutions. The latter, with either H_2O or D_2O
as the solvent, are especially important, since the chemistry of
life takes place in an aqueous medium. However, water the ubi-
quitous biological medium, is an excellent absorber of infrared
radiation, leaving only restricted "windows" for infrared spectroscopy. Raman spectroscopy does not suffer as much from this
limitation, since water is a poor Raman scatterer. Here, the
advantage of Raman over infrared spectroscopy may also be noted.

Lasers now provide the high light power density required
for Raman spectroscopy and allow examination of minute quantities (microliters) of material. The chief obstacle encountered
with biological materials, however, is their complexity. A molecule containing N atoms has 3N-6 (3N-5 for linear molecules)
normal modes of vibration. The macromolecules of biology contain thousands of atoms and have far too many vibrational frequencies to be resolved, let alone assigned in a normal Raman
or infrared spectrum. Fortunately, these frequencies tend to
group themselves into more-or-less discrete bands, which can be
identified with certain classes of structure. These bands can
then be used to monitor changes in gross conformation and this
technique has been fruitfully applied to proteins, nucleic acids
and lipids (1,8). If one is interested in structural features
of a specific site of biological function within a macromolecule
these bands are a serious interference. What is needed is a selective technique that samples only the vibrations of the atoms
in the vicinity of the site. This can be achieved by resonance
enhancement of the Raman spectrum, if the atoms in the site give rise to an isolated electronic absorption band. A normal
Raman spectrum is obtained by illumination of the sample in

a transparent region of its spectrum. In resonance Raman spectroscopy, the illumination is within an absorption band. Most of the Raman bands are attenuated by the absorption, but some bands may be greatly enhanced. This effect is due to a coupling of electronic and vibrational transitions, and the vibrational modes which are subject to enhancement are localized on the grouping of atoms which give rise to the electronic transition, i.e., the chromophore. Resonance Raman spectroscopy therefore provides a means of monitoring vibrational frequencies of a chromophore, with high sensitivity (approaching that of ultraviolet-visible spectrophotometry), and without interference from non-chromophoric components of the sample. Since biological chromophores (hemes, flavins, metal ions. etc...) are usually at sites of biological function, the technique offers high promise as a new probe of biological structure.

The technique is limited by the tunability of laser sources, most of which currently operate in the red and blue regions of the spectrum. Rapid development can be anticipated for the ultraviolet region, which contains the most important biological chromophores, nucleic acid bases and aromatic protein residues.

RESONANCE RAMAN SPECTROSCOPIC TECHNIQUES

CHARACTERISTICS OF RESONANCE RAMAN SCATTERING

I. 1. RAMAN INTENSITIES

The resonance Raman effect was described in considerable detail by Placzek (9). The intensity in the ordinary Raman effect was evaluated theoretically only for the case when the frequency of the exciting line was very different from the frequency of the molecular absorption. The intensity of a Raman band is proportional to the derivative of the polarizability of the sample with respect to the Raman transition. This is a tensor quantity, whose elements are given by the Kramers-Heisenberg-Dirac dispersion equation (Van Vleck, 1929)

$$(\alpha_{ij})_{mn} = \sum_e \frac{(M_j)_{me}(M_i)_{en}}{\nu_e - \nu_0 + i\Gamma_e} + \frac{(M_i)_{me}(M_j)_{en}}{\nu_e + \nu_s + i\Gamma_e}$$

Here m and n are the initial and final states of the molecules, while e is an excited state, and the summation is over all excited states. The quantities $(M_j)_{me}$ and $(M_i)_{en}$ are electric dipole transition moments, along the directions j and i, from m to e and from e to n, and ν_0 and ν_s are the frequencies of the incident and scattered photons.

The Raman intensity is related to the scattering tensor by

$$I_{mn} \propto I_0 \nu_s^4 \sum_{ij} |(\alpha_{ij})_{mn}|^2$$

I_0 is the plane polarized intensity of the incident light.

When ν_0 approaches ν_e the left hand term in the summation can become very large for an allowed electronic transition. It is prevented from reaching infinity by inclusion of the damping term $i\Gamma_e$, which is a measure of the band width of the electronic transition.

For Raman spectra excited in the region of allowed electronic absorption bands, the vibrational modes which are expected to show intensity enhancement are those which lend intensity to the electronic spectrum.

The extension of the theory to account for the resonance Raman effect was attempted by Schorygin (10), Behringer (11), Brandmüller (11), Albrecht (12), Mortensen (13), Peticolas et al. (14), Jacon et al. (15) and Mingardi and Siebrand (16).Raman theory is currently in a state of ferment and development. Certain qualitative results are clearly established, however. There are basically two kinds of scattering mechanisms corresponding to

a) modes which connect the ground state to the excited state through Franck-Condon overlap

b) modes which mix at least two allowed electronic transitions.

The modes of type a) are totally symmetric modes while type b) modes are usually non totally symmetric and of any symmetry that is contained in the direct product of the representations of the two allowed electronic transitions.

1.2. DEPOLARIZATION RATIOS

The polarizability tensor α_{ij} is given by

$$\alpha_{ij} = \begin{vmatrix} \alpha_{xx} & \alpha_{xy} & \alpha_{xz} \\ \alpha_{yx} & \alpha_{yy} & \alpha_{yz} \\ \alpha_{zx} & \alpha_{zy} & \alpha_{zz} \end{vmatrix}$$

In general the $\alpha_{ij,s}$ depend on the frequency of excitation and how close it lies to that of the nearest allowed absorption band. The depolarization ratios in general show a dispersion, or an exact value determined by the symmetry of the mode. The depolarization ratio ρ_ℓ is given by

$$\rho_\ell = \frac{3\gamma_s^2 + 5\gamma_a^2}{45\bar{\alpha}^2 + 4\gamma_s^2}$$

where $45\bar{\alpha}^2 = 5(\alpha_{xx} + \alpha_{yy} + \alpha_{zz})^2$

$$\gamma_s^2 = \frac{1}{2} \left\{ (\alpha_{xx} - \alpha_{yy})^2 + (\alpha_{xx} - \alpha_{zz})^2 + (\alpha_{yy} - \alpha_{zz})^2 \right.$$

$$+ \frac{3}{4} \left. (\alpha_{xy} + \alpha_{yx})^2 + (\alpha_{xz} + \alpha_{zx})^2 + (\alpha_{yz} + \alpha_{zy})^2 \right\}$$

and

$$\gamma_a^2 = \frac{3}{4} \left\{ (\alpha_{xy} - \alpha_{yx})^2 + (\alpha_{xz} - \alpha_{zx})^2 + (\alpha_{yz} - \alpha_{zy})^2 \right\}$$

For the ordinary Raman effect occuring when the excitation frequency is far removed from that of the nearest absorption band the tensor is symmetric and $\alpha_{ij} = \alpha_{ji}$. Hence $\gamma_a^2 = 0$ and ρ_ℓ has its well known form for non resonant Raman spectra of

$$\rho_\ell = \frac{3\gamma_s^2}{45\bar{\alpha}^2 + 4\gamma_s^2}$$

with

$$\gamma_s^2 = \frac{1}{2} \left\{ (\alpha_{xx} - \alpha_{yy})^2 + (\alpha_{xx} - \alpha_{zz})^2 + (\alpha_{yy} - \alpha_{zz})^2 \right\}$$

$$+ 3 \left\{ \alpha_{xy}^2 + \alpha_{zx}^2 + \alpha_{yz}^2 \right\}$$

It is clear then ρ_ℓ can have values from 0 to $\frac{3}{4}$ for totally symmetric modes (class II) but only $\frac{3}{4}$ when $45\bar{\alpha}^2$ is zero and we are concerned with nontotally symmetric modes (class IV). It is readily seen that for resonance Raman bands an antisymmetric part γ_a^2 is introduced to give three extra categories of classes; the first is for totally symmetric modes where ρ_ℓ can be between 0 and ∞ (class I) for not totally symmetric modes for which ρ_ℓ can be between $\frac{3}{4}$ and ∞ (class III); and for nontotally symmetric modes for which $\rho_\ell = \infty$ (class V).

	α^2	γ_s^2	γ_a^2	
class II	+	+	0	$\rho = 0$ to $\dfrac{3}{4}$
class IV	0	+	0	$\rho = \dfrac{3}{4}$
class I	+	+	+	$\rho = 0$ to ∞
class III	0	+	+	$\rho = \dfrac{3}{4}$ to ∞
class V	0	0	+	$\rho = \infty$

Measurements of $I(\perp)$, $I(//)$ and $I(CO)$ or I (contra) lead to individual values of the three polarizability invariants α^2, γ_s^2 and γ_a^2. This information however, is not sufficiently complete to determine the point group symmetry for molecules having modes which give a dispersion of ρ_ℓ . In addition to the measurements at a particular exciting frequency one requires the measurements at various values of ν_o.

The measurement of the reversal coefficients is difficult and prone to considerable error so that if $I(\perp)$ and $I(//)$ are measured at various exciting frequencies one can determine un- ambiguously the class in the table to which the mode belongs but of course no separate evaluation of each of α^2, γ_s^2 and γ_a^2 is possible.

I.3. APPLICATIONS

The information obtained is of two types: one involves the fact that the intensity changes with change of frequency of the exciting line (the so-called excitation profile), and the other involves the change of ρ_ℓ with change of frequency of the exci- ting line (the so-called dispersion of ρ_ℓ). By determining the maximum in the excitation profile one can identify 0-0 and 0-1 bands in the electronic absorption spectrum, for example.

The dispersion of ρ_ℓ can very often give valuable structural information.

SPECIAL INSTRUMENTATION AND TECHNIQUES
FOR RESONANCE RAMAN SPECTROSCOPY

II. 1.   EXPERIMENTAL METHODS

The instrumentation and techniques so far used for resonance Raman work are apart from certain accessories or modifications, identical with those available for ordinary Raman spectroscopy. Schematic illustration of an arrangement for obtaining a Raman spectrum with a laser light source is shown in Fig.I.

In the last few years some tunable dye lasers have become available. The rhodamine 6G argon ion dye laser for example gives a laser line tunable from 550 to 630 nm. Further spectral ranges are available with nitrogen laser pumping a 337 nm and giving a tunable range from $360 \longrightarrow 670$ nm (see Fig. II).

The lasers make it possible to obtain a continuous measurement of intensity excitation profile of a Raman band in the specified spectral regions.

II. 2.   MEASUREMENT OF ALL POLARIZABILITY INVARIANTS

The Raman tensor possesses the three invariants, namely the isotropic part $\bar{\alpha}^2$, the symmetric anisotropy γ_s^2 and the antisymmetric anisotropy γ_a^2 . It is possible to measure the amount of each invariant directly at any particular exciting frequency using both linearly and circularly polarized light. A schematic diagram of the experimental set-up used to measure the polarized spectra is given below. (See Fig.III).

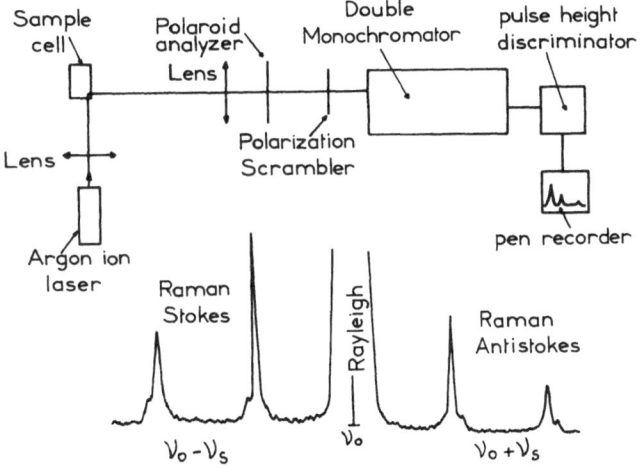

Fig. I Schematic diagram for observation of Raman spectra.

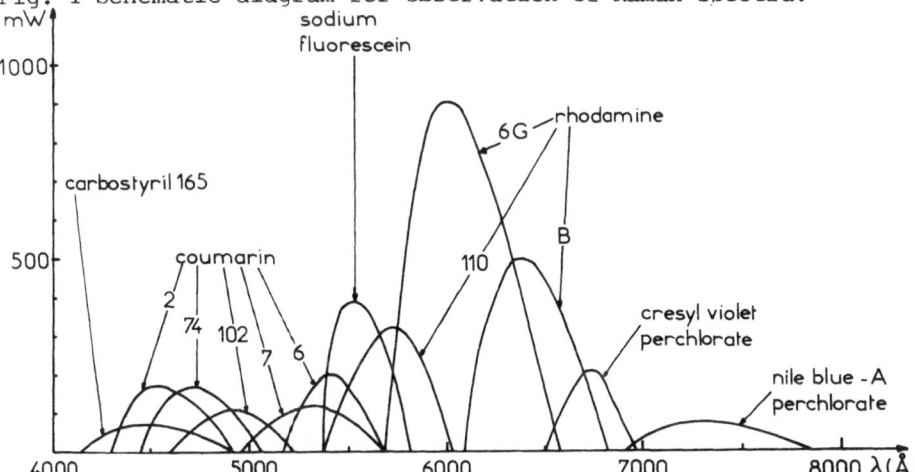

Fig. II The spectral regions covered by various dyes currently used in dye lasers (courtesy of Coherent Radiation, Inc.).

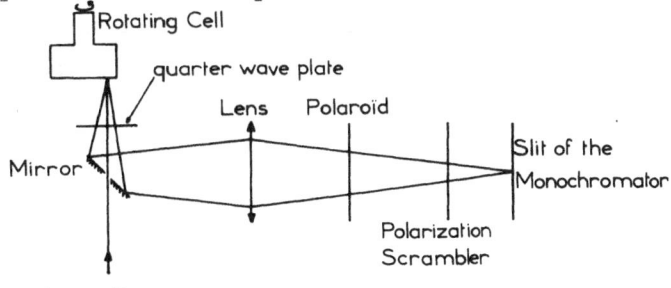

Fig. III Experimental arrangement used to measure the Raman polarizations.

The laser beam which is plane polarized is first passed
through a quarter wave plate that can be rotated to give either
planar or circulary polarized light incident on the sample (see
Fig. III). The back scattered radiation passes through the same
quarter wave plate, is collected by a mirror and a lens and is
analysed with a polaroid disc followed by a polarization scram-
bler.

For the two linearly polarized experiments the optical axis
of the quarter-wave plate was set parallel to the polarization
vector of the incident radiation. In this position the quarter
wave plate merely passes the incident and scattered radiation
without changing the polarization. Thus the spectrum was obtained with the analyser set to transmit the // component of the
scattered light, and that called $\perp$ was obtained by rotating the
analyser by 90°.

For the two circulary polarized experiments, the quarter
wave plate was rotated by 45° to change the polarization of the
incident light from linear to circular. If the scattered radiation is polarized in the opposite sense relative to the incident
polarization (contra-rotating), the $\lambda/4$ plate converts it back
to linearly polarized light with a polarization vector perpendicular to that of the laser beam. The contra-rotating spectrum
can then be observed when the analyser is set to transmit the $\perp$
component of the scattered light. When the scattered radiation
is polarized in the same sense relative to the incident polarization (Co-rotating), the $\lambda/4$ plate converts it to plane polarized
light relative to the laser beam. The co-rotating spectrum may
then be observed by rotating the analyser 90° to pass the // component. Except for same proportionality constant,

$$I \perp = 3\gamma_s^2 + 5\gamma_a^2$$
$$I// = 45\alpha^2 + 4\gamma_s^2$$

$$I_{CO} = 6\gamma_s^2$$

and the three invariants α^2, γ_s^2 and γ_a^2 are obtained from the measurement of $I_\perp$ $I_{//}$ and I_{CO}.

II.3. IMPEDIMENTS TO RESONANCE RAMAN OBSERVATIONS AND THEIR CONTROL

Resonance Raman spectroscopy has special difficulties not found in ordinary Raman spectroscopy. These difficulties and convenient remedies will be discussed below. Furthermore, resonance Raman spectroscopy, unlike ordinary Raman spectroscopy is closely connected not only with I.R. but also with electronic absorption and emission spectroscopy.

The intricacies encountered in experimental resonance Raman spectroscopy can be traced back to three major causes:

1- Absorption
2- Fluorescence
3- Photolysis

We consider them in turn.

II. 3.1. Absorption

Resonance Raman spectra are excited by frequencies near to or even coinciding with absorption frequencies of the sample substance. Using a simplified wording not quite correctly reflecting the exact theory, we can say that the absorption competes with scattering and that in many cases the probability for absorption of the incident light quantum surpasses by many orders of magnitude the probability for scattering. The absorption of the exciting radiation with all its possible consequences thus appears as a process very seriously encroaching upon resonance Raman scattering. Let us consider various aspects of this phenomenon.

II.3.1.1. <u>Absorption losses</u>

First of all, absorption appears to steal intensity from scattering. Depending on the frequency dependence of the absorptivity the sample absorbs the exciting as well as the frequency-unshifted (Rayleigh) or schifted (Raman) scattered light. The ratio of the probabilities for absorption and for scattering of an incident photon, by a sample subunit (atom, molecule, ion, etc...) is an intrinsic property of the sample determined by its chemical composition and its physical state and cannot artificially be varied without more or less significantly changing the substance itself or its state.

II.3.1.2. <u>Practical methods or minimizing the absorption losses</u>

The least incisive practical method consists of optimizing the illumination and observation optics and geometry. Sometimes some change of state (in the strict sense of the word) of the sample must also be admitted in order to obtain good Resonance Raman spectra, e.g., a change of density of the scattering particles or the choice of another phase. There are two typical illumination arrangements:

a) <u>the transmission arrangement</u> (see Figure IV. a)

In the transmission arrangement the laser beam is focused into sample volume close to or grazing the exit surface for observed scattering. Grazing angles 5° are customary. Here clearly only the absorption of the scattered (but not of the exiting) radiation is minimized. Therefore, this very simple and convenient method is useful for not too strongly absorbing samples.

b) <u>the réflection arrangement</u> (see Figure IV. b)

In the reflection arrangement, which is recommendable for strongly absorbing samples (in particular crystals), diminution

A - The transmission arrangement

B - The reflection arrangement

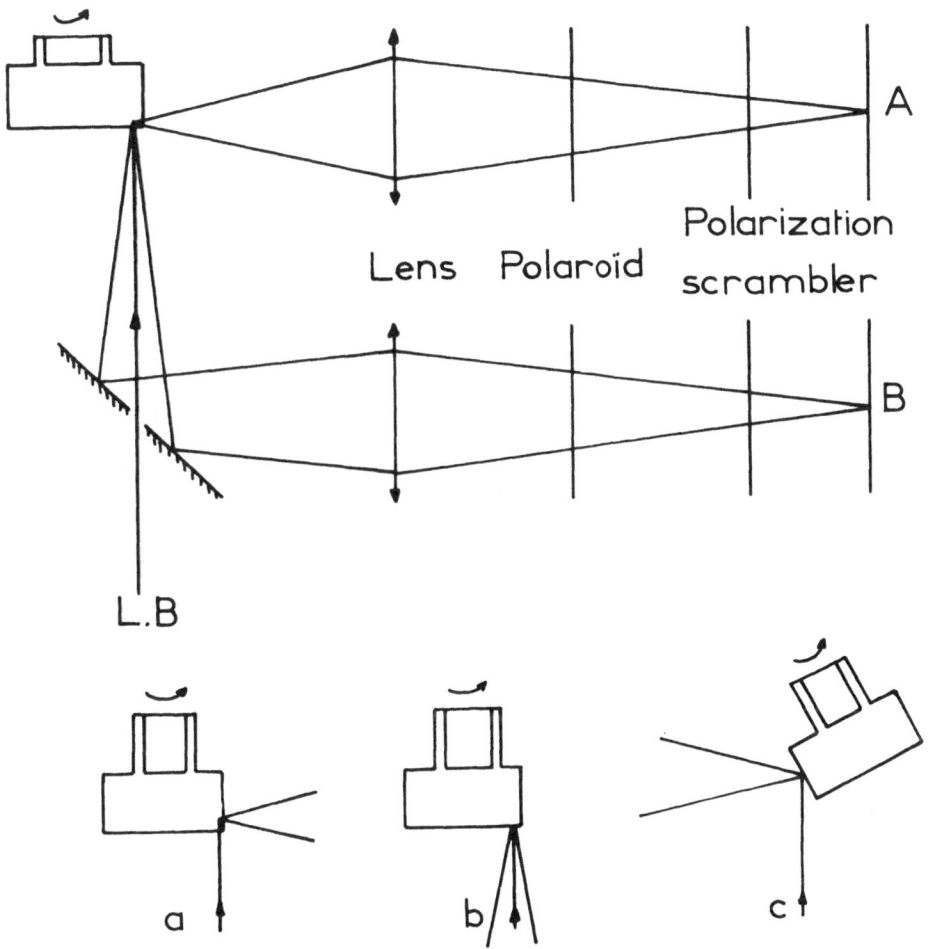

Fig. IV Irradiation of rotating cell for liquids:
(a) which absorb weakly the incident light,
(b) and (c) which strongly absorb the incident light.

of exciting light absorption is also attained by having the beam
impinge from outside into the sample. Often small angles of in-
cidence and observations perpendicular to the incident beam are
used. For gases or liquids this reflection arrangement is often
disadvantageous because the cell material itself may contribute
interfering Raman scattering in certain spectral regions.

II.3.2.2. <u>Sample heating</u>

A consequence of laser light absorption is local heating of
the sample (17). The absorbed radiation energy after a sequence
of radiationless transitions is finally dissipated among neighbouring particles and converted into heat. This local heating
effect, of course, will be all the stronger the higher the gra-
dient of the radiation flux density, i.e., in particular the more
the laser beam diameter is narrowed by focusing with lenses of
short focal lenght into the sample.

Several unpleasant results may ensue from this local temperature enhancement:

- liquids: vaporization (boiling)
- crystals: melting-sublimation-generation of cracks...
- thermal lens's effect.

Furthermore, local heating of the sample as a rule gives
rise to local thermal expansion, which for its part decreases
the refractive index. This decrease is most important near the
focus and along the axis of the laser beam, and leads to what
has been called "the thermal lens effect". This thermal lens
is diverging, i.e., it <u>defocuses</u> the laser beam.

This local heating effect is also a direct connection between temperature and scattering due to the temperature dependence
of the average occupation of possible initial states in the scattering process. In order to understand quantitatively resonance
Raman intensities it should be ascertained whether, and to what

extent, thermal equilibrium within the sample can be assumed and
what the spatial distribution of temperature in the laser-exci-
ted scattering volume looks like. As long as temperature gradients
are very high it will be difficult to give reliable answers to
these questions.

The simplest way out of these difficulties is to reduce the
local heating effect by appropriate experimental measures. Therefore, essentially only three ways for attaining the goal mentioned seem to remain.

a) Diminution of high radiation flux gradients

This can be realized either by lowering the laser beam power
as a whole or by reducing the convergence of the beam;

b) Shortering of the exposure time

For this purpose the use of a focusing lens of longer focal
length or the renunciation of any focusing at all will usually
suffice. The disadvantage of this measure is clearly the simultaneous diminution of the detectable scattering intensity.

c) Diminution of absorbing-particle density or concentration

(only praticable with gases and solutions).

d) Continually moving technique

The reduction of the local heating effect can be realized
simply by continually moving the sample with respect to the laser
beam focus, thus subsequently exposing different parts of the sample to the intense exciting radiation. In particular, gases and
liquids can be investigated when flowing (18). However, this method does not always satisfy the expectations (probably owing to
the relatively small attainable streaming velocities of viscous
media and the immovability of boundary layers at the tube walls).

e) Rotating-cell procedures (19, 20, 21)

In principle, rotating-sample techniques can be devised for

all phases of matter and for transmission as well as reflexion arrangements (see Fig. IV).

The basic idea is to lacate the sample in an approximately uniform distribution along the periphery of a wheelshaped cell, where a small spot is illuminated by the focused laser beam and either to have the cell rotated with respect to the static laser beam or to have the laser beam scanned over the cell kept at rest (22).

The reduction of local heating by the rotating cell or surface-scanning techniques is accompanied by a reduction of the thermal lens effect so that the laser beam can be focused more sharply into the sample with lenses of shorter focal lengths.

II. 3. 2. Fluorescence

Fluorescence emission is also an after-effect of absorption. It militates against the observation of resonance Raman spectra because of its relatively high intensity, which usually surpasses that of scattering by several orders of magnitude. The occurence of fluorescence is more probable in resonance Raman than in ordinary Raman spectra, because fluorescence is excited only by wavelengths short from which fluorescence emission can start.

While in ordinary Raman spectra a fluorescence background usually is caused only by sample impurities with low lying electronic states (even very low impurity concentration may be sufficient to give fluorescence intensities comparable to Raman scattering) the fluorescence concurrent with resonance Raman scattering mostly arises from the sample itself, and therefore cannot be removed by sample purification.

The possibility of fluorescence emission does depend on:
1. the type of the molecule,
2. its interactions with neighbouring molecules,
3. the exact excitation conditions.

II. 3.2.1. The time dependence of fluorescence emission and of Raman scattering.

Fluorescence emission means radiant deactivation of electronically excited particle systems. The emission temporally follows the process of excitation by a statistical low of decay.

In contrast to fluorescence the time dependence of Raman scattering cross-sections is more complicated. It depends not only on the time dependence of the excitation and of the lifetimes of the initial and final states in the scattering process but (particularly in resonance cases) also on the lifetimes of all intermediate states contributing appreciably to the scattering probability. In the Raman scattering in general (except special cases of rigorous resonance) many intermediate states contribute.

There are some quenching mechanisms which practically do not influence the ordinary Raman effect when a large number of intermediary states are involved in the scattering phenomenon. In the case of rigorous Raman effect (R.R.E. or R.F.) this quenching is very efficient because only a few states play a predominant role in the phenomenon.

Reduction of the fluorescence

Fluorescence usually can be lessened appreciably or eliminated by means of the following methods.

1. " Drench-Quench " Method

Fluorescence often is greatly reduced by exposing the sample to 200-500 mW blue laser radiation, significant decay may be attained within a few minutes, but several hours sometimes are required. In fact, some refractory materials must remain in the light path for one or two days before a Raman spectrum can be obtained. If decomposition occurs a red laser source should be utilized.

2. Addition of a Quenching Agent

Suppression of fluorescence may be accomplished by addition of a quenching agent to a liquid sample. This method suffers from the disadvantage that strong bands appear in the spectrum, due to the agent.

3. Distillation, recrystallization, sublimation and extraction

Sample purification by these standard micro and semi-micro techniques often can diminish the fluorescent effect.

4. Suppression of luminescence by instrumental modifications

4.1. F.M. Raman spectroscopy (ref. 23):

The laser light incident on a sample is modulated in frequence between two closely spaced values and the difference of the sample response at the two frequencies is recorded. This quantity is proportional to the derivative of the sample response with respect to input frequency. The F.M. method automatically cancels out any contribution to the detector output which is independent of the laser frequency. Since the spectral distribution of sample luminescence does not vary with the excitation frequency, luminescence peaks do not shift in the experiment and can therefore be made to cancel out.

4.2. Time discrimination:

Since the Raman effect is essentially instantaneous (about 10^{-13} sec), but fluorescence requires a finite rise and decay time (about 10^{-8} sec), the principle of time discrimination to eliminate fluorescence theoretically can be applied to any sample. Therefore, a Raman scatterer illuminated by a ultrashort light pulse of 10^{-13} - 10^{-14} sec will give rise to a Raman pulse approximately as wide as the exciting pulse and the fluorescence spreads over 10^3-10^5 longer time periods. This technique for the rejection of interfering luminescence background signals employing a mode-locked argon ion laser and single photon timing detection electronics is described (24). Only those photons emitted or scattered by the sample during the mode locked laser pulse are passed on to the recording circuitry by the single photon timing detection system. Essentially all of the Raman signals can be recorded and a large fraction of the luminescence background is rejected.

II.3.3. Photolysis

Typical chemical bond energies for single, double, and triple bonds are 50, 100 and 150 Kcal. mol^{-1}, respectively. Neglecting the finer details of photochemical dissociation processes we can qualitatively state that by absorption of photons with wavelengths equal to or smaller than 571, 286 and 190 nm, the respective bonds may be caused to break, provided of course that the necessary electronic vibrational energy transfer takes place. This shows that photochemical instability in principle must always be taken into consideration when R.R.S. of molecules are excited with sufficiently shortwave radiation. The reason is that after the (virtual or real) absorption transition $r \rightarrow m$ from the initial to the intermediate state the molecule acquires an increasing chance to perform a transition into some dissociation continuum state (which subsequently leads to disintegration) instead of going into the scattering final state n.

The transformation of electronic excitation energy during the dissociation process into vibrational and translational energy of the dissociated molecular fragments is one of the causes of local sample heating. In practical resonance Raman spectroscopy, it is advisable to take the following measures in order to exclude photochemical changes:

(1) use fresh samples, (2) avoid unnecessary exposure of the sample to light (use weak intensities or short exposure time, e.g., by applying a shutter or rotating cell) and (3) take appropriate measures to prevent specific photochemical changes (e.g. deoxygenate the sample when there is danger of photo-oxydation, choose favourable temperatures, solvents and concentrations).

Fortunately, such precautionary measures are not always required. Surprisingly, even very complicated biochemical compounds proved to be very stable under long-lasting and very intense irradiation. Photolysis is often reversible and conditions of reversibility should be explored.

APPLICATIONS TO SOME BIOLOGICAL SYSTEMS

INSTRUMENTATION

The Laser Raman schematic is shown in Figure I. The Raman spectra are obtained with a spectra-physics 164-05 argon ion Laser and a Ramanor HG.25 double monochromator using Holographic Gratings.

The laser beam is directed into the sample area by means of dielectric mirrors that maintain the polarization of the laser and reflect more than 99% of the incident light. The radiation is brought to a focus by a simple lens and the sample is accurately placed at the focus in an rotating system to avoid sample decomposition. The focused beam is kept as close as possible to the cell window to minimize self-absorption of the scattered Raman light.

THE RESONANCE RAMAN SPECTRA OF CYTOCHROME C (25, 26, 27)

Horse-heart cytochrome c (Fig. V) was purchased from Sigma (type III) and used without further purification. Sample concentrations ranging from $5\text{-}10^{-3}$ M to 10^{-6} M were tried and the intensity of the Raman spectrum was found to be nearly independant of the concentration (high concentrations fail to give a stronger spectrum because of reabsorption of the scattered light by the sample). At a concentration of $10^{-3} - 10^{-4}$ M the fluorescence of the heme groups is partially self quenched by the protein. The samples are reduced anaerobically by adding a small excess of sodium dithionite or oxidized with excess potassium ferricyanide and the rotating cell technique is used to avoid heating of the sample.

The visible and near-ultraviolet spectra of metalloporphyrins are dominated by two $\pi \rightarrow \pi^*$ transitions. The strong Raman bands are at frequency shifts (1000-1700 cm^{-1}) appropriate for in-plane stetching vibrations.

The appearance of intense bands with anomalous polarization (ρ_ℓ 3/4) is the most remarkable feature of hemoprotein spectra. Their observation (25) was the first experimental confirmation of antisymmetric vibrational scattering, even though the prediction was given by Placzek for point group C_{4V} for which the ρ_ℓ or A_2 modes should be infinity.

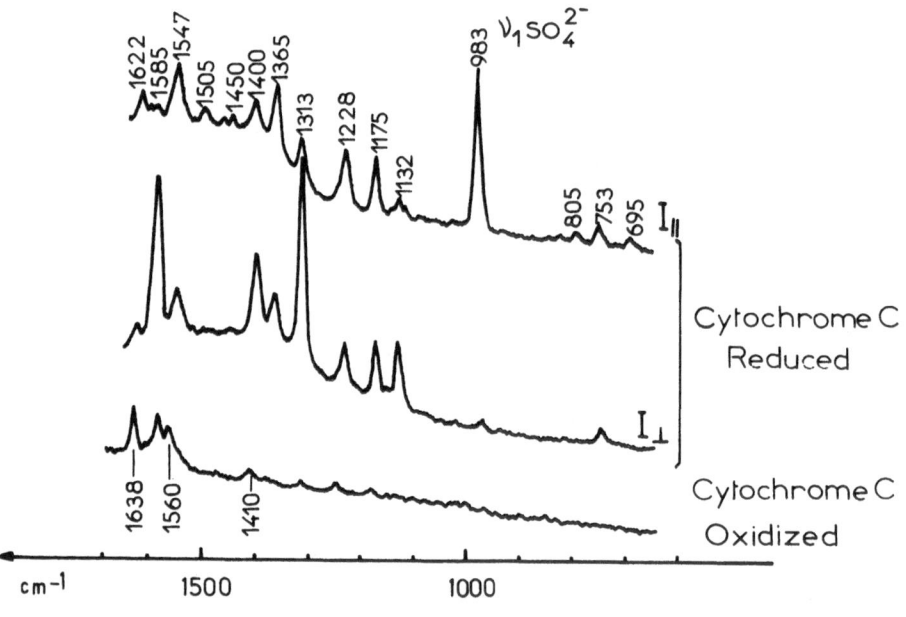

Structure of Cytochrome C

Fig. V Raman spectra for reduced and oxidized horse heart cytochrome C (Sigma type III) in 0.8 mM aqueous solution (ref. 25).

RESONANCE RAMAN SPECTRA OF VITAMIN B 12 (28, 29, 30)

The cyanocobalamin (vitamin B-12) was supplied by Sigma. This was used without further purification since, although serious fluorescence interference with the Raman scattering was encountered at high solution concentration at $1.10^{-4} - 5.10^{-4}$ M concentrations good quality spectra with essentially flot base lines were obtainable directly in the region 0-1800 cm^{-1}. The Figure VI shows Raman spectra obtained from $5-10^{-4}$ M aqueous solutions of cyanocobalamin using approx. 300 mW of focussed 488 nm Ar^+ laser radiation, the beam being held close to the wall of the rotating liquid cell in order to minimise absorption losses.

An intense band at approx. 1500 cm^{-1} is prominent in the spectra and several weaker bands at lower frequencies. All of the bands in both the cobalamin are polarized with depolarization ratios, ρ_ℓ ranging from 0.2 to 0.6. The phenomenon of inverse polarization, exhibited by the resonance Raman spectra of cytochrome c is not observed here. Since the highest symmetry available to the corrin ring is C_{2v} it is not surprising that no inverse polarized bands are found in vitamin B-12 spectra.

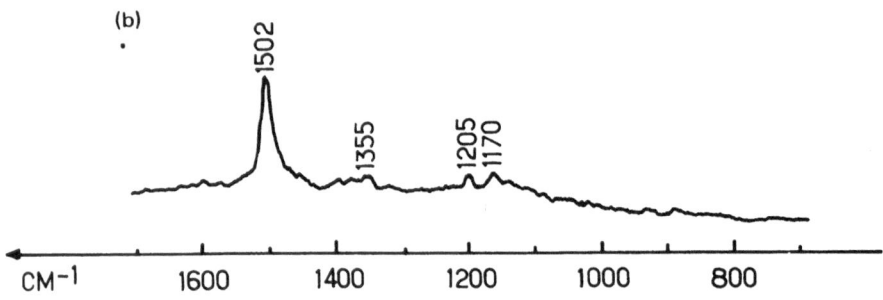

Fig. VI (a) Chemical structure of cobalamins
        (b) Raman spectra of aquocobalamin.

REFERENCES

1. T. G. Spiro. Chemical and Biochemical Applications of Lasers (vol. 1), edited by C. Bradley Moore, Academic Press New York Chapter 2 (1974)

2. A. Lewis and J. Spoonhower; Tunable Laser Resonance Raman Spectroscopy in Biology and Chemistry, Neutron, X-ray, Laser, edited by S. Chen and S. Yip, Academic Press, New York (1974)

3. Applications of Laser Raman Spectroscopy par S.K. Freeman, Wiley-Interscience (1974)

4. T.G. Spiro, Accounts of Chemical Research, 7, 339 (1974)

5. T.G. Spiro and T.M. Loehr, Resonance Raman Spectra of Heme Proteins and other Biological Systems in "Advances in Infrared and Raman Spectroscopy", vol. 1, edited by R.J.H. Clark and R.E. Hester, Heyden and Son, London (1975)

6. J. Behringer, Experimental Resonance Raman Spectroscopy, in "Molecular Spectroscopy " ed. R.F. Barrow, D.A. Long and D.J. Hillen (Specialist Periodical Reports) the Chemical Society, London (1975) vol. 3

7. W. Kiefer, Recent Techniques in Raman Spectroscopy in "Advances in Infrared and Raman Spectroscopy", vol. 3, edited by J.R.H. Clark and R.E. Hester, Heyden and Son, London (1977)

8. J.L. Koenig and B.G. Fruschour, Raman Spectra of Proteins in "Advances in Infrared and Raman Spectroscopy" Vol. 1, edited by J.R.H. Clark and R.E. Hester, Heyden and Son, London (1975)

9. G. Placzek, Handbuck der Radiologie, VI 2, 209 (1934), Leipzig

10. P.P. Shorygin, Dokl Akad. Nauk. SSSR 12, 576 (1948)

11. J. Behringer and J. Brandmüller, Zev. F. Elektrochemie Ber. der Bunsen Ges. für Physik. Chemie, 60, 643 (1956)
                                     62, 544 (1958)
                                       62, 906 (1958)

12. A.C. Albrecht, J. Chem. Phys. $\underline{34}$, 1476 (1961)

13. O.S. Mortensen, J. Mol. Spectry, $\underline{39}$, 48 (1971)

14. W.L. Peticolas, L. Nafie, P. Stein and B. Fanconi, J. Chem. Phys. $\underline{52}$, 1576 (1970)
    L.A. Nafie, M. Pezolet and W.L. Peticolas, Chem. Phys. Lett. $\underline{20}$, 563 (1973)

15. M. Jacon, Advances in Raman Spectroscopy, vol. I, Heyden and Son, London (1972) p. 324
    D. Van Labeke, M. Jacon, M. Berjot and L. Bernard, J. Raman Spectry, $\underline{2}$, 219 (1974)

16. M. Mingardi and W. Siebrand, Chem. Phys. Lett., $\underline{24}$, 492 (1974)
         "   "   "   J. Chem. Phys., $\underline{62}$, 1074 (1975)

17. W. Holzer, W.F. Murphy and H.J. Bernstein, J. Chem. Phys. $\underline{52}$, 399 (1970)

18. T. Kamisuki and S. Maeda, Chem. Phys. Letters, $\underline{19}$, 379 (1973)

19. W. Kiefer and H.J. Bernstein, App. Spectry, $\underline{25}$, 500 (1971)
         "   "   "   , App. Spectry, $\underline{25}$, 609 (1971)

20. H.J. Sloane and R.B. Cook, App. Spectry, $\underline{26}$,589 (1972)

21. J.B.R. Dunn, D.F. Shriver and I.M. Klotz, Proc. Nat. Acad. Sci. U.S.A., $\underline{70}$, 2582 (1973)

22. N. Zimmerer and W. Kiefer, App. Spectry, $\underline{28}$, 279 (1974).

23. F. L. Galeener, Bull. Amer. Phys. Soc., II $\underline{20}$, 44 (1975)
         "   "   , Proceedings of the 5nd Int. Conf. on Raman Spectry Freiburg (1976)

24. R.P. Van Duyne, D.J. Jeanmaire, D.F. Shriver, Anal. Chem. $\underline{46}$, 213 (1974)

25. T.C. Strekas and T.G. Spiro, Biochem. Biophys. Acta (BBA report) $\underline{278}$, 188 (1972)

26. D.W. Collins, D.B. Fitchen and A. Lewis, J. Chem. Phys. $\underline{59}$, n° 10, p. 5714 (1973)

27. M. Pezolet, L.A. Nafie, W.L. Peticolas, J. Raman Spectry, $\underline{1}$, 455 (1973)

28. E. Mayer, D.J. Gardiner and R.E. Hester
   Biochim. et Biophys. Acta, $\underline{297}$, 568 (1973)
   Mol. phys. $\underline{26}$, n° 3, p. 783 (1973)

29. W.T. Wozniak, T.G. Spiro, J. Amer. Chem. Soc. $\underline{95}$, 3402 (1973)

30. F. Galluzzi, M. Garozzo, and F.F. Ricci, J. Raman Spectry $\underline{2}$, 351-362 (1974)

TECHNIQUES OF RAMAN SPECTROSCOPY

by Michel DELHAYE, University of Lille

CNRS-FRANCE

1. - RECENT PROGRESSES IN TECHNIQUES

During the past ten years, new techniques have been developed to improve the study of molecular scattering of light. This
paper will illustrate the efforts that have been concentrated to
take advantages of the progress of LASER light sources, photoelectric detectors, and sophisticated optical systems.

These techniques take benefit of our ability:

- to focus a LASER beam on a sub-micron area
- to deliver a large photon density in a short time by means of
  a pulsed LASER (10^{-3} to 10^{-11} second)
- to discriminate between the intense elastically scattered LASER
  light and the very weak spectral lines which are frequency shifted by Raman effect, by means of holographic gratings
- to measure simultaneously a large number of spectral elements
  by means of multichannel photodetectors and analysers.

Theo M. Theophanides (ed.), Infrared and Raman Spectroscopy of Biological Molecules, 355–365.

The combination of these improvements permitted studying structural and kinetic problems requiring:

- Time resolution

- Spatial resolution.

1.a - <u>Time resolved Raman spectra</u>:

Rapid scanning spectrometers using holographic gratings have been developed by <u>F. WALLART</u> and coworkers. The minimum time which is required to record the spectral range of interest is less than one second. The spectra are treated in real time by a minicomputer system.

Such spectrometers are quite useful for studing a variety of chemical kinetic problems, in the time range from minutes to seconds.

However, many processes occur at much faster rates. For such systems, it is better to utilize a pulsed LASER and a "Multichannel" Spectrometer of the type developped by <u>M. BRIDOUX</u> and coworkers. By means of "imaging photodetectors", all spectral elements are examined simultaneously during a LASER pulse and the signals are stored in an electronic memory. Data analysis is also performed by a minicomputer.

The most evident advantage of the system is the ability of recording the whole range of the Raman spectrum by using a single pulse emitted by the LASER. Data accumulation, averaging or correlation calculation can also be made by recording a series of pulses.

The most recent instrumental advances include the use of Q-switched or modelocked LASERS, which deliver pulses of 10^{-9} to 10^{-11} second duration, in conjunction with concave holographic gratings to reduce the stray light levels in the spectrometer. Some examples of "Spontaneous" Raman spectra recorded in 20 pico-second will be presented. The ability to observe in the same experiment both spontaneous and stimulated effects is interesting.

By using a series of repetitive LASER pulses, the evolution of chemical or physical modifications, or the relaxation phenomena can be explored. As the "multichannel spectrometer" is corrected for astigmatism, the spatial resolution along the slit permits either to observe different parts of the sample, or to improve the time resolution by producing a rapid optical deflexion along the slit.

1.b - "Space Resolved" Raman spectra:

It is well known that the LASER beam can easily be focussed on a spot whose dimensions are only limited by diffraction, to a fraction of micrometer in the visible light. In practice, the limitations of conventional Raman spectrometers, to use this property of the LASER beam for obtaining spectra from microsamples, are determined by:

- the quality of optics

- the thermal degradation of the sample

- the stray light from optical elements.

    With the recent development of holographic optics and sophi-
sticated light detectors, it has been possible to obtain spectroscopic vibrational data from specimen of micrometer dimensions.

2. - PRINCIPLES AND APPLICATIONS OF THE MOLECULAR LASER-RAMAN MICROPROBE M.O.L.E.

    The above described experiments of "space resolved" Raman
spectra have been extended to create a new instrument. which makes
use of the molecular scattering of LASER light to give a precise
identification and an accurate localization of microsized area in
various samples.

    This new analytical instrument has been developed in close
collaboration between our laboratory and french industry - P.
DHAMELINCOURT and M. LECLERQ at C.N.R.S., University of Lille
and E. DA SILVA at I.S.A. LIRINORD have cooperated to create the
Molecular Microprobe, which is called M.O.L.E.

    This instrument offers an attractive complement to other
microanalytical methods. The lack of information on chemical
bonds and the arrangements of atoms in polyatomic molecules ren-
ders most ultra-microanalytical unsuitable for organic or biological samples. Even for inorganic, the unanswered question is
whether or not a given molecular or ionic polyatomic compound
exists in

a small particle or in a microinclusion in an heterogeneous sample.

We feel that molecular inelastic scattering of light, as observed by Raman effect, is rather well suited to the determination of polyatomic compounds, because sharp and characteristic spectra may be obtained from samples in solid, crystalline or amorphous, or in liquid and even gas states.

The M.O.L.E. microprobes resembles an electron microprobe in that a beam of photons, instead of electrons, is focused onto a very small sample area. The photons excite the vibrational transitions of polyatomic species. This results in well define frequency shifts of the scattered light, which characterize unambiguously any molecule or crystal.

The excitation probe is a monochromatic LASER beam, whose power is low enough to avoid thermal degradation of microsamples.

By using the most advanced techniques of optics and electronics, the sensitivity of the M.O.L.E. has been surprisingly increased.

- Most chemical compounds can be identified from subnanogram samples. In many cases, very good vibrational spectra have been obtained from particles of a few cubic microns, typically 10^{-11} grams of matter.

In addition to the Raman spectra the same instrument can give the distribution of the intensity of a characteristic Raman line

in an heterogeneous sample. By isolating this characteristic
Raman radiation, a selective "map" or "image" of the sample can
be rebuilt, which gives a precise distribution of a given polyatomic compound. The MOLE uses a powerful optical microscope, with
a resolving power of the order of 10^{-3} millimetre, and detailed
micrographic "Raman Images" can be obtained from a variety of
samples.

The MOLE combines in a unique instrument, a Raman microspec-
trophotometer and an optical microscope and microprobe. One of
the main advantages is its ability to obtain determinations by
a non destructive method, to study samples in air, or in a con-
trolled atmosphere, or even immersed in a liquid, or under pressure, at low or high temperature.

It has applications in many various fields:

- in geochemistry and geophysics, for the study of rocks, ores
  and minerals, and especially of solid and fluid inclusions

- for identification and characterization of gems and art materials, by a non destructive method, which has been used even
  for large pieces of ancient sculptures

- for the analysis of dust particles and pollution (sulphates,
  silicates, oxides, pesticides)

- for the identification of small particles of impurities (organic or inorganic compounds) in materials (papers, textiles,

plastics, ceramics, semiconductors, food,etc...)

- for studying the evolution of molecular or crystalline composition and of the distribution of species in microscopic samples
during chemical reactions (corrosion, thermal degradation, photochemical processus, electrochemistry, catalysis, etc...)

- for the identification and localization of inorganic or organic
compounds in biology or biochemistry (inclusions, drugs, products
of metabolism, markers, prothesis).

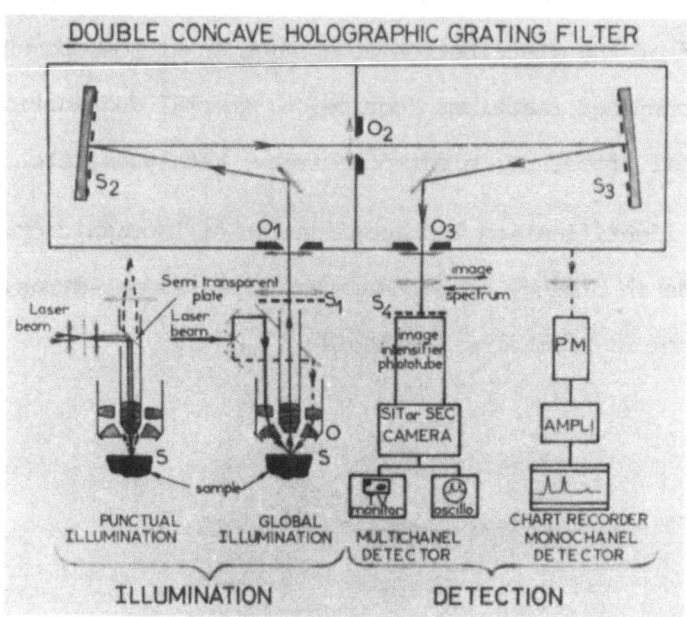

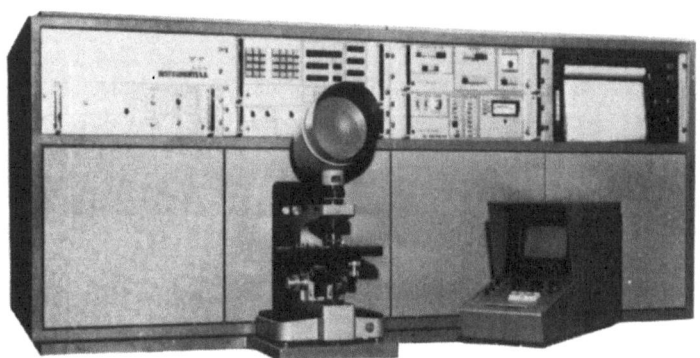

Fig I. The Raman Microprobe and Microscope (M.O.L.E)

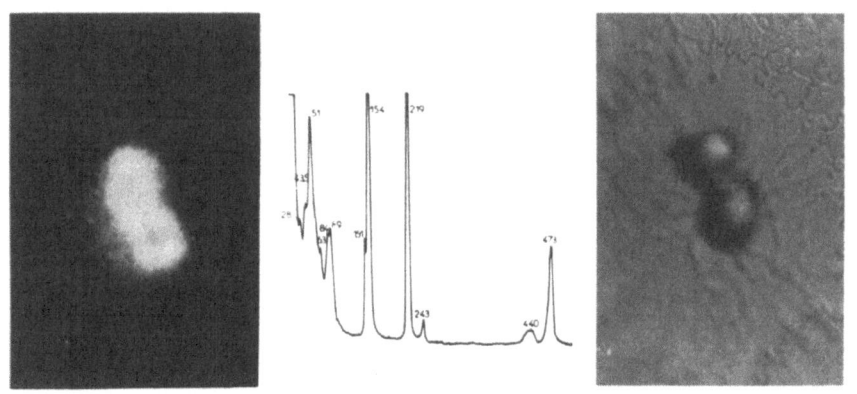

Raman Image($473cm^{-1}$)   RAMAN Spectrum   Micrograph(White light)

Fig 2. Identification of small particles of Sulphur S_8
included in glass bubbles

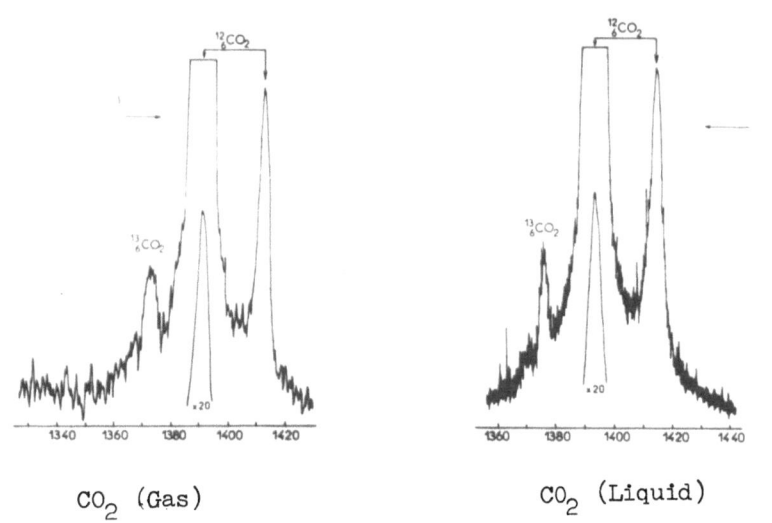

CO_2 (Gas) CO_2 (Liquid)

Fig 3. ^{12}C and ^{13}C components in the Raman spectrum
of inclusions in natural quartz

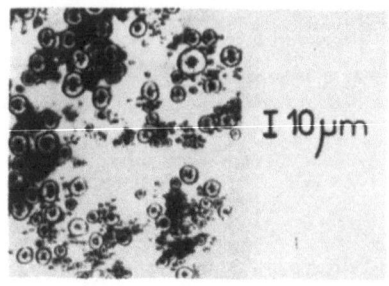

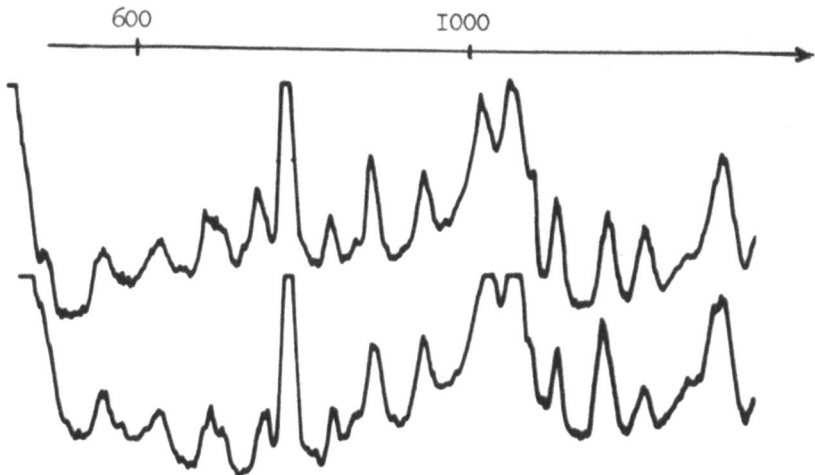

Fig.4. Identification of Urates in biological samples
(see ref.7)

REFERENCES

I. M.DELHAYE and   P.DHAMELINCOURT

      Jnal of Spectroscopy , 3 , 33 , (I975)

2. G.J.ROSASCO , E.S. ETZ and W.A.CASSATT

      Appl. Spectroscopy , 29 , 396, (I975)

3. M.DELHAYE ,P.DHAMELINCOURT and E. DA SILVA

      French Patent . ANVAR 762 I539 (I976)

4. P.DHAMELINCOURT and P.BISSON

      Microscopica   Acta , 79, 267 , (I977)

5. P.DHAMELINCOURT

      Lasers in Chemistry , Elsevier , Amsterdam   (I977)

6. P.DHAMELINCOURT and H.J.SCHUBNELL

      Revue de Gemmologie , 52 , II (I977)

7. C.BALLAN - DU FRANCAIS

      La Cellule , 70 , 3I7 (I974)

I N D E X

367